AF412843

# Separation of Isotopes of Biogenic Elements in Two-phase Systems

# Separation of Isotopes of Biogenic Elements in Two-phase Systems

B.M. Andreev, E.P. Magomedbekov, A.A. Raitman,
M.B. Pozenkevich, Yu.A. Sakharovsky,
A.V. Khoroshilov

*D. Mendeleev University of Chemical Technology*
*Moscow, Russian Federation*

**ELSEVIER**

Amsterdam • Boston • Heidelberg • London • New York • Oxford
Paris • San Diego • San Francisco • Singapore • Sydney • Tokyo

Elsevier
Radarweg 29, PO Box 211, 1000 AE Amsterdam, The Netherlands
The Boulevard, Langford Lane, Kidlington, Oxford OX5 1GB, UK

First edition 2007




**Library of Congress Cataloging-in-Publication Data**
A catalog record for this book is available from the Library of Congress

**British Library Cataloguing in Publication Data**
A catalogue record for this book is available from the British Library

ISBN-13: 978-0-444-52981-7
ISBN-10: 0-444-52981-0


For information on all Elsevier publications
visit our website at books.elsevier.com

Working together to grow
libraries in developing countries

www.elsevier.com | www.bookaid.org | www.sabre.org

ELSEVIER    BOOK AID    Sabre Foundation
            International

Transferred to digital printing 2007

# Contents

# Preface

In the 1940s and 1950s, the *isotopes* of light elements attracted the attention of scientists in the development of nuclear and thermonuclear weapons. This is why enrichment and extraction of such isotopes as $^2$H (deuterium), $^3$H (tritium), $^6$Li, and $^{10}$B were industrially mastered first.

In the 1960s, the peaceful use of nuclear energy, development of new nuclear fuels, and wide application of labelled atoms in various fields of human activities, were favourable for implementing industrial methods of nitrogen, oxygen, and carbon isotope separation.

In recent years, the demand for isotope products used in nuclear medicine has increased sharply. A significant demand relates to the isotopes of biogenic elements (hydrogen, carbon, nitrogen, oxygen). According to the forecasts presented in the monograph *Isotopes: Properties, Production, Application* edited by Yu. V. Baranov (Moscow, IzdAT, 2000, 704 pp.), it is expected that in the coming years demand will increase dramatically for $^{18}$O required for the producion of $^{18}$F used in positron emission tomography, and an increasing use of the isotope breath test leading to a steep rise in demand for $^{13}$C and $^{14}$C isotopes. The use of radiogenic $^3$He in magnetic resonance spectroscopy will spur the production of the radioactive hydrogen isotope tritium.

The above-mentioned monograph discusses all spectrum of problems associated with the technology and application of isotopes, with emphasis placed on the physical methods of separation. The necessity of writing the present book stemmed from two facts. First, the last monograph devoted to the problem of the separation of stable isotopes of light elements, *Separation of Stable Isotopes by Physical–Chemical Methods* by B.M. Andreev, Ya.D. Zelvenskii, and S.G. Katalnikov (Moscow, Energoatomizdat), was published in 1982: in the past 20 years new data on, and novel technologies of, isotope separation processes for these elements have been developed. Secondly, we considered it necessary to more comprehensively describe physical–chemical isotope separation methods for biogenic elements allowing for the development of high-capacity and efficient industrial-scale plants.

The book reflects the present state of research and development, and summarizes both international and Russian experience in the field of separation of isotopes of biogenic elements.

Along with materials gathered by other scientists, the monograph presents the results of practical work done with the participation of the authors.

B.M. Andreev

E.P. Magomedbekov

A.A. Raitman

M.B. Pozenkevich

Yu.A. Sakharovsky

A.V. Khoroshilov

# Introduction

The aim of isotope separation of light elements is an extraction from natural isotope mixtures of less common heavy isotopes, as a rule. This brings about the need for conversion of large masses of raw material flows and the use of cascade schemes, to ensure the required high degree of separation.

To produce stable isotopes of the main biogenic elements (primarily hydrogen, carbon, nitrogen, oxygen) industrial methods of separation are needed. These are based on the physico-chemical process of isotope exchange in two-phase systems: either by rectification or by chemical isotope exchange. The rectification process is well known. The peculiarities of chemical isotope exchange have been investigated to a lesser extent.

Advantages of these methods are connected with the reversibility of single-stage separation. Firstly, unlike methods of separation using inconvertible elementary processes (diffusion, electrolysis, and others), the problem of single-stage isotope effect multiplying can be relatively simply solved by the construction of counter-flow separation columns. Secondly, all power inputs are dependent only on processes of flow reflux at the ends of columns rather than on the elementary act of separation. These advantages allow one to create the high-productivity and economical industrial installations of a rectification and chemical isotope exchange.

In separation columns, where isotope exchange reactions occur, thermal flow reflux (like evaporation or condensation at rectification) or a method with chemical (for instance, electrochemical) conversion can be used. In hydrogen isotope separation, to exclude material expense and shorten energy inputs, the two-temperature method is used in the assembly of inversion of phase flows. This method is based on the dependence of the thermodynamic isotope effect (separation factor) upon the temperature. This allows one to conduct the separation according to the two-column scheme (cold and hot), but without assembly of flow inversion. Here the main expenses of separation are caused by liquid and gas flow circulation and heating (cooling).

The physico-chemical and engineering bases of production of the isotopes of the elements mentioned above in counter-flow columns are considered in this book. The theory of isotope separation in such columns is sufficiently explained in several monographs. So, in chapter 1 only information that is used in subsequent chapters, is given. Besides, in chapter 1 the hydrodynamic features of small packing, used as contact devices in columns for isotope separation of light elements, with the exclusion of hydrogen, are considered. In the last case, because of the large scale of industrial heavy water production, plate columns or columns with regular packing are used.

Hydrogen isotope separation in the past had as its main task the production of heavy water. The main current methods, as in the past, are the chemical isotope exchange, realized according to both dual-temperature schemes and cryogenic hydrogen rectification. At present, interest is moving to the separation of isotope hydrogen mixtures, with radioactive tritium being important in deciding the ecological problems of nuclear energy as well as the development of fuel cycles and systems for radioactive safety of thermonuclear reactors. To solve the tritium problem, it is reasonable to use rectification processes and chemical isotope exchange.

All these questions are considered in chapters 2 and 3, which consider hydrogen isotope separation by rectification and chemical isotope exchange methods in gas–liquid systems. In chapter 3 we show that for tritium entrapping in atomic energy plants the best method is chemical isotope exchange in the $H_2O$–$H_2$ system. When larger volumes should be working over a dual-temperature method in the system, $H_2O$–$H_2S$ can be recommended.

Chapter 4 is devoted to hydrogen isotope separation in systems with a solid phase by methods of chemical isotope exchange of hydrogen with hydride phases of palladium and inter-metallic compounds, as well as by phase isotope exchange in sorption systems (first of all, with zeolites).

At present, no less important are problems of separation of isotopes of the other biogenic elements such as carbon, nitrogen, and oxygen. Heavy stable isotopes of these elements, $^{13}C$, $^{15}N$, $^{17}O$, and $^{18}O$, are indispensable when studying metabolic processes in humans and living organisms. As tagged atoms they are broadly used not only in medical, biological, biochemical, chemical, agricultural, and ecological studies, but also in various technical areas. For instance, interest in the isotope $^{17}O$ is caused by the presence in its atoms of the nuclear magnetic moment, and in the isotope $^{15}N$ for its potential use in the composition of nitride fuel in fast neutron nuclear reactors. Plutonium dioxide, containing only the light isotope $^{16}O$, is used in radioactive sources of electric current (in particular, to ensure the high electrical capacity of implanted artificial valves in the human body, rhythm regulators and heard stimulators).

In the last decennial, world demand for isotopes $^{13}C$ and $^{18}O$ has sharply increased. This is because their use has spread in clinical medicine for the diagnoses of several diseases. Among such diagnostic methods one can note the isotope breath test. It is based on a medical specimen with a high concentration of $^{13}C$; in this method the isotope concentration of $^{13}CO_2$ in exhaled air allows information to be obtained on the condition of internal organs being investigated.

For the diagnosis and evaluation of the efficiency of a treatment for the brain, heart, and different tumors, positron emission tomography (PET) has become widely used throughout the word. It is based on the fact a chemical compound with known biological activity, carrying a short-lived radionuclide, is introduced into the human body, and is disintegrated there with production of positrons; the trace of the emitting positrons allows localization of the region of affected tissue. For targets, the radionuclide $^{18}F$, irradiated beforehand in a cyclotron as $H_2^{18}O$ or $^{18}F_2$, is currently used.

The separation of three stable biogenic isotopes is presented in the last chapters: carbon isotope separation is given in chapter 5; nitrogen in chapter 6; and oxygen in chapter 7. In each chapter the thermodynamic isotope effects in two-phase systems are considered: the mass exchange, the main methods of heavy stable isotope enrichment by rectification and chemical exchange, production of light isotopes of carbon, nitrogen, and oxygen, and perspective processes of separation of these isotopes.

# – 1 –

# Theory of Isotope Separation in Counter-Current Columns: Review

---

### 1.1  SEPARATION FACTOR

Isotope separation in two-phase systems is based on the thermodynamic isotope effect (TDIE), the value of which is conventionally determined by the separation factor of a binary isotopic mixture, $\alpha$, representing the ratio of the relative concentration of isotopes in two different substances or phases in equilibrium:

$$\alpha = \frac{x}{1-x} \Big/ \frac{y}{1-y}, \tag{1.1}$$

where $x$ is the atomic fraction of the target (generally heavy) isotope in one material (X-material), or phase I; and $y$ is the atomic fraction of the same isotope in another material (Y-material), or phase II; $x/(1-x)$ and $y/(1-y)$ is the relative isotope concentrations in X-material and Y-material (phase I and phase II), respectively.

Eq. (1.1), defining a single-stage separation effect, is traditionally written so that the separation factor $\alpha > 1$, and the enrichment factor, $\varepsilon = \alpha - 1$, is positive.

In chemical isotope exchange (CHEX) the aggregative states of working substances (X-material and Y-material) are either the same or different (generally, liquid and gaseous), and phase isotope exchange (PHEX) occurs between the molecules of only one material, forming a two-phase system.

In addition to the separation factor, the isotope exchange reaction can be characterized by an equilibrium constant. In TDIE, the equilibrium constant deviates from a limiting value equal to $K^{\infty}$, with $T \rightarrow \infty$, which signifies an equiprobabilistic isotope distribution between isotope-exchanging molecules.

The values of the separation factor and equilibrium constant coincide only in the event of isotope exchange between molecules with only one exchangeable isotopic atom per molecule, as well as in the case of CHEX reactions of one atom, where $K^{\infty} = 1$. The general forms of these two reactions can be expressed as:

$$AX + BY \leftrightarrow BX + AY, \tag{1.2a}$$

$$A_2X + ABY \leftrightarrow ABX + AY, \tag{1.2b}$$

1

where $A$, $B$ are the light and heavy isotopes, respectively; $X$, $Y$ are the different atoms or groups of atoms, i.e. parts of molecules without the element's exchangeable atoms.

In a general way, if molecules of a material contain $n$ exchangeable isotopic atoms (X-material composed of $A_I B_{n-I} X$ molecules in which $I = 0, 1, 2 \ldots n$), and molecules of another material contain $m$ exchangeable isotopic atoms (Y-material composed of $A_j B_{m-j} Y$ molecules in which $i = 0, 1, 2 \ldots m$), then in such a system the possibility exists of $nm$ isotope exchange reactions with $\alpha_{ij} = K_{ij}/K_{ij}^{\infty}$ separation factors.

On the assumption, that $\alpha_{ij} = \mathrm{const} = \alpha^0$, for a complete isotope exchange reaction,

$$mA_n X + nB_m Y \leftrightarrow mB_n X + nA_m Y, \tag{1.3}$$

the relation between separation factor and equilibrium constant [1] is:

$$\alpha^0 = \sqrt[nm]{K}. \tag{1.4}$$

This expression is sufficient for isotope exchange of all elements, except for hydrogen.

Apart from experimental determination, to compute equilibrium constants for CHEX reactions occurring in gaseous phases, extensive use is made of a quantum-statistical method. Here, for the most interesting case of heterogeneous reactions in liquid–gas systems, when calculating the separation factor ($\alpha_{\mathrm{gas-ql}}$), from the value obtained for the gas reaction ($\alpha_{\mathrm{gas}}$), consideration must be given to the liquid–vapour phase isotope exchange separation factor ($\alpha_{\mathrm{PH}}$) for a substance in its liquid phase:

$$\alpha_{\mathrm{gas-lq}} = \alpha_{\mathrm{gas}} \, \alpha_{\mathrm{PH}} \, . \tag{1.5}$$

In heterogeneous exchange between gaseous and liquid substances, the isotope exchange reaction, as such, occurs, generally, in the liquid phase and is characterized by the separation factor:

$$\alpha_{\mathrm{gas-lq}} = \alpha_{\mathrm{gas}} \, \frac{\alpha_{\mathrm{PH}}}{\alpha_{\mathrm{S}}}, \tag{1.6}$$

where $\alpha_{\mathrm{S}}$ is the separation factor at phase equilibrium of gaseous Y-material with its liquid solution (X-material).

A peculiarity of heterogeneous systems is that several isotope exchange processes occur here simultaneously. Because of such concurrent processes, isotope concentrations in equilibrium phases are established dependent on separation factors of each particular isotope exchange process (CHEX and PHEX reactions). A separation factor calculated using averaged concentrations in each phase is called the effective separation factor.

The simplest case is observed when a single chemical compound in one phase exchanges with an element's several chemical species in another phase (specifically, the chemical species may include the first phase's chemical compound). In this case the effective separation factor ($\bar{\alpha}$) over the area of sparse concentrations of the heavy isotope can be evaluated from separation factors ($\alpha_i$) of all simultaneous processes occurring in the system, using the additive rule, which takes into account the contribution of a particular process to the overall change of isotopic concentrations in equilibrium phases. If the heavy isotope is concentrated in a phase with a complex chemical composition, the effective separation factor is

$$\bar{\alpha} = \sum_{i=1}^{K} M_i \alpha_i, \quad \text{and} \quad \bar{\alpha} - 1 = \bar{\varepsilon} = \sum_{i=1}^{K} M_i \varepsilon_i, \tag{1.7}$$

where $K$ is the number of simultaneous isotope exchange processes, $M_i$ is the element's atomic fraction in a phase with a complex chemical composition, involved in the $i$-th process.

In the second case, when the light isotope is concentrated in a phase with a complex chemical composition, the following equation will be true:

$$\frac{1}{\bar{\alpha}} = \sum_{i=1}^{K} \frac{M_i}{\alpha_i}. \tag{1.8}$$

Of wide occurrence are the CHEX reactions between gases and liquids, complicated by either PHEX reactions between a gaseous phase substance and its liquid solution, or PHEX reactions between a liquid and its vapor in the gaseous phase, or, again, by both PHEX processes simultaneously. In the first case, for example, in the isotope exchange between water and hydrogen sulphide,

$$H_2O + HTS \leftrightarrow HTO + H_2S \tag{1.9}$$

the effective separation factor at low temperature, when the water vapor concentration in the gaseous phase may be ignored, and in the region of low-tritium content, will equal

$$\bar{\alpha}_{\text{gas-lq}} = \frac{1}{1+S}(\alpha_{\text{gas-lq}} + S\alpha_S), \tag{1.10}$$

where $S$ is the hydrogen sulphide water solubility, $H_2S$ mol/$H_2O$ mol.

The second case is characteristic for poorly soluble gas systems, such as in isotope exchange between water and hydrogen,

$$H_2O + HT \leftrightarrow HTO + H_2, \tag{1.11}$$

occurring, as well, in the region of low content of the heavy isotope. Here, in line with equation (1.8), the following relation will be true:

$$\frac{1}{\bar{\alpha}} = \frac{1}{1+h}\left(\frac{1}{\alpha_{\text{gas-lq}}} + \frac{h}{\alpha_{\text{PH}}}\right), \tag{1.12}$$

where $h$ is water vapor content in hydrogen, $H_2S$ mol/$H_2$ mol.

Finally, in the last case, expressing $\overline{\alpha}$ in terms of $\overline{x} = (x + Sy_S)/(1 + S)$ (for the liquid phase) and $\overline{y} = (y + hy_{PH})/(1 + h)$ (for the gaseous phase), we obtain the following expression true for the region of low heavy isotope concentrations [2,3]:

$$\overline{\alpha}_{gas-lq} = \frac{\alpha_{gas-lq} + S\alpha_S}{1 + h\alpha_{gas-lq}/\alpha_{PH}} \times \frac{1+h}{1+S}, \tag{1.13}$$

where $\alpha_{gas-lq} = x/y$; $\alpha_S = y_S/y$; $\alpha_{PH} = x/x_{PH}$, with $y_S$ and $x_{PH}$ being the heavy isotope concentration in the dissolved gas and liquid's vapor, respectively.

Unlike $\alpha$, the effective separation factor $\overline{\alpha}$ depends on pressure due to the pressure influence on the phases' chemical composition.

It is the temperature that makes the greatest impact on the CHEX equilibrium constant, and thus on the separation factor. Also, the change in temperature affects the isotope effect direction as well, that is, results in the inversion of the isotope effect.

Over limited temperature ranges, the separation factor's temperature dependence may generally be represented as

$$\ln \alpha = a + \frac{B}{T} \text{ or } \alpha = Ae^{\frac{B}{T}}. \tag{1.14}$$

Taking into account the relation between the reaction's equilibrium constant and variations in isobaric–isothermal potential

$$RT \ln K = -\Delta G = -\Delta H + T\Delta S, \tag{1.15}$$

and considering the discussed above relation between $K$ and $\alpha$, we have

$$\ln \alpha = \frac{\Delta S}{R} - \frac{\Delta H}{RT}, \tag{1.16}$$

where $\Delta H$ and $\Delta S$ are enthalpy and entropy changes in the course of one atom displacement in the CHEX reaction, where products of symmetry numbers of parent materials' molecules, and those of reaction products' symmetry numbers, are equal.

If in one atom's CHEX reaction $K^\infty \neq 1$, then $S$ will be related to the constant of equation (1.14) by

$$\Delta S = (\alpha + \ln K^\infty)R. \tag{1.17}$$

Hence eq. (1.14) is valid for such temperature ranges where $\Delta H$ and $\Delta S$ values remain constant.

For the most part, the mixtures of one element's isotopes may be considered ideal regardless of the substance aggregative state. This allows calculation of the separation factor of the PHEX process from properties of individual substances (monoisotope compounds); that is, to relate $\alpha$ and isotope effects in the substances' properties.

By this means, using pressures of saturated vapors, $P^0_{AX}$ and $P^0_{BX}$, of material X pure components comprising molecules $AX$ and $BX$ with a single isotope substitution degree, the ideal separation factor in liquid–vapor phase equilibrium can be determined:

$$\alpha_{id} = \alpha^0 = P^0_{AX}/P^0_{BX}. \tag{1.18}$$

For a substance of which the molecules contain several exchanging atoms (e.g. $n$), the ideal separation factor's relation with the ratio of pressures of saturated vapors of monoisotope substances comprising molecules $A_nX$ and $B_nX$ is:

$$\alpha^0 = \sqrt[n]{\frac{P^0_{A_nX}}{P^0_{B_nX}}}. \tag{1.19}$$

A PHEX special case is the separation factor determination at sorption equilibrium. In this case, as distinct from the liquid–vapor system discussed above with a single degree of freedom ($T$ or $P$), temperature and pressure are independent parameters of sorption equilibrium, and the sorption isotherms of the mixture's individual components at a corresponding temperature are required to calculate the separation factor.

Since the separation factor can depend on the sorbed gas amount, the concept of a differential separation factor [4] (characterizing isotope effect on a given portion of the sorption isotherm) is introduced:

$$\alpha_{\mathrm{diff}} = (P^0_{AX}/P^0_{BX})\alpha_H, \tag{1.20}$$

where $P^0_{AX}$ and $P^0_{BX}$ are the equilibrium pressure of the pure components over a sorbent at its equal filling $\alpha_H$.

The most important isotope effects are seen in molecular hydrogen sorption marked by the largest relative mass difference between isotope species. For the filling of all the previous isotherm portions, the following expression for the separation factor of the $A_2$ and $B_2$ molecules mixture (e.g. $H_2$ and $T_2$) can be derived:

$$\ln \alpha_{A_2-B_2} = \frac{1}{a_H}\int_0^{a_H} \ln \alpha_{\mathrm{diff}}\, da_H = \frac{1}{a_H}\int_0^{a_H} \ln \left(\frac{P^0_{A_2}}{P^0_{B_2}}\right)_{a_H} da_H. \tag{1.21}$$

When hydrogen sorption is accompanied by its dissociation into atoms, the separation factor will be equal:

$$\ln \alpha^0_{A-B} = \frac{1}{2a_H}\int_0^{a_H} \ln \alpha_{\mathrm{diff}}\, da_H = \frac{1}{2a_H}\int_0^{a_H} \ln \left(\frac{P^0_{A_2}}{P^0_{B_2}}\right)_{a_H} da_H, \tag{1.22}$$

where $\alpha^0_{A-B}$ is the separation factor at equal ratio of heavy and light hydrogen isotopes in the gas phase.

## 1.2 KINETICS OF CHEX REACTIONS AND MASS EXCHANGE IN COUNTER-CURRENT PHASE MOVEMENT

A peculiarity of CHEX reactions of virtually all elements, excluding hydrogen, is that the reactions can be described by a kinetics equation with a single constant, overall exchange rate, $R$. The reason is that if an insignificant TDIE in these reactions is ignored, then kinetics of commonly termed "ideal" isotope exchange obeys the unified exponential equation [5,6] (not appropriate for complicated isotope exchange, e.g. with diffusion process during the transport of substances, or with more than two exchanging chemical species):

$$-\ln(1-F) = R\left(\frac{1}{nn_x} + \frac{1}{mn_y}\right)\tau = r\tau, \tag{1.23}$$

where $n_X$ and $n_Y$ are the number of X-material and Y-material moles, respectively; $n$ and $m$ are the number of exchanging atoms of X-material and Y-material; and $r$ is the observed rate constant.

Exchange degree, $F$, is defined by the relation

$$F = \frac{x - x_0}{x_\infty - x_0} = \frac{y_0 - y}{y_\infty - y}, \tag{1.24}$$

where $x_0$ and $y_0$ are the initial concentrations of the isotope $B$ in X-material and Y-material, respectively; $x$ and $y$ are isotope concentrations at the $\tau$ instant of time; $x_\infty$ and $y_\infty$ are equilibrium concentrations determined by the separation factor (in the case under study, $x_\infty = y_\infty$).

A simple exponential kinetics equation will also govern isotope exchange in hydrogen isotope exchange reactions with significant thermodynamic isotope effect, if they occur in the region of low concentration of one of the isotopes, or with a small amount of one of the reagents [7]:

$$-\ln(1-F) = \vec{R}\left(\frac{1}{\alpha nn_x} + \frac{1}{mn_y}\right)\tau \quad (\text{at } x, y, \ll 1); \tag{1.25}$$

$$-\ln(1-F) = \vec{R}\frac{\varepsilon y_0 + 1}{\alpha nn_x}\tau \quad (\text{at } n_y \gg n_x), \tag{1.26}$$

where $\vec{R}$ is the initial rate of direct exchange reaction [7].

Like any chemical reaction, the rate of direct or reverse exchange reaction $R$ depends, apart from temperature, on the reagents' concentrations [5–8]

$$R = kC_x^p C_y^q, \tag{1.27}$$

where $k$ is the rate constant; $C_x$ and $C_y$ are the concentration of X-material and Y-material; $p$ and $q$ are the reaction order of X-material and Y-material, respectively.

The half-exchange time $\tau_{0.5}$, with the exchange degree $F = 0.5$, is commonly taken as a characteristic of the exchange kinetics:

$$\tau_{0.5} = \frac{-\ln(0.5)}{r} = \frac{0.693}{r} = \frac{0.693}{R\left(\dfrac{1}{\alpha n n_x} + \dfrac{1}{m n_y}\right)}. \tag{1.28}$$

Depending on the exchange conditions (number of moles $n_x$ and $n_y$), even at a constant rate of exchange $R$, the half-exchange time may vary over a wide range. That is why consideration must be given to the relative nature of this isotope exchange kinetics characteristic.

The equations discussed above refer to isotope exchange reactions both in homogeneous and heterogeneous systems. In the former case, X-material and Y-material are in the same reaction volume. That is why in this case instead of number of moles $n_x$ and $n_y$, the reactants' molar concentration is generally used, so the exchange rate $R$ is expressed in mol/(l·s).

If the heterogeneous isotope exchange occurs on the interphase boundary surface, the exchange rate $R$ is related to the surface unit $S$, then the kinetics eqs. (1.23, 1.25, 1.26) will involve the product $R_{Sp}S$ (the dimension of $R_{Sp}$ is mol/(m²·s)). The most representative example of chemical exchange systems with fixed contact surface is systems with a solid phase, discussed in chapter 4.

In counter-current separation in columns, of the greatest interest are the CHEX reactions in gas–liquid systems. A distinguishing feature of the kinetics study in such systems is that, unlike systems with a solid phase, the surface of phase contact here is not strictly fixed. Moreover, to eliminate the influence of diffusion processes in the contacting phases on chemical kinetics, it is necessary to intensively mix the phases, which is generally difficult to realize with the surface unchanged and constant. In addition, the pattern may be complicated by the fact that the reaction occurs not on the interphase boundary surface, but in the liquid phase, between the phase substance and the gas dissolved in the substance. This is why the isotope exchange rate is often related to the liquid phase's volume unit, resulting in appropriate changes in the kinetics equation's notation. To illustrate, when the exchange occurs between a liquid substance X and a gas Y in a system with thermodynamic isotope effect at $x, y \ll 1$, and $n = m = 1$, the kinetics equation will be written as

$$-\ln(1-F) = \vec{R}V\left(\frac{1}{\alpha n_x} + \frac{1}{n_y}\right)\tau, \tag{1.29}$$

where $V$ is the amount of liquid in l; and $\vec{R}$ is expressed in mol/(l·s).

The isotope exchange in gas–liquid systems is frequently performed in such conditions where isotope concentration change in the liquid phase may be disregarded (i.e. $n_x \ll n_y$). Then the observed rate constant $\vec{r} = \vec{R}V/n_y$, that is, the smaller the gas phase proportion in the system, the higher the rate of isotope equilibrium establishment in the system. If the isotope exchange rate is unaffected by the gas pressure (zero-order with respect to

Y-material), then $r$ will be inversely related to the pressure. Conversely, the coincidence of kinetic curves derived for different pressure values points to the first order of the CHEX reaction with respect to the gaseous substance Y.

For reactions performed with heterogeneous catalysis (a solid catalyst in the liquid phase), the isotope exchange rate is generally related to the catalyst surface unit, or to the catalyst mass unit (if the catalyst surface area is unknown). In the former case, at $x, y \ll 1$, and $n = m = 1$, the calculations are performed by the equation:

$$-\ln(1-F) = \vec{R}_{SP} S_{SP} g \left( \frac{1}{\alpha n_x} + \frac{1}{n_y} \right) \tau, \tag{1.30}$$

where $S_{SP}$ is the catalyst surface area, m²/g; $R_{SP}$ is exchange rate related to the surface unit, mol/(m²·s); $g$ is the catalyst mass, g.

In the latter case, the following kinetic equation form is used:

$$-\ln(1-F) = \vec{R}_{SP} g \left( \frac{1}{\alpha n_x} + \frac{1}{n_y} \right) \tau, \tag{1.31}$$

where $R_{SP}$ is the exchange rate related to the catalyst mass unit, mol/(g·s).

Mass-exchange in a counter-current column may be represented as composed of the following stages: isotope mass-transport in each phase, and isotope mass-transfer, caused by the CHEX reaction, from one phase into another. Hence the equation of mass-transfer resistance additivity [9] will be as follows [10–12]:

$$\frac{1}{K_{OY}} = \frac{1}{\beta_Y} + \frac{1}{m\beta_X} + \frac{1}{m'\beta_{IE}} \qquad \text{or} \tag{1.32}$$

$$\frac{1}{K_{OX}} = \frac{1}{\beta_X} + \frac{m}{\beta_Y} + \frac{m}{m'\beta_{IE}}, \tag{1.33}$$

where $K_{OY}$ and $K_{OX}$ are the mass-transfer coefficients for the phase of Y-material and X-material, respectively, mol/(m²·s); $\beta_Y$ and $\beta_X$ are the diffusion coefficients of mass-exchange for the phase of Y-material and X-material, respectively, mol/(m²·s); $\beta_{IE}$ is the chemical component of the mass-transport coefficient due to the CHEX reaction; $m$ is a coefficient equal to the equilibrium line slope ratio, taken to be constant within narrow interval of isotope concentrations.

In accordance with eq. (1.1) the equilibrium line slope ratio equals $m = dx/dy = (\alpha - \varepsilon x)^2/\alpha$ (as is commonly assumed in the literature on mass-transfer, $m = dy/dx$). It follows that $m$ varies from $\alpha$ in the range of low concentrations of heavy isotope to $1/\alpha$ in the region of the high isotope concentrations.

As indicated above, the liquid–gas CHEX reactions occur, as a rule, between molecules of liquid phase substance and those of the solute gas. That is why the coefficient $m'$ accounts for the isotope effect on gas dissolving ($\alpha_S$).

As applied to the gas–liquid systems, eqs. (1.32) and (1.33) refer to the instance when the X-material, in liquid or gaseous phase, respectively, is enriched in the heavy isotope.

If X-material is in solid phase, then the equation of mass-transfer resistance additivity [11–13] will be:

$$\frac{1}{K_{OY}} = \frac{1}{\beta_Y} + \frac{1}{m\beta_X} + \frac{1}{m\beta_{IE}} \quad \text{or} \tag{1.34}$$

$$\frac{1}{K_{OX}} = \frac{1}{\beta_X} + \frac{m}{\beta_Y} + \frac{1}{\beta_{IE}}. \tag{1.35}$$

According to eqs. (1.32) and (1.33), the height of the transfer unit (HTU) may be represented as a sum of constituents [11–13]:

$$h_{OY} = h_Y + \frac{\lambda}{m} h_X + \frac{\lambda}{m'} h_{IE} \quad \text{or} \tag{1.36}$$

$$h_{OX} = h_X + \frac{m}{\lambda} h_Y + \frac{m}{m'} h_{IE}, \tag{1.37}$$

where $h_{OY}=G/SK_{OY}a=G_{SP}/K_{OY}a$; $h_{OX}=L/SK_{OX}a=L_{SP}/K_{OX}a$; $h_Y=G_{SP}/\beta_Y a$; $h_X=L_{SP}/\beta_X a$; $h_{IE}=L_{SP}/\beta_{IE}a$; $G$ and $L$ are flows of Y-material and X-material respectively, mol/s; $S$ is the column cross section, m$^2$; $a$ is the specific surface of phase contact, m$^2$/m$^3$; $\lambda$ is the flow ratio, $\lambda = G/L$.

So, the efficiency of mass-transfer in the column is specified by the mass-transfer coefficients ($K_{OY}$ and $K_{OX}$), by HTU values ($h_{OX} = m/\lambda\, h_{OY}$), and by the height equivalent for the theoretical plate of separation (HETP) related to HTU in a wide range of isotope concentrations by

$$h_E = h_{OY} \left\{ \frac{m \ln m}{m-1} - \ln[1+(m-1)x] \right\}. \tag{1.38}$$

In the low heavy isotope content the relation between HETP and HTU is expressed as follows:

$$h_E = h_{OY} \frac{\ln(\alpha/\lambda)}{1-\lambda/\alpha} = h_{OX} \frac{\ln(\alpha/\lambda)}{\alpha/\lambda-1}. \tag{1.39}$$

In hydrogen isotope separation (especially the separation of tritium-containing isotope mixtures), HETP and HTU may differ significantly ($h_{OX} > h_E > h_{OY}$), whereas in the separation of other elements' isotopes, separation factors and flow ratios are close to 1, therefore it is generally assumed that $h_{OX} \approx h_E \approx h_{OY}$.

From the HTU expression it follows that the volume mass-transfer coefficient $K_{OY,V} = K_{OY} a = G_{SP}/h_{OY}$ or $K_{OX,V} = K_{OX} a = L_{SP}/h_{OX}$, incorporates both the column specific flows of gaseous (vapor) or liquid phases, and the height of transfer unit. Consequently, the volume mass-transfer coefficient comprises both hydrodynamic and mass-transfer characteristics of the packing layer volume unit (in terms of moles or mass units per unit volume of the column), that is, represents the most indicative integrated quantitative characteristic of the separation column packing layer.

Since the value $\beta_{IE}$ is independent of the column hydrodynamic environment, HTU increase with a rise in loading is associated, above all, with the decisive contribution of the chemical constituent in the overall mass-transfer resistance. This can be exemplified by the gas–liquid isotope exchange in a column with a fine effective packing, for which the mass-transfer resistance from the gas can be ignored. In this case, in the criterion equation for the coefficient of mass-transfer in the liquid phase

$$Nu_X = ARe_X^m Pr_X^n \tag{1.40}$$

the exponential order $m$ is close to 1 ($m = 0.8–1.0$) [14–18], i.e. $Nu_X$ is approximately proportional to $Re_X$, and thus $L_{SP}/\beta_X = const$ and $h_X = const$ ($Nu_X = \beta^C d/D$, where $D$ is diffusion coefficient; $d$ is the determining size of the packing material particles; $\beta^C$ is the coefficient of mass-transfer in the liquid, m/s; and $\beta_X = \beta^C (\rho_X/M_X)$; $\rho_X$ and $M_X$ are the liquid density and molecular mass).

At a high wet ability of the packing, $a = const$, and eqs. (1.36) and (1.37) present virtually the straight line equations:

$$h_{OY} = h_Y + \frac{\lambda}{m} h_X + \frac{\lambda L_{SP}}{m'\beta_{IE} a} = const + \frac{G_{SP}}{m'\beta_{IE} a} \quad \text{or} \tag{1.41}$$

$$h_{OX} = h_X + \frac{m}{\lambda} h_Y + \frac{m}{m'} \frac{L_{SP}}{\beta_{IE} a} = const + \frac{m L_{SP}}{m'\beta_{IE} a}. \tag{1.42}$$

In the first case, the straight-line slope ratio on the dependence of HTU on the column loading (Figure 1.1) equals $1/(m' \cdot \beta_{IE} a)$, and in the second case – $m /(m' \cdot \beta_{IE} a)$. With increase in $\beta_{IE}$ and $a$, the dependence of HTU on the column loading weakens, and with the decisive contribution of diffusion resistance may practically vanish. Figure 1.1 presents the dependence of HTU on the column loading. As the CHEX reaction rate increases, the straight line slope diminishes. If the CHEX reaction is catalytic, Figure 1.1 can represent the influence of the catalyst influence on HTU at $T, P = const$. The same pattern will be observed with increase in temperature (at $P = const$) or in pressure (at $T = const$), since the exchange rate rises with increasing temperature, or, as a rule, with an increasing amount of solute gas.

The next figure illustrates how the packing specific surface influences the HTU dependence on the loading. Figure 1.2 demonstrates that with a decrease in the size of packing particles (increase of $a$), and all other factors being the same, the HTU dependence on the loading becomes less distinct and the line segment, intercepted on the y-axis by the straight line and determined by the value $h$, shortens. For the relation $h_{OY} = f(G)$, such a line segment will equal $\lambda h_X/m$. Examples of similar relations in systems with solid phase are given in references [13, 17, 18].

Now we focus upon the fact that even when $\beta_{IE}$ and $a$ remain invariant, the contribution of the chemical constituent in $h_{OY}$ and $h_{OX}$ will vary if CHEX reactions are accompanied by significant thermodynamic isotope effects which, as indicated above, are characteristic of hydrogen isotope exchange. Even with the absence of kinetic isotope

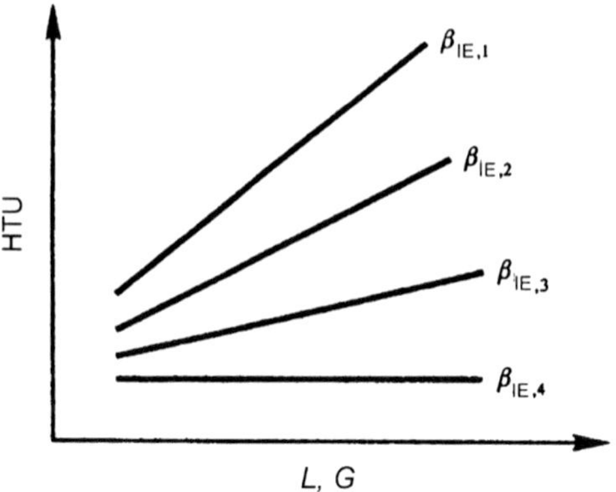


**Figure 1.1**   Dependence of HTU on the column loading at different values of $\beta_{IE}$ ($\beta_{IE,1} < \beta_{IE,2} < \beta_{IE,3} < \beta_{IE,4}$).

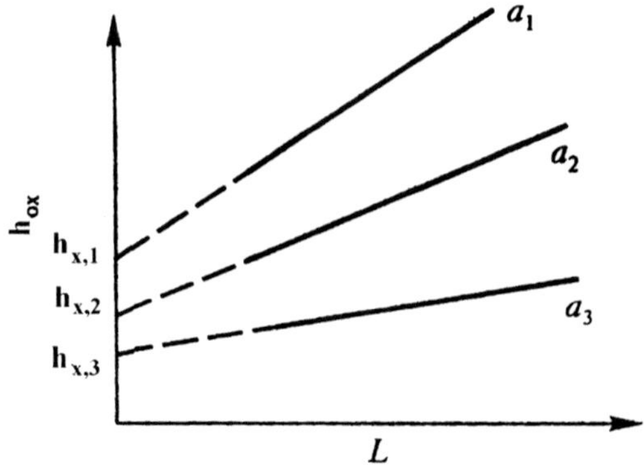


**Figure 1.2**   Influence of the packing size on the dependence of $h_{OX}$ on the liquid flow ($a_1 < a_2 < a_3$).

effects, the relation $h_{OX} = f(L)$ in the H–T isotope exchange, with all other factors being the same ($\beta_{IE}$ = const), will be more pronounced than that in the H–D isotope exchange. At the same time the $h_{OX} = f(L)$ relations may coincide due to a minor thermodynamic isotope effect during the gas dissolving ($\alpha_{S,HT} \approx \alpha_{S,HD} \approx 1$).

The above consideration is true with the assumption that the isotope concentrations in the gas and water vary only with the column height and are invariant across the column lateral section, which is typical only for small-diameter columns. In the general case proper allowance must be made for deviations from the model of ideal (complete) displacement. The simplest way to do so is to supplement the additivity eqs. (1.36) and (1.37) with addends specified by longitudinal mixing ($h_{LM}$) and transverse irregularity ($h_{TI}$) [9, 19]. Since the last addend depends on the column diameter (the effective radial diffusion coefficient), it is this addend that is mainly responsible for the departure of the scale-up factor (SF), allowing for HTU (HETP) increase in packed columns of greater diameters, from one; that is, such an approach assumes the absence of influences of the real structure of flows in the separation column on $h_{IE}$.

In the above examination we did not dwell on the $\beta_{IE}$ value calculations and on possible influence of $\beta_{IE}$ on the diffusion components of HTU. The reason is that the general theory of heterogeneous mass transfer for the CHEX reactions in columns has yet to be developed. Some studies [20, 21] suggest that the isotope exchange in gas–liquid systems be considered as identical with a chemical gas adsorption process. With a set of assumptions [22], a different mathematical model is advanced describing the isotope exchange process in packed columns for low isotope concentrations in the CHEX reactions occurring both on interphase boundary surface and between liquid and solute gas.


## 1.3   STATIONARY STATE OF THE COLUMN WITH FLOW REFLUX

In isotope separation, multistage counter-current separation processes (generally continuous) are used, which allow isotopes to be produced at any required concentration. Such processes are done in cascades of separation elements (or stages). In separation elements of the first type (Figure 1.3a) the inlet flow is divided into two: enriched with the target isotope flow and waste flow. In a cascade of such separation elements, the enriched flow enters the next separation stage, and the waste one is fed to a preceding stage with the same isotope content.

To multiply a single isotope effect, two-phase systems make use of counter-current columns incorporating separation elements of the second type, and flow-conversion units.

The separation elements of the second type (Figure 1.3b) have, with two outlet flows (e.g. a liquid enriched with required isotope, and a gas (vapor) depleted of the same isotope), two inlet flows (liquid flow from the upper stage and a gas or vapour flow from the lower stage). In such elements the redistribution of isotopes between moving counter-current phases occurs due to the phase- or chemical isotope exchange.

If the gas (vapor) phase leaving a stage is in isotope equilibrium with the liquid phase leaving the stage, this is referred to as the theoretical plate (TP) of separation.

Figure 1.4 shows the principle of continuous separation column plants. The 'open' scheme, or the scheme with a reservoir of infinite size (Figs. 1.4a and b) represents a

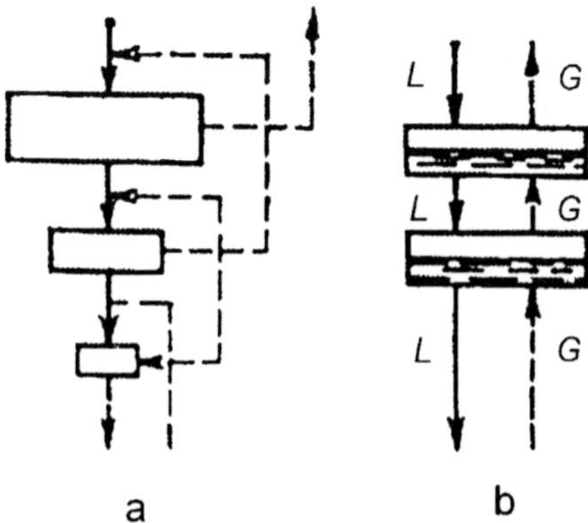

**Figure 1.3**  Types of separation element: *a*, with one inlet flow; *b*, with two inlet flows.

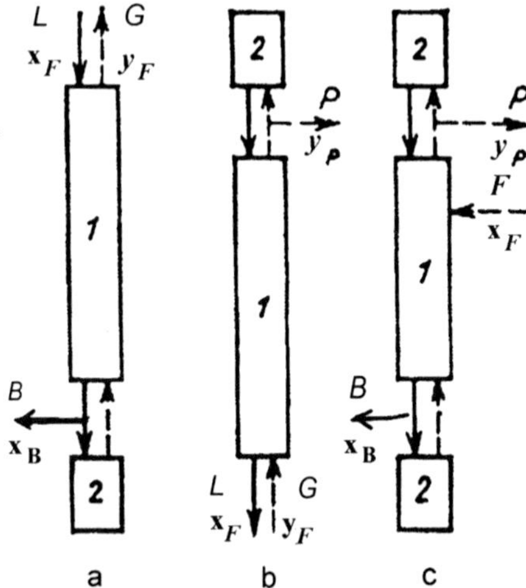

**Figure 1.4**  Principle of continuous separation column plant: *a*, open scheme of the heavy component concentration; *b*, open scheme of the heavy component depletion; *c*, scheme with depletion; *1*, separation column; *2*, flow conversion unit.

column with the concentration section only, where the isotope concentration is increased from $x_F(y_F)$ to $x_B(x_P)$, and one enriched flow reflux unit.

The concentration and depletion scheme (Figure 1.4c) incorporates, in addition to the column, two flow conversion units for the flows enriched with, and depleted of, the same isotope. The flow conversion units provide for the realization of counter-current movement of liquid and gas (vapor) phase ($L$ and $G$, respectively) in the column.

In the absence of losses and without interconversion of phases, $L$ and $G$ flows remain constant along the whole column height up to the feeding point (flow ratio $\lambda = G/L = \mathrm{const}$).

Inlet and outlet (or external) flows (Figure 1.4c) are related by the material balance equations with respect to flows and isotope

$$F = B + P; \tag{1.43}$$

$$Fx_F = Bx_B + Px_P. \tag{1.44}$$

For the given separation task (product withdrawal $B$ and concentrations $x_F$, $x_B$, and $x_P$) the feed flow rate can be determined from the joint solution of eqs. (1.43) and (1.44)

$$F = B\frac{x_B - x_P}{x_F - x_P}. \tag{1.45}$$

From the material balance equation with respect to the target isotope for concentration or depletion column sections

$$Lx_{i-1} = Gy_i + Bx_B; \tag{1.46}$$

$$G_D y_{i+1} + L_D x_i + Px_P \tag{1.47}$$

with regard to the relation (1.43) we obtain the following expressions, referred to as operating line equations relating $x$ and $y$ in any cross-section of concentration or depletion column sections:

$$x = \lambda y + (1 - \lambda)x_B; \tag{1.48}$$

$$x = \lambda_D y - (\lambda_D - 1)x_P; \tag{1.49}$$

where $\lambda_D = G_D/L_D$ ($L_D$ and $G_D$ is the liquid and gas (vapor) phase flows in the column depletion section).

At $\lambda$, $\lambda_D = \mathrm{const}$, the operating lines eqs. (1.48) and (1.49) in $xy$-coordinates (McCabe–Thiele diagram) represent the equations of straight lines with slope ratios equal to $\lambda$ and $\lambda_D$, respectively, reaching the diagonal at the points with $x_B$ and $x_P$ ordinates, and converging at the point with $x_F$ ordinate (Figure 1.5).

In the absence of withdrawal function ($\lambda = \lambda_D = 1$), the isotopic composition of liquid and gas (vapor) will coincide in any column cross-section, and operating lines will agree with the diagonal segment between $x_P$ and $x_B$.

As indicated above, the feed flow rate can be determined from the material balance using eq. (1.45) (the complexity is associated only with the justification of the target isotope concentration in the waste flow $P$).

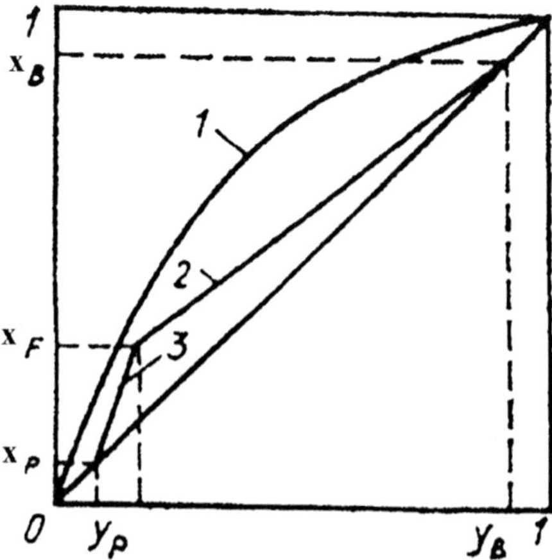

**Figure 1.5**    Graphical representation of the separation process in the column, in *xy*-coordinates: *1*, equilibrium line; *2*, operating line of the concentration section; *3*, operating line of the depletion section.

For the 'open' scheme, the problem of the determination of the feed flow ratio is not simple since the feed flow ratio will depend on the value (degree) of the target isotope extraction from the feed, which may vary from zero (non-extraction mode) to a maximum value determined, first of all, by the value of $\alpha$.

The extraction degree ($\Gamma$) is the amount of the target isotope in the enriched product as related to the amount of the same isotope entering the separation plant with the source (feed) flow:

$$\Gamma = \frac{Bx_B}{Fx_F}. \tag{1.50}$$

From the material balance of an 'open' scheme column with respect to the target isotope, it follows that

$$\Gamma = \frac{Fx_F - Gy_F}{Fx_F} = 1 - \lambda\frac{y_F}{x_F} = (1-\lambda)K. \tag{1.51}$$

The extraction degree is conventionally determined in terms of the maximum extraction degree ($\Gamma_m$) representing an extraction degree for infinite number of theoretical separation plates (NTP), and relative withdrawal $\theta$. If a separation column is infinitely high, isotopic concentrations at its upper end will be interrelated by the separation factor $\alpha$ (see Figure 1.4a), i.e. the operating line lower point (with $x_F$ ordinate) will be resting on the equilibrium line.

For the 'open' scheme, there are two variations of the process realization for the infinite NTP: (1) at $K$ = const and the flow ratio decreasing to the minimum value $\lambda_m$ (Figure 1.6a); and (2) at $\lambda$ = const and the separation factor increasing to $K' = x'_B/x_F$ (Figure 1.6b).

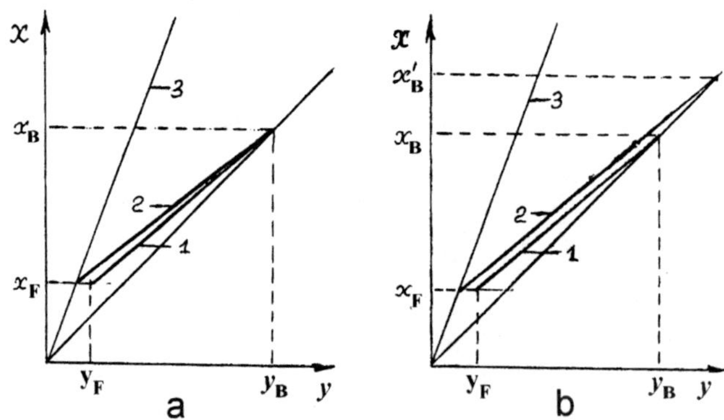

**Figure 1.6**   *xy*- diagram of the 'Open'-scheme column at *x*, *y* ≪ 1, with the withdrawal of first kind: *1*, the column operating line; *2*, operating line for the maximum extraction degree ($n = \infty$); *3*, equilibrium line; *a*, the maximum extraction degree at $K = $ const; *b*, the maximum extraction degree at $\lambda = $ const.

In the initial concentration region (at $x_F \ll 1$), from eq. (1.51) it follows that

$$\Gamma_{1,m} = 1 - \frac{\lambda_m}{\alpha}, \tag{1.52}$$

$$\Gamma_{1,m} = (1 - \lambda_m)K, \tag{1.53}$$

$$\lambda_m = \frac{x_B - x_F}{x_B - \dfrac{x_F}{\alpha}} = \frac{K - 1}{K - 1/\alpha}, \tag{1.54}$$

and for the second variation we obtain

$$\Gamma'_{1,m} = 1 - \frac{\lambda}{\alpha}, \tag{1.55}$$

$$\Gamma'_{1,m} = (1 - \lambda)K', \tag{1.56}$$

$$K' = \frac{\alpha - \lambda}{\alpha(1 - \lambda)}. \tag{1.57}$$

From the correlation between eqs. (1.52) and (1.54) it follows that $\Gamma_{1,m} > \Gamma'_{1,m}$. But in this case $\lambda_m$ is still negligibly less than 1 at $\varepsilon \ll 1$ (i.e. with separation of all elements except for hydrogen). The result is that for the 'open' scheme $\Gamma_m$, $\Gamma'_m \ll 1$, and the major part of the target isotope contained in the source material is found in the waste flow *G*.

If the column is a stage of a separation cascade and operates at $x_F \ll 1$, then the value $B \ll L$, and flow ratio $\lambda$ is close to 1 ($\lambda = G/L = 1 - B/L \approx 1$). **In** this case, termed the withdrawal of second kind,

$$\Gamma_m = 1 - \frac{1}{\alpha}. \tag{1.58}$$

Inserting $\lambda_m$ from eqs. (1.54) to Eq. (1.52), one obtains

$$\Gamma_{1,m} = \frac{\alpha - 1}{\alpha - \dfrac{1}{K}}, \tag{1.59}$$

from which it follows that at a moderate separation degree and at $\varepsilon \ll 1$, the value $\Gamma_{1,m}$ may considerably exceed the maximum extraction degree in the withdrawal of second kind $\Gamma_m$. Notice that the relation between $\Gamma_m$ and the maximum extraction degree $\Gamma'_{1,m}$ will be as follows:

$$\Gamma'_{1,m} = \Gamma_m + \frac{1 - \lambda}{\alpha}. \tag{1.60}$$

After inserting $y_1 = x_F/(\alpha - \varepsilon x_F)$, which results from eq. (1.1), in eq. (1.51) we obtain, that in a wide range of concentrations,

$$\Gamma_m = 1 - \frac{\lambda_m}{\alpha - \varepsilon x_F}. \tag{1.61}$$

Considering that $\lambda_m$ can be determined from the coordinates of points $x_F$, $y_1$ and $x_B = y_B$, we have

$$\lambda_m = \frac{x_B - x_F}{x_B - x_F/(\alpha - \varepsilon x_F)}, \tag{1.62}$$

then substituting eqs. (1.62) in eq. (1.54), we will find the value $\Gamma_m$.

With infinite NTP, the withdrawal flow $B_m$ will be maximal (at $x_B = $ const) since $\Gamma = \Gamma_m$. For a real plant with a limited NTP the extraction degree equals

$$\Gamma = \Gamma_m \theta, \tag{1.63}$$

where $\theta$ is the relative withdrawal ($\theta = B/B_m$).

It is evident that according to eqs. (1.51) and (1.53), the relative withdrawal $\theta$ for the first variation (Figure 1.6a) will be

$$\theta = \frac{1 - \lambda}{1 - \lambda_m}, \tag{1.64}$$

and for the second variation (Figure 1.6b) from eqs. (1.51) and (1.56) we obtain

$$\theta = \frac{K}{K'}. \tag{1.65}$$

Then the value of $\lambda$ can be found from

$$\lambda = \frac{K'-1}{K'-1/\alpha}. \tag{1.66}$$

Unlike the 'open' scheme discussed above, the plant with a depletion section (see Figure 1.4c) can provide for a high extraction degree that equals:

$$\Gamma = 1 - \frac{Px_P}{Fx_F} = 1 - \frac{P}{F} \times \frac{1}{K_D}, \tag{1.67}$$

where $K_D = x_F/x_P$.

If $B \ll F$, the extraction degree is determined by the value $K_D$

$$\Gamma \approx 1 - \frac{1}{K_D}. \tag{1.68}$$

The prime objective of the column stationary state calculation is limited to determining the column NTP and flows. Given NTP and HETP, the height of the column separation section can be defined, and knowing flows and their velocities, we can determine the column diameter.

For the 'open' scheme the flow $L = F$ can be found from eq. (1.50), after prior determination of $\Gamma$ from eqs. (1.63) and (1.52–1.57), or (1.61).

For the withdrawal of second kind $G = L$, and to determine the flow $G$ for the withdrawal of first kind, we need to calculate the value of $\lambda$, which equals:

$$\lambda = \frac{L-B}{L} = 1 - \frac{B}{L}, \tag{1.69}$$

where

$$\lambda = 1 - \theta(1 - \lambda_{\mathrm{m}}). \tag{1.70}$$

In the column depletion section, the flow ratio $\lambda_D = G_D/L_D$ equals

$$\lambda_D = \frac{L_D + P}{L_D} = 1 + \frac{P}{L_D}, \tag{1.71}$$

where $L_{\mathrm{D}} = L - F$.

The maximum value of $\lambda_{Dm}$ for the initial concentration region is

$$\lambda_{Em} = \frac{x_F - x_P}{x_F/\alpha - x_P} = \frac{K_D - 1}{K_D/\alpha - 1} = 1 + \frac{P}{L_{Em}}, \tag{1.72}$$

where $L_{Em}$ is the the minimum liquid flow in the column depletion section.

From the relation

$$\frac{G_{Dm}}{G_D} = \frac{L_{Dm} + P}{L_D + P} = \theta \tag{1.73}$$

and eqs. (1.71) and (1.72), we can obtain the expression relating $\lambda_D$ to $\lambda_{Dm}$ by the relative withdrawal

$$\frac{(\lambda_D - 1)\lambda_{Dm}}{(\lambda_{Dm} - 1)\lambda_D} = \theta. \tag{1.74}$$

HETP can be found by joint solution of the equilibrium equation and material balance equation for each separation stage. For cases of a prime practical significance, appropriate analytical dependencies have been deduced, of which the simplest is the Fenske equation for non-withdrawal mode:

$$n = \frac{\ln K}{\ln \alpha}, \tag{1.75}$$

where $K = x_B/(1 - x_B) \times (1 - x_F)/x_F$ (for concentration column); $K = x_F/(1 - x_F) \times (1 - x_P)/x_P$ (for depletion column).

For the column operating with withdrawal mode, the equations interrelating $n$, $K$, and $\alpha$ will comprise one more parameter which characterizes withdrawal: relative withdrawal $\theta$ and/or flow ratio $\lambda$.

The simplest equations are derived in special cases for the linear form of the equilibrium equation, i.e. in the region of sparse concentrations of one of isotopes. These equations can be obtained primarily with the use of the so-called graphico-analytic method. The essence of the method is graphical plotting of NTP between operating and equilibrium lines in the column with the use of $xy$-diagram. The process motive force is determined by concentration pressure or the distance between operating and equilibrium lines. The shorter distance, the greater TP is needed to achieve the required separation degree. The distance is characterized by segments $a_T$ and $a_B$ between operating and equilibrium lines at the upper and lower ends of the column

$$n = \frac{\ln(a_T/a_B)}{\ln(\alpha/\lambda)}. \tag{1.76}$$

From the above equation, and eqs. (1.51), (1.54), and (1.57), for the column with the withdrawal of second kind ($\lambda = 1$), the following expression [23] can be obtained:

$$n = \ln \frac{K - \theta}{1 - \theta} \bigg/ \ln \alpha, \tag{1.77}$$

where $K = x_B/x_F$.

The following are the equations analogously derived for a concentration column with the withdrawal of second kind when the column is an individual plant or the last stage of a cascade:

$$n = \ln \frac{K(\alpha - 1)\lambda}{\alpha - \lambda - K(1 - \lambda)\alpha} \bigg/ \ln \alpha\!\big/\!\lambda = \ln \frac{K(\alpha - 1)\lambda}{(1 - \theta)(\alpha - \lambda)} \bigg/ \ln \frac{\alpha}{\lambda}, \tag{1.78}$$

and when the column is a cascade intermediate stage ($\lambda \neq 1$) with the enriched product return from the succeeding stage, because of which the upper end of operating line does not reach the diagonal line (e.g. column II in Figure 1.11):

$$n = \ln \frac{K - \theta}{1 - \theta} \bigg/ \ln \alpha/\lambda. \tag{1.79}$$

Eqs. (1.78) and (1.79) have been derived in reference [23], where eq. (1.54) at $\lambda = \mathrm{const}$ was used for the maximum extraction degree. As a consequence, in the NTP calculation by eq. (1.78), first the $\theta$ value is specified, and $K'$ is determined from eq. (1.65); then the value of flow ratio $\lambda$ is derived from eq. (1.66).

When using eq. (1.79), the $K'$ value is also determined from eq. (1.65), where by the separation degree is meant not the column $K$ separation degree appearing in eq. (1.79), but the separation degree $K_\Sigma$ accounting for the succeeding cascade stages (i.e. column II in Figure 1.11 $K_\Sigma = x_B/x_B^I$).

For the depletion column the following expression was obtained [10]:

$$n_D = \frac{\ln[K_D(1 - \theta)]}{\ln \alpha/\lambda_D}. \tag{1.80}$$

In a wide range of concentrations the equilibrium equation cannot be approximated by the straight-line equation, and the cross point method [10] can be utilized for the NTP calculation. And yet it is the Cohen equation [24] that gained the widest acceptance in isotope practice:

$$n = \frac{1}{\varepsilon \Delta \psi} \ln \frac{x_B + x_F - 2x_B x_F - (x_B - x_F)\theta \psi + (x_B - x_F)\Delta \psi}{x_B + x_F - 2x_B x_F - (x_B - x_F)\theta \psi - (x_B - x_F)\Delta \psi}, \tag{1.81}$$

where

$$\psi = \frac{B}{\varepsilon L\theta}; \Delta\psi = \left[1 + 2\psi\theta(1-2x_B) + (\psi\theta)^2\right]^{1/2}.$$

The calculation of the column separation stage height can be performed as well by using the number of transfer units (NTU), both by analytical and graphical [9] methods. The simplest analytical solution is obtained for non-withdrawal mode, within the region of low concentrations of one of isotopes.

If the target isotope in the equilibrium conditions concentrates in the liquid phase (X-material), then NTU calculated in terms of the mass-transfer coefficients $K_{OY}$ and $K_{OX}$ will equal:

$$N_y = \frac{\alpha \ln K}{\alpha - 1}, \tag{1.82}$$

$$N_x = \frac{\ln K}{\alpha - 1}. \tag{1.83}$$

When $\alpha$ is little different from 1, eq. (1.83) is practically identical to the Fenske equation (1.75), since $\ln \alpha \approx \alpha - 1$.

In a wide range of concentrations in non-withdrawal mode, instead of eq. (1.82) the following expression is obtained:

$$N_y = \frac{\alpha \ln K}{\alpha - 1} + \ln \frac{1 - x_F}{1 - x_B}, \tag{1.84}$$

where  $K = \dfrac{x_B}{x_F} \times \dfrac{1 - x_F}{1 - x_B}.$

For the withdrawal of first kind at $x, y \ll 1$, the following equations will be true:

$$N_y = \frac{\alpha}{\alpha - \lambda} \ln \frac{K\lambda(\alpha - 1)}{(\alpha - \lambda) - K\alpha(\lambda - 1)} \tag{1.85}$$

for the concentration column, and

$$N_y = \frac{\alpha}{\alpha - \lambda_D} \ln \frac{\alpha(\lambda_D - 1) + K_D(\alpha - \lambda_D)}{\lambda_D(\alpha - 1)} \tag{1.86}$$

for the depletion column.

In the general case, that is, for the column operating with withdrawal in a wide range of concentrations, V.L. Pebalk [25], having accepted the equilibrium equation $y'=x/\alpha-(\alpha-1)x$, obtained the following solution of the integral determining NTU:

$$N_y = \frac{1}{1-\beta}\left[\ln\frac{x_2-d}{x_1-d}+\beta\ln\frac{c-x_1}{c-x_2}\right],\tag{1.87}$$

where $\beta=1+(\alpha-1)c/[1+(\alpha-1)d]$, and $c$ and $d$ are the roots of quadratic equation $c=A+\sqrt{A^2-B}$ and $d=A-\sqrt{A^2-B}$, $A=1/2[\alpha\lambda/(\alpha-1)-1/(\alpha-1)+\hat{a}]$   $B=b/(\alpha-1)$; $\hat{a}=+(1-\lambda)x_B$ for the concentration column, and $\hat{a}=(1-\lambda_D)x_P$ for the depletion column ($b$ is the $y$-axis line segment intercepted by the operating line).

The optimum separation conditions (first of all, temperature, pressure, and loading in the column) depend on physicochemical properties of the operating system and characteristics of contactors. At the same time, a compromise between the column height and diameter (to be specific, between NTP and flows) is determined by the $\theta_{opt}$ value, which at $x \ll 1$ depends only on the separation degree $K$.

Considering the minimum column-specific volume $V_{SP}=V/P$ (column volume per unit of product) as an optimality criterion, C. Marchetty [26], for the concentration column at $x \ll 1$, obtained dependence of $\theta_{opt}$ on the column separation degree at $b = 0$, shown in Figure 1.7, and depicted below by eq. (1.89).

The optimization of columns with flow conversion by specific volume, however, is not strict, since the costs proportional to flow $L$ $(G)$ make a significant contribution to the general costs of isotope production unit $(C_{SP})$:

$$C_{SP} = \left(\frac{C_{CV}}{\tau}+C_{OV}\right)\frac{V}{P_I}+\left(\frac{C_{CL}}{\tau}+C_{OL}\right)\frac{L}{P_I}=C_V\frac{V}{P_I}+C_L\frac{L}{P_I},\tag{1.88}$$

where $C_{CV}$ is the costs per column unit volume (i.e. cost of the column with packing); $C_{OV}$ is the operating costs per column unit volume (i.e. energy costs of gas circulation); $C_{CF}$ is the capital costs per unit flow (i.e. flow conversion unit development costs); $C_{OF}$ is the operating costs per unit flow (i.e. power costs of electrochemical flow conversion, or thermal energy costs of thermal flow conversion); $\tau$ is the service life; $P_I$ is the productivity of pure (100%) isotope.

A. Rozen [10], at $x \ll 1$, with the extremum $\partial C_{SP}/\partial\theta = 0$, obtained the following dependence of $\theta_{opt}$ on $K$, and on $b$ characterizing the relation of costs proportional to the column volume and flow:

$$\ln = \frac{K-\theta_{opt}}{1-\theta_{opt}}=\frac{\theta_{opt}}{1-\theta_{opt}}\frac{K-1}{K-\theta_{opt}}-b,\tag{1.89}$$

where $b = \ln \alpha\,(C_L/C_V)$.

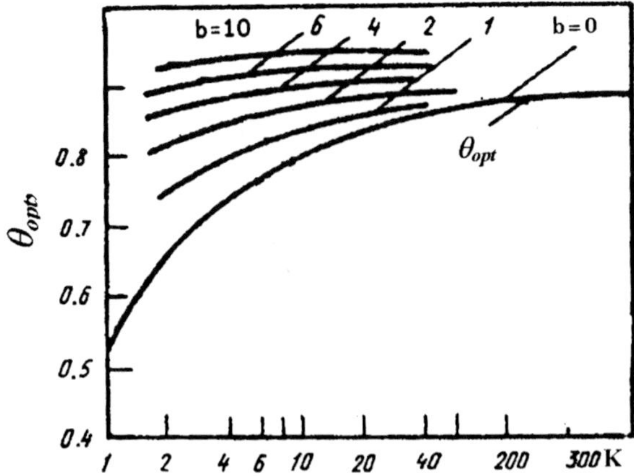

**Figure 1.7**    Optimum value of relative withdrawal degree at various separation degree values.

In Fig. 1.7 are also shown the dependences of $\theta_{opt}$ on $K$ for a variety of $b$ values, calculated by cq. (1.89). With a rise in costs proportional to the flow, the $\theta$ optimum value increases, bringing about an enhancement of the extraction degree $\Gamma$ and, at $B$ = const, a diminution of flows.

Over a wide range of concentrations, $\theta_{opt}$ depends not only on $K$, but also on the concentration $x_F$ since with an increase in concentration in source material, $\theta_{opt}$ decreases [27]. The presence of the depletion stage leads as well to a reduction in $\theta_{opt}$, which in this case depends both on $K$ and on $K_D$ [12, 28].


## 1.4    UNSTEADY STATE OF THE COLUMN AND CASCADES OF COLUMNS

The preceding section discussed the steady state of the column. This state is characterized by time-invariant concentrations in each column cross-section. However, it is not established immediately but over a more or less long period that is referred to as start-up period, or equilibrium (accumulation) time (period).

During the start-up period, the column generally functions without withdrawal (at $\lambda = 1$), and most equations for the equilibrium time $\tau$, known in the literature, apply to the 'open' scheme, with the following assumptions: isotope accumulation in gas (vapor) phase is disregarded, and the liquid amount on all TP (TP holdup, or delay) is taken to be uniform ($\Delta H_i = \Delta H$ = const).

The target isotope accumulation in a time unit is termed isotope transfer $j$ and is determined by the material balance equation:

$$j = Lx_F - G\tilde{y}_1 = L(x_F - \tilde{y}_1) = Lx_F - \tilde{x}_1/(\alpha - \varepsilon\tilde{x}_1),$$

where $\tilde{y}$ is the unsteady concentration of isotope in the gas (vapor) phase flow leaving the column.

The transfer is maximum ($j_0$) at the initial instant of time $\tau = 0$, when $\tilde{x}_1 = x_F$

$$j_0 = Lx_F\left[1 - \frac{1}{\alpha - \varepsilon x_F}\right] = L\varepsilon\frac{x_F(1 - x_F)}{\alpha - \varepsilon x_F}. \tag{1.90}$$

According to the above equation the maximum value $j_0$ corresponds to the equal content of both isotopes ($x_F = 1 - x_F$), and at their peak concentrations ($x_F$, $1 - x_F \to 0$), the $j_0$ value becomes infinitesimal. It is evident that the $j_0$ dependence on $x_F$ will be

$$j_0 = Lx_F(\alpha - 1)/\alpha. \tag{1.91}$$

As an illustration, Figure 1.8 presents the dependence of maximum isotope transfer per flow unit on the concentration $x_F$ at $\alpha = 1.006$. Hydrogen isotope separation by water vacuum rectification is generally done at this value of separation factor (see chapter 7).

On the attainment of the column steady state determined by the change to the withdrawal mode, the target isotope accumulation at $i$-th theoretical stage is $\Delta H(x_i - x_F)$. The aggregate accumulation, referred to as equilibrium accumulation, in the $M_E$ plant, can be expressed by the equation:

$$M_E = \Delta H\sum_{i=1}^{n}(x_i - x_F) + \Delta H_R(x_n - x_F), \tag{1.92}$$

where $\Delta H_R$ is the amount of liquid in the flow reflux system (the isotope composition of the liquid is taken to be identical to that in the lower TP).

If the transfer were constant and equal to the initial transfer $j_0$, then, at $x_F \ll 1$ the time $\tau_0$, termed the relaxation time, would equal [29]:

$$\tau_0 = \frac{\alpha}{\alpha - 1}\left\{\frac{\Delta H}{L}\left[\frac{\alpha(\alpha^n - 1)}{\alpha - 1} - n\right] + \frac{\Delta H_R}{L}(\alpha^n - 1)\right\}. \tag{1.93}$$

On the assumption that the degree of approximation to the steady state $\varphi$ is uniform in all TP at any instant of time, the authors of reference [29] obtained the equation relating time $\tau$ and the concentration in the column lower cross-section $\tilde{x}_{n,\tau} = x_B$

$$\tau = \frac{\alpha}{\alpha - 1}\left\{\frac{\Delta H}{L}\left[\frac{\alpha(\alpha^n - 1)}{\alpha - 1} - n\right] + \frac{\Delta H_R}{L}(\alpha_n - 1)\right\}\ln\frac{1}{1 - \varphi}, \tag{1.94}$$

where, $\varphi = \dfrac{\tilde{x}_{n,\tau} - x_F}{x_{n,m} - x_F} = \dfrac{K_\tau - 1}{K_m - 1}$, $K_m = \dfrac{x_{n,m}}{x_F} = \alpha^n$, $K_\tau = \dfrac{\tilde{x}_{n,\tau}}{x_F}.$

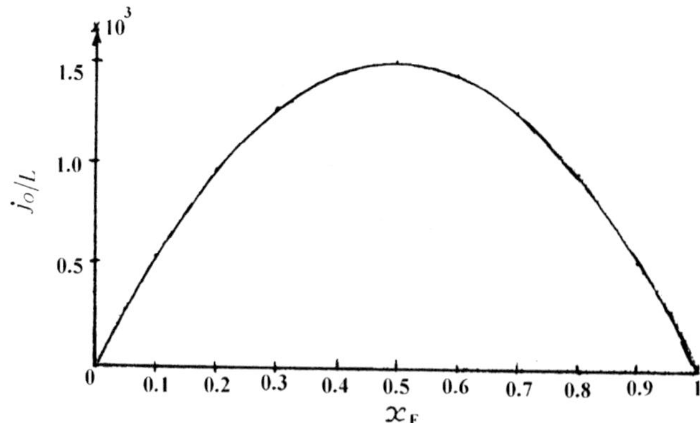

**Figure 1.8**   Dependence of maximum isotope transfer on the concentration $x_F$.

Similarly, A. Rozen [10] derived the following equation differing insignificantly from eq. (1.94):

$$\tau = \frac{\alpha}{\alpha-1}\left\{\left[\frac{\Delta H(\alpha^n - 1)}{\ln\alpha} - n\right] + \frac{\Delta H_R}{L}(\alpha^n - 1)\right\}\ln\frac{1}{1-\varphi}. \qquad (1.95)$$

The difference between the equations is due to the methods of equilibrium accumulation calculation. On deducing eq. (1.95) it was assumed that isotope concentration varied continuously through the column height, whereas eq. (1.94) was derived with discrete changes of concentration at theoretical stages of separation. This difference, however, becomes significant only at high values of $\alpha$.

From eqs. (1.94) and (1.95) it follows that

$$\tau = \tau_0 \ln\left[\frac{1}{(1-\varphi)}\right]. \qquad (1.96)$$

Thus, it is evident that at $\varphi = 1$, $\tau = \infty$, i.e. fundamentally, to completely attain the steady state in non-withdrawal mode, an infinite time is required. Figure 1.9 represents the nature of the time-dependence of $K_\tau$. Since $\theta = 1 - \varphi$ [29], then $\tau < \tau_0$, considering that, generally, $\theta_{opt} = 0.7 - 0.85$ (at $\tau = \tau_0, \varphi = 0.631$). Figure 1.9 shows as well the dependence

$$j = j_0 e^{-\frac{\tau}{\tau_0}} = j_0(1-\varphi), \qquad (1.97)$$

whence it follows that the transfer $j$ varies from maximum value $j_0$ (at $\tau = 0$) to zero (at $\tau = \infty$).

In estimating the accumulation time it is convenient to use the mean transfer value [30], which, at $x \ll 1$, equals:

$$j_M = Lx_F \frac{\alpha - 1}{\alpha} \frac{\varphi}{\ln\left[1/(1-\varphi)\right]} = j_0 \frac{\varphi}{\ln\left[1/(1-\varphi)\right]}. \tag{1.98}$$

From the above equation it follows that during the start-up period up to $\varphi = 1 - \theta_{\text{opt}}$, a decrease in the mean transfer compared with maximum transfer, does not exceed 10–15%.

To calculate $\tau$ over a wide range of concentrations, the following equation was suggested [31]:

$$
\begin{aligned}
\tau = \frac{1}{L} &\left\{ \frac{\varepsilon}{1+\varepsilon x_F} + \frac{\alpha}{\varepsilon(1-x_F)(1+\varepsilon x_F)^2}\left[(\alpha^n - 1)\left(\frac{1-\tilde{x}_{n\tau}}{\tilde{x}_{n\tau}-x_F}\right)+\varepsilon\right] \right. \\[2mm]
&\times \ln\left[1+\frac{1+\varepsilon\tilde{x}_F}{x_F(\alpha^n-1)\dfrac{1-\tilde{x}_{n\tau}}{\tilde{x}_{n\tau}-x_F}-1}\right] \\[2mm]
&\times \left[(\tilde{x}_{n\tau}-x_F)\left(\Delta H_R - \frac{n\Delta H}{\left(\dfrac{1-\tilde{x}_{n\tau}}{1-x_F}\right)\alpha^n-1}\right)\right. \\[2mm]
&\left.\left. + \frac{\Delta H(\alpha^n-1)}{\ln\alpha}\frac{1-\tilde{x}_{n\tau}}{\dfrac{1-\tilde{x}_{n\tau}}{1-x_F}\alpha^n-1}\ln\left(\frac{1-x_F}{1-\tilde{x}_{n\tau}}\right)\right]\right\}. 
\end{aligned}
\tag{1.99}
$$

The problem of determining $\tau$ for the non-withdrawal mode of the column with depletion stage was considered as well [10]. Thus it was accepted that isotope composition in the column cross-section with concentration $x_F$ is kept invariant. However, it was shown later that the assumption is not strict [32].

From the above equations it is evident that the time of the target concentration achievement increases with reducing $\varepsilon$ (in proportion to $\sim \varepsilon^2$), and with increasing holdup in the column and flow conversion system. Because of the distinct dependence of $\tau$ on NTP, and for the purpose of diminishing the separation plant size, separation column cascades with staged flow reduction and comparatively small NTP are used.

Such cascades allow for reducing both the separation degree in a single column, and the isotope holdup in the flow conversion system, which results in shortening the time of the target concentration achievement.

At $\varepsilon \ll 1$ it is practically impossible to obtain highly enriched product (from source material with low content of target isotope) without recourse to the cascade scheme.

To take an example, in the cited method of isotope separation by water vacuum rectification in a single column, owing to the small value of $\varepsilon \approx 0.006$ the time of achievement

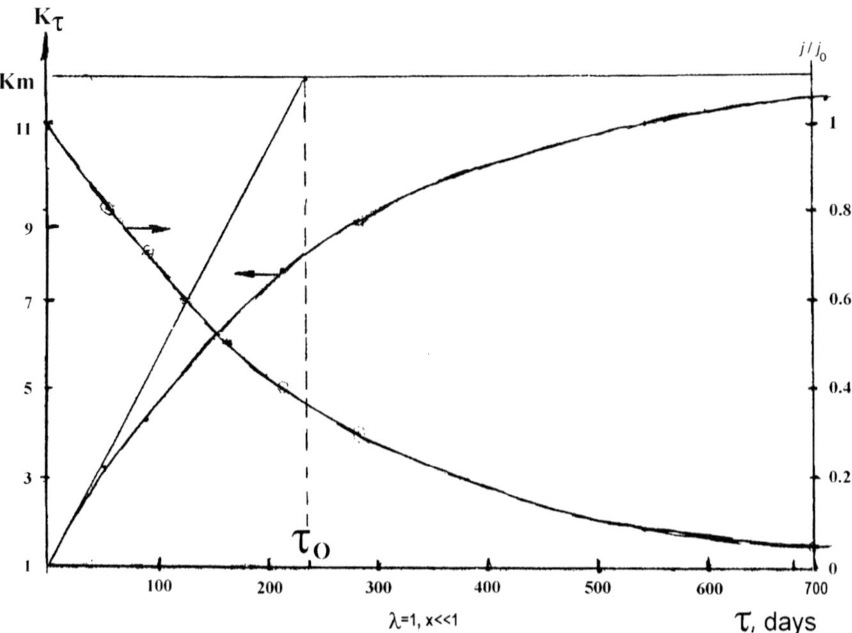

**Figure 1.9**   Time-dependence of the separation degree $K$ and relative transfer in the column at $\lambda = 1$, $x \ll 1$, $\alpha = 1.005$, $n = 500$, $L/\Delta H = 100$ h$^{-1}$, $L/\Delta H_K = 1$ h$^{-1}$.

of $^{18}$O concentration at a level of 95 at. % would have amounted to tens of years, exceeding the plant depreciation period.

Fig. 1.10 shows the scheme of a three-stage concentration column cascade. The cascade is fed with liquid at an extent of $F = L^I$ supplied to the stage I column as reflux. The liquid with a concentration $x^I_B$ leaving the column is divided into two parts, of which the smaller in the amount of $L^{II}$ is supplied to the stage II column as reflux, and the larger in the amount of $L^I - L^{II}$ is supplied to the bottom flow conversion unit. Gas (vapor) produced by the flow conversion unit, together with the stage II gas (vapor) flow containing the heavy component with concentration $y^{II}_F$, is supplied to the bottom section of the stage I column. Similarly, the liquid with the concentration $x^{II}_B$ in the amount of $L^{II}$ from the stage II column is fed to the stage III column as reflux, and the rest is supplied to the stage II bottom flow conversion unit. The product with a concentration $x^{III}_B = x_B$ is withdrawn from the last stage. The column of the cascade last stage operates with withdrawal of first kind, and the columns of remaining stages with withdrawal of second kind realized owing to the difference in target component amounts between flows drawn off from the columns and those returned from the succeeding stage.

The separation degree value at each stage of the cascade is generally chosen with regard to the minimization of the column's total volume. Since direct costs are determined by the initial stages, cascades with progressive stages, in which the separation degree rises with an increase in the stage number, are commonly utilized.

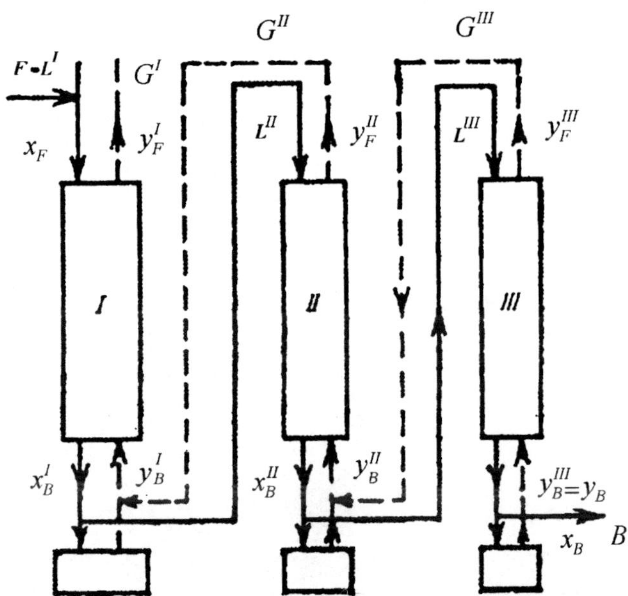



**Figure 1.10**   Scheme of three-stage concentration column cascade.

In the absence of losses the target isotope transfer for all cascade stages is uniform and equal to the withdrawal $Bx_B$. The cascade separation degree is determined by the stage I separation degree $\Gamma = Bx_B/L^{I}x_F$. With the extraction degree of stages taken as uniform, the flow reduction rate is $\sigma^{i} = L^{i}/L^{i+1} = K^{i}$.

An $xy$-diagram of the cascade functioning over a wide range of concentrations is shown in Figure 1.11. As is evident from the figure, operating lines of columns II and III are closer to the equilibrium curve than the operating line of one column equivalent to the cascade by separation degree and production rate. Because of this, the overall NTP of the cascade scheme is higher than that of a column without flow reduction.

But because the flow reduction in cascades, a substantial gain in the overall volume of the columns is obtained. When compared with the volume of the column without flow reduction, the gain amounts to 30–40% for the two-stage scheme, and to 40–50% for the three-stage scheme.

At the sequential start, the total start-up time $\tau_\Sigma$ of the cascade up to the commencement of withdrawal of product with a concentration $x_B$ is determined by the sum:

$$\tau_\Sigma = \tau_{I} + \tau^{I} + \tau_{II} + \tau^{II} + \tau_{III} + \dots . \tag{1.100}$$

Since in such start-up mode non-stationary processes in the cascade stages will carry in the same way as in individual columns, the time of non-withdrawal mode operation of the columns I, II, and III ($\tau_{I}$, $\tau_{II}$ and $\tau_{III}$) can be obtained using the above equations.

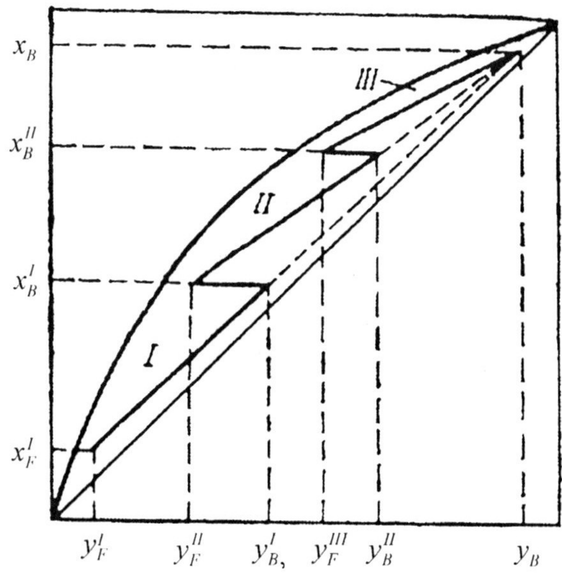



**Figure 1.11**   $xy$-Diagram of three-stage concentrating cascade of columns.

It is not difficult to find the stage II fill-up time from the equation:

$$(n\Delta H + \Delta H_R)^{II} x_B^{I} = L^{I} x_F \Gamma \tau^{I}, \qquad (1.101)$$

where indexes I and II are cascade stage numbers.

Similarly, the stage III fill-up time $\tau^{II}$ is found.

## 1.5   SEPARATION COLUMN CONTACTORS

### 1.5.1   Types and characteristics of packing

Packed contactors have found the widest application in isotope separation columns, whereas tray columns are currently used only for the large-scale heavy-water production. That is why the present section covers only packed contactors, and the use of tray contactors will be discussed further in the sections dedicated to specific methods of heavy-water production.

Packed columns differ generally in the utilized types of packing. The packing materials, in turn, varying in shape and size of its elements, are made of metal (plate, mesh, wire), silicate materials (ceramics, glass, quartz), or polymeric materials (fluoroplastic, caprone).

Geometric and mass characteristics of packing are expressed in the following quantities: the nominal size of particle (diameter or width, height and thickness), mass per unit volume (bulk weight) $\gamma$ (kg/m$^3$), void space $V_S$ (m$^3$/m$^3$), total geometric surface area of the packing elements per unit volume (specific surface) $a$ (m$^2$/m$^3$).

The bulk weight and void space are related by the equation:

$$V_S = 1 - \frac{\gamma}{\delta},$$

(1.102)

where $\delta$ is the packing material density.

In isotope separation both random (dump) and regular packing materials are used. Depending on particle size, the random packing materials can be subdivided into two groups: industrial packing with a particle size of 10mm and over (e.g. Raschig rings, Pall rings, Berl saddles, and Intalox saddles), and fine, or high-efficiency packing with a particle size of under 10mm extensively used in isotope separation and high-clean materials production.

In Russia high-efficiency spiral-prismatic packing (SPP or Levin's packing) is used, made of wire coil sections wound round a trihedral frame; and abroad Dixon rings, a woven stainless steel mesh material formed into a cylindrical shape (with a height equal to diameter). Data obtained in the D. Mendeleev University of Chemical Technology of Russia (MUCTR) is indicative of the high efficiency of the packing in the shape of rings with a vertical partition made of brass or steel mesh (metal mesh rings or MMR).

Table 1.1 presents the characteristics of fine packing. Among regular packings, it is Sulzer packing that has found the widest application in isotope separation. Sulzer packing is made of vertical stainless-steel gauze strips placed at a fixed distance in parallel with one another. The packing characteristics are shown in Table 1.2.

Structured packing is made of corrugated thin gauge metal or gauze coiled in the form of a double Archimedes spiral with a diameter approaching the column diameter. The packing is manufactured with oblique or straight-type corrugation. Table 1.3 presents the characteristics of mesh structured packing (with 45° corrugation angle) developed by MUCTR.

An important advantage of the structured packing is that, with a properly arranged reflux distribution, an increase in the column diameter does not result in a notable deterioration in performance.

In random packed columns, even with the use of efficient reflux distributors, HETP rises with an increase in the column diameter. That is why the diameter of columns with fine high-efficiency packing generally does not exceed 150mm.


### 1.5.2  Hydrodynamics of countercurrent gas (vapor)–liquid two-phase flows in the packing material layer

*Flow rate in the packing layer*

Gas or liquid moves at a varying velocity in the void space between packing material's particles through variable-section channels. So, in the packed column calculation,

**Table 1.1**

Characteristics of Fine High-Efficiency Packing Materials

| Packing type | Material | Particle size[a], mm | Void space $V_S$ (m³/m³) | Specific surface $a$, m²/m³ | Bulk weight $\gamma$, kg/m³ |
|---|---|---|---|---|---|
| Raschig rings | Porcelain | 5.2 × 5.4 × 1.3 | 0.64 | 896 | 1,060 |
| | Ceramic | 4.4 × 3.7 × 1.1 | 0.60 | 1,000 | 1,060 |
| | Quartz | 4.1 × 5.2 × 0.6 | 0.65 | 1,380 | 780 |
| | Glass | 4.1 × 4.1 × 0.5 | 0.71 | 1,300 | 710 |
| | | 6.2 × 6.2 × 0.6 | 0.77 | 850 | 560 |
| | PVC | 5 × 5 × 0.5 | 0.74 | 1,150 | 350 |
| | Fluoroplastic | 5.6 × 5.9 × 0.9 | 0.64 | 920 | 785 |
| | Nickel | 3 × 3 × 0.5 | 0.66 | 1,470 | |
| Single coils | Glass | 4.0 × 0.5 | 0.74 | 1,645 | 610 |
| | Quartz | 5.0 × 0.8 | 0.71 | 1,725 | 890 |
| | | 5.8 × 1.6 | 0.64 | 960 | 670 |
| | Steel | 2.9 × 0.4 | 0.78 | 2,130 | 1,670 |
| | | 4.0 × 0.5 | 0.82 | 1,440 | 1,370 |
| Double-coil spirals | Quartz | 5.6 × 4.4 × 1.5 | 0.72 | 760 | 500 |
| Spiral-prismatic packing | Nickel wire | 3 × 3 × 0.5 | 0.85 | 2,640 | |
| | Stainless steel wire | 2 × 2 × 0.2 | 0.82 | 3,500 | 1,400 |
| | | 3 × 3 × 0.2 | 0.86 | 2,580 | 1,114 |
| | | 4 × 4 × 0.2 | 0.89 | 2,140 | 855 |
| | Caprone | 2 × 2 × 0.22 | 0.81 | 2,700 | 220 |
| | | 3 × 3 × 0.22 | 0.85 | 2,440 | 185 |
| | | 4 × 4 × 0.22 | 0.89 | 1,610 | 122 |
| | | 5 × 5 × 0.22 | 0.90 | 1,450 | 110 |
| | Polypropylene | 4 × 4 × 0.3 | 0.83 | 2,140 | 148 |
| Gauze rings with mesh Partition (MMR) [b] | Brass gauze, 400 holes per cm² | 3.4 × 3.1 × 0.3 | 0.82 | 2,500[b] | 1,530 |
| | | 4.3 × 3.3 × 0.3 | 0.86 | 1,600 | 1,230 |
| | | 5.1 × 5.1 × 0.3 | 0.89 | 1,100 | 970 |
| | Brass gauze 100 holes per cm² | 5.2 × 5.2 × 0.3 | 0.93 | 1,050 | 600 |
| | Stainless steel gauze, 400 holes per cm² | 5.1 × 5.2 × 0.2 | 0.904 | 1,180[c] | 760 |
| | Stainless steel gauze, 03X18H9T, No. 014, 2500 holes per cm² | 5.0 × 5.0 × 0.1 | 0.965 | 1,260 | 280 |

[a] Diameter × height × wire diameter.
[b] MUCTR clone of Dixon rings.
[c] Element surface is defined as the ring and partition surfaces.

**Table 1.2**

Sulzer Packing Materials Characteristics

| Packing type | Material | Hydraulic diameter[a] $d_h$, mm | Void space $V_S$ (m³/m³) | Specific surface, $a$, m²/m³ | Bulk weight $\gamma$, kg/m³ | Corrugation angle, degrees |
|---|---|---|---|---|---|---|
| AX | Stainless steel wire mesh. Wire of 0.16mm in diameter | 15 | 0.95 | 250 | 125 | 30° |
| BX | Stainless steel wire mesh. Wire of 0.16mm in diameter | 7.5 | 0.90 | 500 | 250 | 30° |
| CY | Stainless steel wire mesh. Wire of 0.16mm in diameter | 5.0 | 0.85 | 700 | 350 | 45° |
| BX | Polymeric Mesh | 7.5 | 0.85 | 450 | 120 | 30° |

[a] Hydraulic diameter $d_h = 4/\alpha - 2\delta$, where $\delta$ is the wall thickness, mm.

**Table 1.3**

Gauze Structured Packing

| Material | Bulk weight $\gamma$, kg/m³ | Void space $V_S$, m³/m³ | Specific surface $a$, m²/m³ | Equivalent channel diameter $D_E = 4V_S/a$, mm |
|---|---|---|---|---|
| Screen cloth with brass-wire warp, $\varnothing$ 0.3mm, twisted weft 0.01 ×8.96 holes per cm² | 700 | 0.92 | 800 | 4.5 |
| Brass-wire screen cloth $\varnothing$ 0.1mm, 900 holes per cm² | 290 | 0.96 | 980 | 3.9 |
| Stainless steel wire screen cloth $\varnothing$ 0.1 mm, 1200 holes per cm² | 360 | 0.97 | 1,220 | 3.2 |

a conventional quantity of the gas (vapor) velocity related to the total column cross-section $w_0 = G/S$ (where $G$ is the gas or vapour flow, m³s), and average velocity in the packing material $w_y$, is used. Void (pore) space in the column with a cross-sectional surface $S$ and a packing bed height $H$ equals

$$V_{\mathrm{P}} = SHV_{\mathrm{S}}, \qquad (1.103)$$

where $V_{\mathrm{S}}$ is the column void space.
The mean pore cross-section then will be

$$S_{\mathrm{P}} = V_{\mathrm{P}}/H = SV_{\mathrm{S}}. \qquad (1.104)$$

The gas (vapor) average velocity is

$$w_y = G/S_{\mathrm{P}} = G/SV_{\mathrm{S}} = w_0/V_{\mathrm{S}}. \qquad (1.105)$$

The equivalent pore diameter [9] is equal to:

$$d_{\mathrm{E}} = 4S_{\mathrm{P}}/P = 4SV_{\mathrm{S}}/Sa = 4V_{\mathrm{S}}/a, \qquad (1.106)$$

where $P$ is the overall perimeter of all pores ($P = Sa$), and $a$ is the packing geometric specific surface, m²/m³ (assuming that the entire packing surface is irrigated with liquid).
In the calculation of the gas (vapor) Reynolds number, $Re_{\mathrm{Y}} = w_{\mathrm{Y}}l\rho_{\mathrm{Y}}/\mu_{\mathrm{Y}}$ (where $\rho_{\mathrm{Y}}$ is density and $\mu_{\mathrm{Y}}$ is dynamic viscosity), the equivalent diameter is taken as linear dimension $l$. Expressing $w_{\mathrm{Y}}$ in terms of eq. (1.105) we obtain:

$$Re_y = \frac{4G_{SP}}{a\mu_y}, \qquad (1.107)$$

where $G_{\mathrm{SP}}$ is specific flow ($G_{\mathrm{SP}} = G/S$); $G$, kg/s.
If the liquid flow section is expressed as the product of wetted perimeter $P$ by mean liquid film thickness $\delta$, then the average liquid flow velocity across the bed will be:

$$w_x = \frac{L}{P\delta\rho_x},$$

where $L$ is liquid mass flow, and $\rho_{\mathrm{X}}$ the liquid density.
With regard to the fact that $d_{\mathrm{E}} = 4\delta$ and $P = Sa$, the liquid-phase Reynolds number will equal:

$$Re_x = \frac{4L_{SP}}{a\mu_x}, \qquad (1.108)$$

where $L_{\mathrm{SP}}$ is liquid specific flow ($L_{\mathrm{SP}} = L/S$).

*Load limit (throughput rate)*

The counter-current movement of two phases in the packed column with a film-type inter-phase contact is possible only within a certain load range. As the load (flow rate) varies, the hydrodynamic state in the counter-current two-phase column changes, which is externally manifested in the nature of the dependence of the irrigated packing hydraulic resistance on the flow rate. At low gas (vapor) and liquid throughputs, the interactions between the phases is moderate, the curve of the pressure drop dependence on the load is similar to, but higher than, an analogous curve for non-irrigated packing. This state changes into a state characterized by the gas flow retarding liquid, which leads to a decrease in void space and a faster rise in hydraulic resistance. The transfer from the first to the second state is referred to as the point of column flooding. The liquid accumulates on the packing bed until the whole void space is overflowed with liquid, and then the flooding bed state, termed also phase inversion state, sets in. Gas (vapor) in this state ceases to be a uniform phase and begins bubbling through the liquid flooding the packing bed, and the hydraulic resistance rises all the steeper. With further increase in load, the flooding bed state passes into inverted movement of the liquid phase which is entrained by the flowing gas (vapor), i.e. the countercurrent flow is upset.

The load (linear or mass velocity of gas and liquid) in which a minor increase results in the onset of bed flooding is referred to as load limit, or packing throughput capacity.

According to several authors, the conditional gas (vapor) velocity value representing the inversion point $w_o^*$ can be obtained from

$$\lg\left[\frac{(w_o^*)^2 \cdot a \cdot \rho_y}{g \cdot V_S^3 \cdot \rho_x}\right] = A - 1.75\left(\frac{\rho_y}{\rho_x}\right)^{1/8}\left(\frac{L}{G}\right)^{1/4}. \tag{1.109}$$

The mass load limit $G_{SP}^* = 3600 w_o^* \cdot \rho_y$, where $G_{SP}^*$, kg/(m².h).

For industrial ring packing materials in gas–liquid systems the coefficient $A = 0.022$, in vapour–liquid systems $A = -0.125$ [9, 15], and for fine packing materials $A = -0.4$ [15]. At a gas–liquid flow ratio close to one, the load limit for fine packing materials and for regular structured packing materials can be derived from a simpler equation taking into account the influence of packing element size, and gas and liquid properties [12]:

$$L_{SP}^* = A + B\left(d_E \cdot \rho_y \cdot \rho_x\right)^{0.5}, \quad \text{kg/(m}^2 \cdot \text{h)}. \tag{1.110}$$

Values of factors (or coefficients) $A$ and $B$ in the equation (1.110) are presented in Table 1.4.

Experiments on liquefied gas rectification at a pressure of up to 1 MPa with a structured packing made of a denser mesh ($V_S = 0.92$, unit volume mass 700 kg/m³, and $d_E = 4$ mm) gave a similar load limit value $B = 4,800$.

As follows from the above equations, when gas or vapor phase approximates an ideal gas, the mass throughput capacity of the packed column varies in proportion with pressure to the power of about 0.5.

*Holdup*

An important characteristic of counter-current columns is the packing liquid holdup $\Delta H$ (holding power, retention), while the gas holdup is generally disregarded. The holdup affects the hydraulic resistance and the target concentration achievement time (see above).

The holdup is the liquid amount on the packing bed falling at the packing unit volume, and most commonly expressed in $m^3/m^3$. In the packed columns, the total holdup is formed with static holdup $\Delta H_S$ and dynamic holdup $\Delta H_D$:

$$\Delta H = \Delta H_S + \Delta H_D. \tag{1.111}$$

By the holdup is meant a liquid amount held on the packing by capillary forces. The holdup dynamic component results from the liquid amount moving along the packing with gas (vapor) flow. Dynamic holdup and is determined by hydrodynamic conditions and does not depend on the packing material, while static holdup depends on the packing shape and material, and on the reflux liquid properties. In general, the liquid holdup amount is obtained experimentally. Experimental data on the fine packing holdup (in $m^3/m^3$) are adequately described [15] by

$$\Delta H = 1.83 \cdot Ga_x^{-0.3}\left(\frac{We}{Fr}\right)^{-0.14}\left(\frac{d_i}{d_e}\right)^{-0.5} \cdot Re_x^{0.35}, \tag{1.112}$$

where $d_i$ и $d_e$ are the packing element's internal and external diameter, respectively, Galilei criterion $Ga_x = d^3_x \cdot \rho^2_x/\mu^2_x$; $We/Fr = \rho_x d^2_E/\delta$.

To calculate holdup of fine spiral-prismatic packing materials, the following equation [15] is also advised:

$$\Delta H = K_l Re_x^{1/3}\left(\frac{a \cdot \mu_x}{\rho_x}\right)^{2/3}. \tag{1.113}$$

**Table 1.4**

Values of Coefficients $A$ and $B$ from eq. (1.110)

| Packing type | $A$ | $B$ |
|---|---|---|
| Spiral-prismatic (SPP) | 0 | 1,900 |
| Mesh rings (MMR) | 700 | 2,400 |
| Quartz spirals or rings | 700 | 1,400 |
| Fluoroplastic rings | 500 | 950 |
| Mesh structured packing (MSP) | 0 | 5,300 |

In rectification over a temperature range of 270–370K, coefficient $K_1 = 5.0$, in rectification of liquefied gases (80–200K), $K_1 = 5.3$.

Eq. (1.113) adequately describes as well gauze ring packing material at $K_1 = 10.0$ and structured packing material at $K_1 = 4.6\rho_x^{0.3}$ (in rectification over a wide pressure range).

The holdup of the above packing materials within recommended load range is most commonly 0.08–0.12m³/m³.


*Hydraulic resistance*

Contactors' hydraulic resistance determines the gas (vapor) flow pressure drop, and, consequently, the pressure variation in the column. In rectification, the pressure changes imply temperature variation, and the lower absolute pressure, the wider the temperature variation. The variations in parameters $P$ and $T$, in their turn, affect equilibrium and kinetics of separation processes, as well as energy consumption.

From the aforesaid it is evident that taking into account hydraulic resistance is essential in deciding on a contactor type, and in column design calculation.

Hydraulic resistance of dry (non-irrigated) packing to the gas (vapor) flow, as a case of granular bed resistance, can be derived from

$$\left(\frac{\Delta p}{H}\right)_y = \zeta \frac{w_0^2 \cdot a \cdot \rho_y}{8 \cdot V_S^3}, \tag{1.114}$$

where $(\Delta P/H)$ is the resistance of a packing bed with a thickness of 1 m (Pa/m).

According to experimental data [33], for industrial ring packing, resistance coefficient $\zeta = 140/Re_y$ at $Re_y < 40$, and $\zeta = 16/Re_y^{0.2}$ at $Re_y > 40$. For fine packing materials within the range $Re_y = 10 - 200$, it is more advisable to take the resistance coefficient $\zeta = 80/Re_y^{0.64}$ [34].

Hydraulic resistance of fine and regular packing materials in vapor–liquid counter-current flow can be obtained from the equation accounting for the void space decrease [12, 15]:

$$\left(\frac{\Delta p}{H}\right)_{xy} = \zeta \frac{w_o^2 a \cdot \rho_y}{8(V_S - \Delta H)^3}. \tag{1.115}$$

Experiment shows that the irrigated packing resistance coefficient is generally expressed by an equation of the type:

$$\zeta = \frac{A}{Re^B}. \tag{1.116}$$

Values of factors (or coefficients) $A$ and $B$ of the equation (1.116) are presented in Table 1.5.

*Real flows pattern and scale factor*

Mass-transfer process calculation is generally based on the assumption that gas (vapor) and liquid concentrations remain constant across the column and vary only along the column height. This assumption conforms to an idealized piston model, under which all particles of each phase flow move without mixing in parallel at an equal velocity, with gas and liquid flows evenly distributed over the apparatus cross section. In this case, all particles of a phase flow reside in the apparatus for a uniform time. Another idealized model of flows pattern assumes complete intermixing in a single phase or in both phases, when a phase concentration remains constant throughout the volume.

For a variety of reasons, the real movement of gas (vapor) and liquid flows in mass-transfer apparatus mostly differs from these idealized models. Among the reasons are the velocity variability and non-uniform flow distribution across the apparatus, including the packing bed; complexion of the particles trajectories; stagnation areas and partial flow bypass.

Alternatively, owing to a concentration gradient, component transfer by turbulent and molecular diffusion is present along the apparatus axis in a sense opposite to the separation direction (longitudinal diffusion).

All these phenomena, conventionally referred to as axial mixing, lead to variations in the concentration field and to a decrease in motive force compared with the idealized piston model.

To quantitatively evaluate the axial mixing, various simplified models are used. According to a diffusion model, it is assumed that the axial mixing can be described by the diffusion equation where the diffusion coefficient is replaced by the axial mixing coefficient $E_x$ with the same dimensionality (m$^2$/s).

The axial mixing in gas (vapor) phase can usually be neglected since it is the liquid phase flow variation from the perfect displacement model that has the strongest influence on the mass transfer. In a dimensionless form, the axial mixing is characterized by Pecle's criterion, $Pe_x = w_x H/E_x$, where $w_x$ is the liquid flow average velocity, $H$ is the apparatus length (packing bed height). In the idealized piston model ($E_x = 0$), Pecle's criterion tends to infinity, and at complete intermixing $Pe$ approaches zero.

**Table 1.5**

Values of Coefficients $A$ and $B$ from eq. (1.116)

| Packing type | Pressure | Coefficient $\zeta$ from eq. (1.116) | |
|---|---|---|---|
| | | $A$ | $B$ |
| Spiral-prismatic | 80–1,000 hPa | 210 | 0.77 |
| | 3–65 hPa | 175 | 0.5 |
| Mesh rings | 3–65 hPa | 90 | 0.5 |
| | 65–1,500 hPa | 12 | 0.2 |
| | 0.1–1.0 MPa | 16 | 0.2 |
| Mesh structured | 100–400 hPa | 100 | 0.86 |
| | 0.1–1.0 MPa | 50 | 0.4 |

The variations of the column flows' real pattern from the idealized piston model, which bring about a change in concentration field and a decrease in motive force, are more simply taken into account in calculations by a corresponding increase in HTU or HETP; that is, by retaining the above methods and equations for NTP (NTU) and/but introducing supplementary summands into the additivity eqs. (1.36) and (1.37), thus accordingly increasing HTU or HETP. The axial mixing contribution is allowed for by a supplementary summand:

$$h_{LM} = \frac{E_x}{w_x} + \frac{E_y}{w_y}.$$ (1.117)

To correct for an increase in HTU or HETP on transition to packing layers of larger cross-sections compared with the model discussed above, a scale factor (SF) is introduced in the following form:

$$SF = 1 + \frac{h_{LM}}{h_{\text{model}}}.$$ (1.118)

The quantity $h_{LM}$ allows for the axial mixing effective contribution and cannot be defined from reduced-scale experiments.

For a diffusion model:

$$h_{LM} = \left(\frac{E_{LM}}{w}\right)_x + \left(\frac{E_{LM}}{w}\right)_y,$$ (1.119)

where $E_{LM}$ is a constituent of the effective coefficient of axial mixing in a corresponding environment.

The deviation from the idealized piston model in the vapor phase, that is the quantity $E_y/w_y$, is normally ignored as small compared with $E_x/w_x$.

The available experimental data indicate that even in packed columns with a diameter of up to 150mm, an increase in diameter leads to a decline in separation efficiency, that is to a rise in HTU.

To lower HTU, it is advisable to augment the load, i.e. flow rate, to improve the reflux distribution system, to laterally and longitudinally sectionalize the packing layer. Regular packings are characterized by a lower scale factor.

No sufficiently reliable equations exist to calculate flow scale factor. As a result of experimental data generalization [35], the following equation was suggested:

$$SF = 1 + 2\lg\frac{D_I}{D_0} \times \frac{1-\eta}{\eta},$$ (1.120)

where $D_I/D_0$ is the ratio of industrial and model column diameters, $\eta$ is the quantity characterizing the process order degree. It has been suggested that for irregular dump packing

$\eta = 0.6$–$0.8$ (complex packing with multipoint reflux), and $\eta = 0.4$–$0.6$ (with inadequately uniform reflux), and for regular packing $\eta = 0.8$–$1.0$ [34].

For columns with spiral-prismatic and fine ring packing, it may be provisionally thought that, with an increase in the column diameter from 30mm to 150mm, HTU rises in proportion to the diameters ratio to the 0.8 power with single-point reflux distribution, and to the 0.2 power with multipoint distribution. Data on fine-packed columns with a diameter of over 150mm are not yet available.

*Phase contact surface*

Since the mass-transfer rate is directly proportional to the phase contact surface, it is of fundamental importance to know it. In packed columns, the active surface of phase contact $a_C$ (m$^2$/m$^3$) participating in mass transfer differs generally from the packing geometric surface $a$. In the general case:

$$a = a_C + a_1 + a_2, \qquad\qquad (1.121)$$

where $a_1$ is the specific surface part not irrigated by liquid but washed by gas (vapor), $a_2$ is the surface part wetted by liquid but not washed by gas (vapor).

The ratio $a_C/a = \psi_C$ characterizes the packing geometric surface efficiency. In normal counter-current column operation $\psi_C < 1$ (for regular packing, $\psi_C$ approaches one), in bubbling mode with flooded packing it is possible that $\psi_C = 1$. It is a matter of difficulty to reliably define the quantity $\psi_C$ since it is dependent on many diverse factors.

According to several authors, in industrial packing materials the surface efficiency $\psi_C$ increases with a rise in specific load (reflux density), but even at high densities the surface efficiency does not exceed 0.6–0.7.

At the flow rates that are of practical interest, for small packing materials characterized by capillarity, the degree of surface efficiency $\psi_C$ is independent almost of the liquid flow rate and equals 0.8–0.9 [15].

# REFERENCES


1. A. I. Brodskii *Chemistry of Isotopes,* Izd. Akad. Nauk SSSR, Moscow, **1957**, 594.
2. G. G. Philippov, K. N. Sakodynskii, Ya. D. Zelvenskii *Khim. Prom.*, **1965**, 1, 10.
3. B. M. Andreev, S. G. Katalnikov *Khim. Prom.*, **1965**, 4, 28.
4. V. E. Kotchurikhin, Ya.D. Zelvenskii *Zh. Fiz. Khim.*, **1964**, 38, 2594.
5. S. V. Roginskii *Theoretical Fundamentals of Isotopic Methods of Chemical Reactions Research,* Izd. Akad. Nauk SSSR, **1957**, 611.
6. Yu. A. Sakharovskii, In: *Proc. Mendeleev Univ. Chem. Technol.* Russia, **1983**, 89.
7. B. M. Andreev, E. P. Magomedbekov, M. B. Rozenkevich, Yu. A. Sakharovskii *Heterogenous Reactions of Tritium Isotope Exchange*, editorial URSS, **1999**, 208.
8. A. N. Murin, V. D. Nefedov, V. P. Shvedov *Radiochemistry an Chemistry of Nuclear Processes,* Izd. Khim. Lit., **1960**, 784.
9. V. V. Kafarov *Mass Transfer Fundamentals,* Vyshaya Shkola, **1972**, 496.
10. A. M. Rozen *Theory of Isotope Separation in Columns,* Atomizdat, Moscow **1960**, 436.

11. B. M. Andreev, Ya. D. Zelvenskii, S. G. Katalnikov, V. V. Uborskii *Isotopenpraxis*, **1977**, 13, 440.
12. B. M. Andreev, Ya. D. Zelvenskii, S. G. Katalnikov *Heavy Isotopes of Hydrogen in Nuclear Technology*, Energoatomizdat, **1987**, 456.
13. B. M. Andreev, E. P. Magomedbekov, G. H. Sicking *Interaction of Hydrogen Isotopes with Transition Metals and Intermetallic Compounds*, Springer Verlag, Heidelberg, **1996**, 168.
14. V. M. Ramm *Gas Adsorption*, Khimiya, **1966**, 768.
15. Ya. D. Zelvenskii, A. A. Titov, V. A. Shalygin *Rectification of Diluted Solutions*, Khimiya, **1974**, 216.
16. P. G. Romankov, N. B. Rashkovskaya, V. F. Frolov *Mass Transfer Processes in Chemical Technology*, Khimiya, **1975**, 334.
17. B. M. Andreev, A. S. Polevoi *TOKhT*, **1995**, 29, 261.
18. B. M. Andreev, A. S. Polevoi *TOKhT*, **1995**, 29 373.
19. A. M. Rozen (ed) *Scaling in Chemical Technology*, Khimiya, **1980**, 320.
20. H. K. Rae In: *Separation of Hydrogen Isotopes* (ed. R. F. Gould), In: Am. Chem. Soc. Symp. Ser. Washington, **1978**, 68.
21. Yu. A. Sakharovskii *Utilization of Catalytic Reactions of Hydrogen Isotope Exchange for Hydrogen Isotope Separation* – RCTU im. D. I. Mendeleeva, **1983**, 84.
22. A. V. Kaminskii, B. Sh. Dzhandzhgava, Yu. K. Ledyakov *Isotopenpraxis*, **1984**, 20, 183.
23. N. M. Zhavoronkov, Ya. D. Zelvenskii In: *Processes and Apparatus of Chemical Technology* (ed. A. G. Kasatkin), Goskhimisdat, **1953**, 7.
24. K. Cohen In: *Theory of Isotope Separation as Applied to Large Scale Production of $^{235}U$*, (ed. George M. Murphy) McGraw-Hill Book Co., New York, **1951**.
25. V. L. Pebalk *Zh. VKhO*, **1961**, 6, 589.
26. E. Cerrai, C. Marchetty, R. Renzomi, *Chem. Engng. Progr.*, **1954**, 50, 271.
27. S. G. Katalnikov, A. V. Khoroshilov, A. S. Sobolev, A. P. Timashev *Isotopenpraxis*, **1982**, 18, 208.
28. B. M. Andreev, V. V. Uborskii *TOKhT*, **1981**, 15, 664.
29. S. I. Babkov, N. M. Zhavoronkov *Rep. USSR Acad. Sci.*, **1956**, 106, 877.
30. B. M. Andreev, Ya. D. Zelvenskii, S. G. Katalnikov *Separation of Isotopes by Physical–Chemical Methods*, Energoatomizdat, **1982**, 208.
31. S. I. Babkov, N. M. Zavoronkov, *Kernenenergie*, **1962**, 5, 219.
32. O. A. Fedorenko, I. A. Alekseev *V. D. TOKhT*, 1997.
33. N. M. Zhavoronkov *Hydraulic Fundamentals of Scrubber Process* **1944**, Goskhimizdat, 223.
34. V. A. Kaminskii *et al. At. Energ.*, **1971**, 30, 48.
35. E. D. Vertuzaev *Khim. Prom.*, 1982, 8, 458.

# – 2 –

# Hydrogen Isotope Separation by Rectification

## 2.1 D$_2$O PRODUCTION BY WATER RECTIFICATION

Water rectification is the simplest method of heavy water production. This method offers a variety of apparent advantages such as limitless raw material resources and the possibility to carry out the process on the simplest non-depletion principle, absence of corrosion, toxicity, inflammability and explosion hazards, freedom from chemicals, possibility of using natural and waste low-temperature heat sources, and simplicity of utilized apparatus.

With the assumption that the appropriate ideal laws are adequately obeyed by mixtures of water species in liquid and vapour phases, and water homomolecular isotope exchange reac-tion (HMEX) proceeds sufficiently fast with equilibrium constant $K_{HDO} = [HDO]^2/[H_2O][D_2O] = 4$ and equilibrium pressure of deuterium–protium water (HDO) vapour $P^0_{HDO} = \sqrt{P^0_{H_2O} P^0_{D_2O}}$ (in conformity with a well-known geometric mean rule), then hydrogen isotope separation factor in water rectification is $\alpha = \sqrt{P^0_{H_2O}/P^0_{D_2O}}$, which agrees well with experimental data.

The values of the separation factor for hydrogen isotopes (deuterium and tritium) in the liquid–vapor system, that is in water rectification over the most important temperature range are represented in Table 2.1 [1].

The ratio of light and heavy water vapor pressures within a temperature range from 4 to 114°C can be calculated by the Jones equation [2] generalizing experimental data.

$$\ln\left(\frac{P^0_{H_2O}}{P^0_{D_2O}}\right) = -\frac{70.87 \pm 0.43}{T} + \frac{33,630 \pm 160}{T^2} \tag{2.1}$$

### Table 2.1

Hydrogen Isotope Separation Factor $\alpha_{HD}$ in Water Rectification

| $T$, K | 303.16 | 313.16 | 323.16 | 333.16 | 343.16 | 353.16 | 363.16 | 373.16 | 383.16 |
|---|---|---|---|---|---|---|---|---|---|
| $\alpha$ | 1.069 | 1.061 | 1.053 | 1.047 | 1.041 | 1.035 | 1.030 | 1.026 | 1.022 |

According to W. Jones, the normal boiling point of heavy water is 101.42°C, and the evaporation heat is 2–3 % higher than that of natural water.

The rectification optimum conditions (pressure and temperature) should be selected with regard to their influence both on the separation factor, and on the reaction kinetics and hydrodynamics, which requires taking into account the adopted type of column contactors. Being lower than atmospheric pressure, the optimum water rectification pressure for columns with irregular (dump) packing is in the range 220–250kPa, and is somewhat lower for columns with regular packing. It seems unreasonable to use tray columns with their high hydraulic resistance and holdup for water vacuum rectification.

A practical implementation of the water rectification method for industrial production of $D_2O$ began in 1943–1945 in the U.S.A., where three plants were equipped with similar water rectification systems with overall monthly potential output of 2,400kg [3].

In consequence of several poor engineering solutions (specifically, tray bubble-cap columns with a diameter of up to 4.5m were adopted for the first stages), design limitations, and installation faults, the actual average output was found to be approximately half as high as the potential one, and the heating steam consumption amounted to 275–316tons per kilogram of heavy water.

Owing to unsatisfactory performance characteristics, the US plants ceased to operate at the end of 1945.

The estimates show that even in optimum conditions, when using water rectification technique at the initial concentration (IC) stage, the natural water flow to be processed amounts to 170–180tons per kilogram of $D_2O$. Thus, at least 180tons of heating water vapour is required to heat the rectification column vaporizers, which is unacceptable even with regard to the possibility of utilizing low-grade vapor.

Developed in the U.K. in 1950–1955, a project of heavy water production by water rectification in Spraypack regular-packing columns based on the use of natural geothermal steam in Wairakei (New Zealand) has never been implemented. For the second time, in 1972 Japanese companies addressed to the New Zealand government and put forward a suggestion to produce heavy water with the use of natural steam. There is no evidence about the project realization. In Russia similar geothermal fields can be found in Kamchatka.

Both Russian and foreign experts thoroughly considered possible ways of power saving in the rectification process, including vapour recompression rectification (heat pump) and multiple-unit rectification. Although such energy-saving schemes allow for reducing power consumption multiple times, the use of water rectification for heavy water production is not yet competitive with other well-proven initial concentration techniques.

Of particular interest are methods of water rectification facilities combined with steam-turbine plants of thermal and nuclear power stations. The idea of such a combination consists in installing rectification columns between the steam turbine and its condenser; that is, the turbine switches to the counter pressure mode and dead (waste) steam is used for the rectification column heating, and dump (deuterium-depleted) steam from the column head is fed to the condenser acting as a reflux unit where a pressure is maintained optimum for rectification processes. In this approach, the rectification power consumption is determined by reduced power generation associated with the turbine switching to the operation mode with a minor counter pressure (0.2–0.3MPa). According to American scientists

[4, 5], a water rectification plant based on a thermal power station with an output of 1GW is capable of producing 150tons of heavy water per year, and that based on a nuclear power station of the same output 225tons per year. There is no information on the practical implementation of the above scheme.

As is evident from the above brief overview, water rectification at the heavy water IC stage, as opposed to other well-proven initial concentration methods, is not justified from the standpoint of power consumption.

A different situation arises with water rectification at the heavy water final concentration stage (FC), when processed flows are several hundred times smaller, which is why energy consumption is not of decisive importance here. Nowadays, the vacuum water rectification is the main EC technique used by plants (including the most powerful Canadian plants) carrying out the initial concentration with the use of two-temperature hydrosulfuric method (see section 3.2).

At the Savannah River site (U.S.A.), the final concentration (from 10%) was performed in a five-stage cascade of bubble-cup tray columns with a height of 24.4m each, and with an absolute pressure of 135hPa in each column head. At a performance of 25kg per hour of the end product containing 99.75% of heavy water, the column diameters were 2.28, 1.73, 1.52, and 1.22 meters, and the vapour consumption for rectification about 500kg per kilogram of product [6].

It is more advisable to carry out final concentration in a system of packed rectification columns. Heavy water of the required final concentration was obtained at a Soviet chemical plant with the use of Rashig steel rings with dimensions of $10 \times 10$mm and $15$mm $\times 15$mm as a packing material in a three-stage cascade of columns with a height of 33m each. The use of columns with effective packing is the most efficient FC method which allows to achieve a required NTP (250–300) in a single column with a relatively small height. This makes it possible to considerably simplify the production concept, diminish the product loss and reduce start-up time (owing to the exclusion of intermediate reservoirs).

Tubular columns packed with Dixon rings made of metal gauze were suggested by the Sulzer Corporation for final concentration of heavy water. The column with a potential output of 30tons of heavy water per year features a 2m diameter, incorporates 112 rectification tubes with a diameter of 100mm filled with packing elements of 2–3mm in diameter, and provides a concentration strengthening from 10% to 99.8% D$_2$O[7] with the packing bed of 12m in height. At a specific flow rate of 900kg/(h·m$^2$), HETP amounts to 2.5–3cm.

Later the Sulzer Corporation developed CY-type effective regular packing for water rectification. In water rectification, HETP equals 10cm; pressure drop at one TP is $\Delta P/H = 60$–70Pa [8]. The packing advantage is the HETP independence from the column diameter in the conditions of efficient reflux distribution (up to 300 reflux points per square meter of cross-sectional surface are recommended) which makes it possible to construct high-performance columns completely filled with packing material. Sulzer columns are installed at the most powerful Canadian heavy-water plants, as well as at nuclear sites in many countries.

The vacuum water rectification in packed columns is as well successfully used in many countries for the diluted heavy water processing into standard product.

Rectification columns with spiral-prismatic packing materials made both of stainless steel and of caprone have been used for years for substandard heavy-water processing at several Russian plants.

In India, the Heavy Water Board of Bhabha Atomic Research Center (Trombay)[9] has developed and, at present supplies, regular packed rectification columns for final concentration and substandard heavy water processing. Cylindrical packing elements are assembled from parallel corrugated strips of deoxidized bronze gauze. The corrugations are inclined to the column axis, with adjacent strips inclined in opposite direction. To improve wet ability, the packing is pretreated to coat the material with a cupric oxide layer. The packing specific surface is $800 m^2/m^3$ with a void space of 0.9 to 0.95. The packing elements are superimposed on one another. Liquid collectors and reflux distributors are placed every 3m throughout the packing bed height (Figure. 2.1).

The packing layer of a required height is composed of such sections and compressed between upper and lower supporting rings. H. Sudhukhan *et al.*[9] have presented the results of the packing tests in water rectification columns with diameters of 100–1,500mm

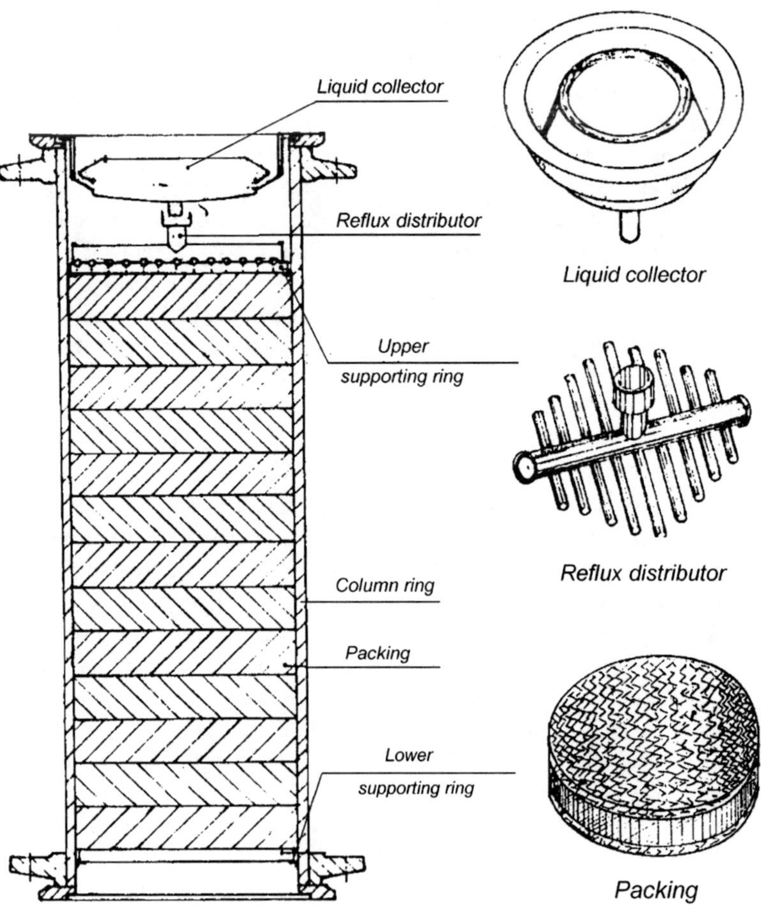


**Figure 2.1**    Rectification column with regular gauze packing and reflux distributor (Atomic Research Center, India).

at a specific flow rate of 1,500kg/(h·m²) to 2,000kg/(h·m²). At 1,600kg/(h·m²), NTP ranged from 16 to 18 per meter with pressure drop equal to 5.3–5.6hPa per meter.

Such a load range seems to be too low for regular packing (unfortunately, there is no information on the packing's limit load [9]). India exploits 11 such rectification columns for reprocessing of discharge heavy water, and eight columns were prepared for sale [9].

## 2.2   HEAVY WATER PRODUCTION BY AMMONIA RECTIFICATION

Heavy water can be obtained not only by water rectification, but also by rectification of other hydrogen compounds, of which ammonia is of chief interest. The virtues of ammonia as a working substance for deuterium extraction are: (1) a higher hydrogen molecular content, and, consequently, higher deuterium content, which is 1.59 times as high as that of water; (2) ammonia evaporation heat is almost two times lower than that of water; and (3) ammonia is produced on a large scale, with individual plants of high production capacity (a million tons per year and over).

One ton of ammonia contains a deuterium amount equivalent to 0.25kg of $D_2O$. Hence it follows that with an extraction degree of 80%, 1 million tons of $NH_3$ allows for production of 200tons of heavy water.

The triple point temperature of ammonia is 195.42K, normal boiling point 239.76K, critical temperature 405.6K. These constants determine conditions that make it possible to carry out rectification. In ammonia rectification at atmospheric pressure and below, artificial cooling is required to form wet reflux. With natural water cooling, the rectification column head pressure will range from 0.8 to 1.0MPa.

The difference in boiling point between $NH_3$ and $ND_3$ equals 2.37K at atmospheric pressure ($ND_3$ is less volatile). Ammonia species with different deuterium–protium substitution degree are related by the homomolecular isotope exchange reactions:

$$NH_3 + NHD_2 \leftrightarrow 2NH_2D \qquad \text{and} \tag{2.2}$$

$$NH_2D + ND_3 \leftrightarrow 2NHD_2. \tag{2.3}$$

With an equiprobabilistic isotope distribution, equilibrium constants of both reactions equal 3, and this value differs little from the constants' design thermodynamic quantities [10].

Ammonia of natural isotope composition contains deuterium in the form of monodeuterium ammonia $NH_2D$. The ideal hydrogen-isotope separation factor in a liquid–vapor system is

$$\alpha^0 = \sqrt[3]{P^0_{NH_3} \big/ P^0_{ND_3}}. \tag{2.4}$$

Within the temperature range from $-55$ to 0°C, the $NH_3$–$ND_3$ vapor pressure ratio varies from 1.183 to 1.0834, and, at atmospheric pressure, makes up 1.132 [11]. This value, according to eq. (2.4), conforms to the value of ideal separation factor for the $NH_3$–$NH_2D$

mixture at atmospheric pressure $\alpha^0 = 1.042$. As indicated by other sources [12], at the normal boiling point, $\alpha^0 = 1.044$ which approaches the above value.

Within the temperature range from 0 to 20°C, the $NH_3$–$ND_3$ vapor pressure ratio varies from 1. 07503 to 1. 0568 [13], and, at 20°C $\alpha^0$ $\sqrt[3]{P^0_{NH_3}/P^0_{ND_3}} = 1.019$.

Table 2.2 presents the results of $\alpha$ measurements made in a $NH_3$–$NH_2D$ system at above-atmospheric pressure [14–16] As Table 2.2 indicates, experimental data of different authors vary noticeably. Since the data of the French researchers [15] are interposed between the data reported in sources [14, 16], they can be used in calculations as more reliable.

Considering that the ammonia rectification requires lower-grade heat than water vacuum rectification, these two rectification processes can be combined into a multiple-unit rectification scheme where water vacuum rectification exhaust gas is utilized for heating the ammonia column evaporator.

Even though ammonia resources, as mentioned above, are abundant, to enhance the deuterium extraction degree it is expedient to perform ammonia rectification by a scheme with depletion. After deuterium extraction, ammonia is returned to the ammonia plant for conventional reprocessing into nitric acid or nitrogen fertilizers. Such a transit scheme whereby rectification stage intervenes between ammonia reception and reprocessing, does not involve ammonia consumption, except for inevitable minor losses.

A high-rate reaction of hydrogen isotope exchange between ammonia and water takes place

$$NH_3 + HDO \leftrightarrow NH_2D + H_2O. \tag{2.5}$$

The relation between separation factor and equilibrium constant is $\alpha = (2/3)K$. The $\alpha$ values are shown below [17]:

| Temperature, °C | 0 | 25 | 50 | 100 | 150 |
|---|---|---|---|---|---|
| $\alpha$ | 1.09 | 1.07 | 1.06 | 1.04 | 1.03 |

Thus, contact of ammonia with water results in a certain enrichment of the former with deuterium.

**Table 2.2**

Separation Factor $\alpha_{HD}$ in Ammonia Rectification

| $t$,°C | $\alpha^{15}$ | $\alpha^{16}$ | $t$,°C | $\alpha^{14}$ |
|---|---|---|---|---|
| −20 | 1.036 | | | |
| −10 | 1.031 | | | |
| 0 | 1.027 | | 0 | 1.0245 |
| 10 | 1.023 | 1.0246 | 9 | 1.0216 |
| 20 | 1.020 | 1.0216 | 17 | 1.0189 |
| 30 | 1.017 | 1.0187 | 27.3 | 1.0163 |
| 40 | 1.015 | 1.0161 | 38 | 1.0135 |
| 50 | | 1.0136 | | |
| 60 | | 1.011 | | |
| 80 | | 1.007 | | |

The process of ammonia–water isotope exchange can be used for the creation of a circulation scheme of ammonia rectification (Figure 2.2). Depleted of deuterium, ammonia leaves the rectification plant and enters the isotope exchange column (IE) where it contacts with oncoming water flow. Owing to the ammonia absorption and isotope exchange reaction (eq. 2.5), ammonia restores near-natural deuterium content.

Ammonia–water interaction occurs in the middle part of the IE column; in the lower part water is stripped of $NH_3$ on water vapour heating, and the upper part is used for ammonia dehydration by refluxing with ammonia wet reflux. A portion of liquid ammonia leaving the IE column condenser is returned as a reflux, and another portion is fed into rectification plant (columns I and II). The water depleted of deuterium is withdrawn from the lower part of the IE column. It is evident that water is the deuterium source in this scheme, and ammonia circulates in a closed circuit between the rectification column and the isotope exchange column.

Isotope exchange reaction (eq. 2.5) can also be performed in the opposite direction to extract deuterium from enriched ammonia and to enrich water with the extracted deuterium. For this purpose, deuterium-enriched ammonia withdrawn from the bottom of rectification plant is supplied into a similar but smaller isotope exchange column fed by water at a rate conforming to the rectification plant output. Ammonia from the upper part of this column is returned to the lower part of the rectification column.

The circulation scheme of ammonia rectification together with ammonia–water isotope exchange has the advantage of independence from the ammonia source. But it has as well a serious drawback: additional power consumption in the IE column. According to the calculations presented by Barr and Drew [18], a product cost for the isotope-exchange scheme is 1.5 times higher than that for the transit scheme of ammonia rectification.

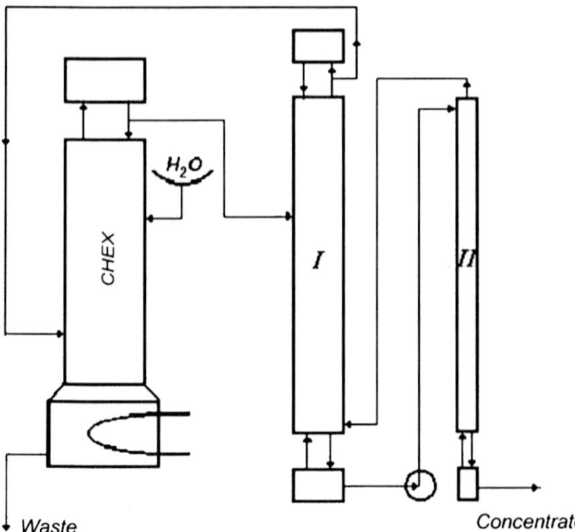


**Figure 2.2**   Scheme of ammonia rectification with ammonia–water isotope exchange: CHEX, isotope exchange column; I, II, columns of the first and second stage of the distillation plant.

The scheme of ammonia rectification with the use of a heat pump holds the greatest promise. A British company CJB (Contractors John Brown) has developed a five-stage ammonia rectification scheme to concentrate deuterium within a concentration range from 0.4 to 75% as translated into $D_2O$ (pre-enrichment from natural concentration to 0.4% is designed to perform by isotope exchange with hydrogen, and final concentration, by rectification of water obtained from isotope exchange with enriched ammonia). The ammonia rectification operating pressure is 110kPa (upper part of the column) and 140kPa (lower part). Exiting from the upper part of the first-stage column, ammonia is compressed from 110 to 290kPa by a radial flow compressor and supplied to all five stages as a heat-carrier for evaporator heating. Ammonia condensate is accumulated in a reservoir and pumped to the first-stage column as a reflux, with a certain portion withdrawn from the rectification system and returned to the pre-enrichment section. All rectification stages operate at a uniform pressure for which purpose ammonia from the successive stage head is supplied to the previous stage bottom with the use of blower. To reduce hydraulic resistance and power consumption, the project uses Spraypack-type columns with regular packing (the project is described more fully in reference [3]).

In 1955, a full-scale plant for heavy water production by ammonia rectification with heat pump was put into service in Norilsk, Russia. The plant was developed on the basis of engineering and scientific research done in 1946–1952. The plant and its commissioning are described by A.M. Rozen [19, 20].

The plant represented a cascade of five rectification columns with 100m in height each and with diameters of 4.7, 1.7, 0.8, 0.3, and 0.3m. The plant is schematically shown in Figure 2.3. Column 1 was provided with depletion section. The Column 1 main section with the next columns (2, 3, 4, 5) formed a five-stage concentration cascade with flow reduction.

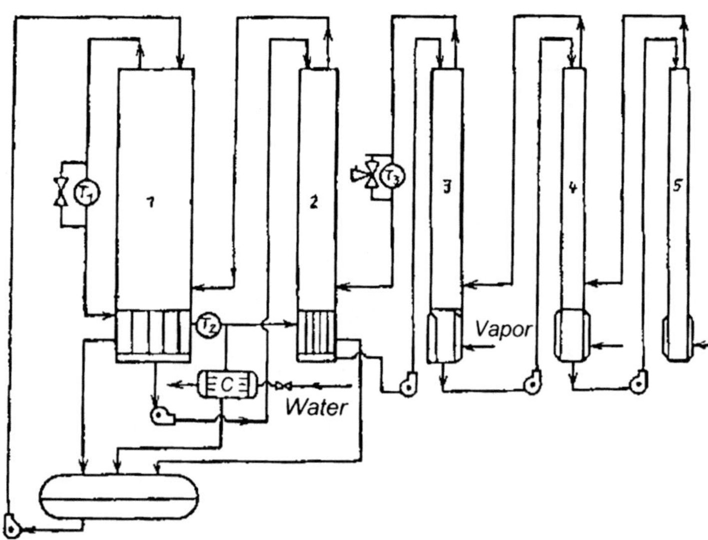


**Figure 2.3**  Flowchart of the plant for heavy water production by ammonia rectification with heat pump in Norilsk, Russia.

At deuterium concentrations in the ammonia source flow close to natural concentration (after ammonia–water isotope exchange), the concentration in the still of the last (fifth) stage amounted to 99.85%. The degree of deuterium extraction from the source flow was 27.5%. The first three tray columns were fitted with Kittel perforated plates with ordered liquid movement around the plate. The plates were overlain by fenders. Determined from operational data, performance factors of depletion section, and of the first, second, and third stages of concentration section were 11.5%, 13%, 20%, and 28% respectively.

The fourth column was filled with packing elements in the shape of perforated stainless steel semi-rings with a diameter of 12mm (Cannon packing), and the fifth with Rashig porcelain rings measuring 15mm $\times$ 15mm $\times$ 2mm.

With multipoint reflux distribution, HETP of the columns with a diameter of 0.3m is estimated at 250mm. It turned out that in rectification, isotope exchange reactions (2.2) and (2.3) between different deuterium ammonia forms proceed sufficiently fast, which is presumably due to the ammonia admixtures with catalytic effect.

The pressure in the column 1 head is 0.6MPa (which corresponds to a temperature of 10°C), and in the succeeding columns the pressure increases according to the columns' hydraulic resistance.

From the column 1 head, vaporous ammonia arrives at the heat pump, the function of which is performed by a steam turbine-driven turbocharger $T_1$ compressing 240tons of ammonia per hour to 0.9MPa. The ammonia is utilized as heating steam in the column 1 evaporators where it condenses and accumulates in the liquid ammonia collector (see Figure 2.3). The turbine drive consumes 20tons of water vapour per hour. The second turbocharger $T_2$ (second stage heat pump) boosts 40tons of ammonia per hour to a pressure allowing the use of compressed steam as heater of the second rectification stage column's evaporators. The condensate is sent to the aforementioned liquid ammonia collector. A portion of the steam may be passed through a water condenser $K_W$ to level off the heat balance, thus ensuring complete condensation of ammonia in the frost-free season.

The third turbocharger $T_3$ with a capacity of 5tons per hour serves to return the ammonia vapour from the third stage top to the second rectification stage to ensure against undesirable pressure increase in the last stages. The end product $D_2O$ was obtained by the two-stage combustion of 99.9% $ND_3$ withdrawn from the bottom of the fifth stage, with air over a nickel catalyst.

The ammonia feed flow accounting for 7.1tons per hour circulated between the top of the column 1 depletion section and the unit of isotope exchange with water, as shown in Figure 2.2. The water flow representing the deuterium source made up 16tons per hour, and the ammonia wet reflux in the isotope exchange column 4tons per hour. The ammonia stripping from the waste depleted water was performed at a temperature of 190°C with heating by water vapour with a pressure of 3MPa.

The withdrawn enriched flow was equal to 0.47kg per hour. Actual output with respect to $D_2O$ made up 45% of the potential output. The main reason is poor performance of the first- and second-stage tray columns. As an example, in the first-stage concentration section, the separation degree accounted for 4 instead of the projected 8, while the flow supplied to the second stage as a reflux was kept at the projected level, that is, was cut down by a factor of eight compared with the first-stage flow.

The plant consumption rates per 1kg of $D_2O$ were as follows: electric power 6,000kWh; water steam exhaust 3.5 tons; high-pressure steam (3MPa) for isotope exchange 10 tons; water 5,500–6,000m$^3$.

The author of publications [19, 20] believes that the process performance characteristics could be drastically improved by switching to the transit scheme (with a plant installed near a large-scale ammonia production facility) since it dispenses with vapor consumption for isotope exchange, or by changing from tray columns to regular packed columns which halves the electric power consumption.

## 2.3　HEAVY WATER PRODUCTION BY CRYOGENIC RECTIFICATION OF HYDROGEN

### 2.3.1　Fundamentals

Liquid hydrogen rectification for the purpose of extracting deuterium and producing heavy water is a major industrial method of heavy water production. Hydrogen rectification offers such advantages as high separation factor, low molecular weight of the working substance, sufficiently large hydrogen resources (worldwide hydrogen production exceeds 90 million tons per year). Electrolytic hydrogen, nitrogen–hydrogen mixture (NHM) to synthesize ammonia, and hydrogenous gases from petroleum refineries can be used as raw materials sources. Of prime importance is the deuterium extraction from NHM owing to the ever-expanding scale of NHM production. The deuterium extraction from hydrogen may be combined with the production of liquid hydrogen as a fuel component for rocket and space technologies. All the more important, increase of hydrogen resources in the future is due to the progress of nuclear-power engineering wherein hydrogen will act as energy carrier and deoxidizing agent for process technology. To industrially implement hydrogen liquefaction and rectification, it is necessary to counter several problems associated with hydrogen properties, discussed below.

The hydrogen critical temperature $T_C$ is 33.24K; the normal boiling point temperature is 20.39K. At the liquid hydrogen temperature all admixtures that can be found in liquefied gas, except for helium,–water vapours, carbon dioxide, argon, nitrogen, oxygen, methane, carbon oxide, – are transformed into a solid state and hence deposited in apparatus which upsets the production process.

Furthermore, solidified oxygen impurity in the hydrogen environment (hydrogen solubility of oxygen is modest, $S = 2 \times 10^{-9}$ molar fractions at a temperature of 22K) represents an extremely dangerous high explosive. In consequence, hydrogen liquefaction must necessarily be preceded by the removal of all impurities down to a level of under $10^{-9}$ volume fraction of oxygen, and $10^{-7}$ to $10^{-8}$ volume fraction of all the remaining admixtures. The finer the purification, the longer the non-stop run of equipment. Thereafter, reliable and sufficiently sensitive impurity content control becomes a necessity.

In the construction of refrigeration cycles for hydrogen liquefaction, account must be taken of the fact that under normal conditions, when throttled, hydrogen is not cooled but warmed up, and a positive throttle effect is observed only in a temperature range $-76$ to $-92°C$ under inversion point (depending on the pressure). That is why hydrogen must be pre-cooled well below the inversion point to obtain a sufficiently high throttle effect value.

Equipment operating at the liquid hydrogen temperature cannot be protected from refrigeration loss by conventional bulk insulation to avoid air solidification between insulating particles. It is necessary to use vacuum insulation, or to fill insulation jacket with non-freezing hydrogen or helium with a pressure maintained at a somewhat higher level than atmospheric.

To reduce refrigeration losses (heat penetration) through radiation which are significant due to a large temperature difference, it is advisable to place a nitrogen-cooled shield in the insulation bed.

Molecular hydrogen (all three hydrogen isotopes) may be in the ortho or the para state differing in nuclear spins: ortho-hydrogen has parallel (of like sign) spins of its atoms' nuclei, and para-hydrogen antiparallel (of unlike sign) spins. In normal conditions, hydrogen contains 25% para and 75% orthomodifications; deuterium 33% para- and 67% ortho-deuterium, (they are normal hydrogen $nH_2$ and normal deuterium $nD_2$, respectively).

Properties of hydrogen ortho- and paramodifications differ significantly: the boiling point of para-hydrogen, as an example, is 20.26K, and that of ortho-hydrogen $-20.45K$.

At the hydrogen normal boiling point 20.39K, para-hydrogen is more volatile than ortho-hydrogen, and the separation factor of these modifications is $\alpha_{O-P} = P^0_{para}/P^0_{ortho} = 1.055$.

Ortho-hydrogen can turn into para-hydrogen with heat release o-$H_2 \rightarrow$ p-$H_2$ + 1,418J/mol.

Hydrogen with equilibrium ortho-para composition is referred to as equilibrium hydrogen and designated by e-$H_2$. At the liquid hydrogen temperature, para-hydrogen and ortho-deuterium are thermodynamically stable. The quantity of heat $Q$ liberated in the normal hydrogen transformation into equilibrium hydrogen at various temperatures is:

| $T$, K | 20 | 30 | 40 | 50 | 100 | 150 | 200 | 273 |
|---|---|---|---|---|---|---|---|---|
| $Q$, J/mol | 1056 | 1016 | 897 | 731 | 177.5 | 30.1 | 4.15 | 0.29. |

So, the normal hydrogen transformation into para-hydrogen involves a significant heat release which exceeds evaporation heat and, consequently, leads to extra cold and refrigerating energy consumption.

Spontaneous ortho–para transformation of hydrogen at low temperatures proceeds very slowly, which makes it possible to produce liquid hydrogen of the approximate ortho–para composition of normal hydrogen.

Broadly speaking, the ortho-hydrogen transformation into para-hydrogen proceeds in liquid hydrogen at a rate of about 0.5% per hour. In the deuterium extraction by liquid hydrogen rectification, it allows prevention of a significant ortho-para transformation and the ensuing undesirable heat release.

Natural hydrogen contains deuterium in the form of HD molecules. Hence there is a need to consider the homomolecular isotope exchange (HMEX) $H_2 + D_2 \leftrightarrow 2$ HD $-$ 650J/mol, with equilibrium constant $K_{HD}$, of which the values are given in the literature [1]. Without catalysts, the reaction rate at the liquid hydrogen temperature is low, and therefore it is possible to extract thermodynamically unstable liquid HD from natural hydrogen. Liquid HD, mixed with $H_2$ molecules, acts in these conditions as independent component.

With catalysts, HD decomposition can be realized even at low temperatures, but in this case a necessity arises for an extra cold consumption to remove the reaction heat. In actual practice, HD decomposition required for the concentrated deuterium production is carried out with catalysts at an above-zero temperature.

The hydrogen isotopes vary widely in their properties. Data for vapour pressure of all hydrogen isotope species are known [21]. The separation factor values for $H_2$–HD mixture with liquid–vapor equilibrium for the HD low concentrations are shown below [1]:

| Pressure $P$, MPa | 0.101 | 0.121 | 0.142 | 0.155 | 0.162 | 0.172 | 0.183 | 0.202 |
|---|---|---|---|---|---|---|---|---|
| $T$, K | 20.39 | 21.02 | 21.18 | 21.84 | 22.08 | 22.32 | 22.55 | 22.97 |
| $\alpha$ | 1.63 | 1.60 | 1.55 | 1.54 | 1.52 | 1.51 | 1.49 | 1.47 |

### 2.3.2   Hydrogen rectification for deuterium extraction

Since hydrogen of natural isotopic composition contains deuterium in the HD form, and the HD decomposition reaction does not occur at a low temperature, it is evident that, at best, only concentrated HD can be separated at a single stage. To reduce the accumulation time it is worthwhile to divide HD concentration into two stages with appropriate flow reduction. In this case, as evident from calculations, with concentration at the first stage up to 10% of HD, accumulation time is reduced by a factor of about 12 compared with the production of 90% HD. On this basis, the process technology of heavy water production by hydrogen cryogenic rectification is composed of five stages following purification and refrigeration processes:

(1)   initial concentration to produce 5–10% HD;
(2)   the second rectification stage to produce concentrated HD (90% and over);
(3)   HD decomposition to produce $H_2$ + HD + $D_2$ mixture;
(4)   extraction of concentrated deuterium;
(5)   combustion of deuterium in oxygen to produce heavy water.

Owing to a low natural deuterium concentration in hydrogen, the highest energy consumption and largest dimensions of equipment are associated with the initial concentration stage. To make the best use of the available resources, at the first stage it is preferable to employ a rectification column with depletion section. To achieve a deuterium extraction degree of 0.9 (or 90%), in the column depletion section it is necessary to lower the deuterium concentration in the waste flow $x_P$ by a factor of 10 compared with the concentration in the feed flow. To form wet reflux, it is economically feasible to utilize the heat pump concept (thermo compression of hydrogen leaving the column). From the above reasoning, the scheme of the stage of HD initial concentration and extraction from hydrogen will take the form shown in Figure 2.4 [1].

Compressor 1 delivers hydrogen feed flow through heat exchange and purification system 2 into rectification column 3 consisting of depletion and concentration sections.

When leaving the column top, hydrogen, depleted of deuterium, is divided into two portions. The smaller portion is the waste flow delivered counter-currently through the heat exchange system to the feed flow and returned to the hydrogen user. The larger portion is the wet reflux supplied through system 4 gas exchangers into compressor 5 and compressed to a pressure of about 0.5 MPa required to produce a temperature difference allowing the wet reflux condensation heat to be transferred to the boiling still liquid. Then the wet reflux is purified and delivered, as a heat-transfer medium, to the coil in the column still. Losing heat to the still liquid evaporation, the wet reflux is liquefied and, after throttling, delivered to column 3 as a reflux unit.

Withdrawn from the column still, HD-enriched concentrate arrives at stage II for further concentration, and the waste flow from stage II returns to the relevant column section (not shown in Figure 2.4).

Considering that the hydrogen waste flow must pass through the purification and heat exchange system offering a substantial hydraulic resistance, the column head minimum pressure can be estimated as no less than 0.15MPa which corresponds to a temperature of 22K and hydrogen isotope separation factor $\alpha = 1.52$, and the column still pressure will be about 0.2MPa, $T = 23K$, and $\alpha = 1.47$. Setting the mean separation factor at $\alpha = 1.5$, and with extraction degree of 90%, there is no difficulty in obtaining the minimum wet reflux number $R_{min} = 1.7$.

The world's first industrial plant for deuterium extraction from hydrogen by cryogenic rectification was developed and put into service in the Soviet Union [22, 23]. The use of electrolytic hydrogen as a source material makes it possible to simplify the purification process. Hydrogen pre-cooling and refrigeration loss compensation at temperatures over 80K is performed by ammonia refrigeration cycle and liquid nitrogen. Cooling at temperatures under 80K is performed by a two-pressure throttling cycle with nitrogen pre-cooling. The wet reflux flow is divided into two portions: cooling flow with a pressure of

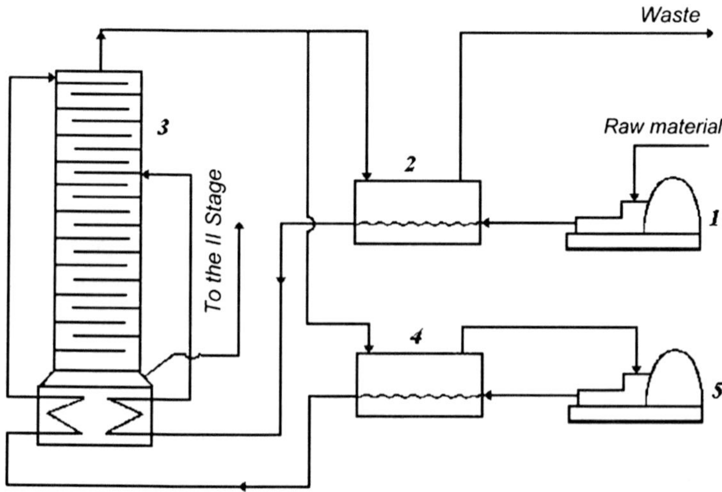


**Figure 2.4**   Scheme of heavy water initial concentration by hydrogen rectification with thermal compression of hydrogen issuing from the column: 1, 5, compressors; 2, 4, purification and heat exchange system; 3, rectification column.

up to 15.0MPa, and low-pressure flow with a pressure of 0.4–0.5MPa. The plant concept has been detailed in some reports [1, 23, 24].

The plant with a column of 1,050mm in diameter processes about 4,000m$^3$ stp of hydrogen feed flow per hour. The wet reflux number with a reserve is set equal to 2.5, so the circulating flow accounts for 10,000m$^3$ stp per hour. Provided with 80 bubble-cap trays of 35% efficiency factor, the column ensures a HD extraction factor of about 90%. A long-term operating experience confirmed the plant's reliability and safety.

The plant is serviced by a nitrogenous liquefying refrigeration cycle (80K) and an ammonia refrigerating unit with a cooling performance at a level of minus 45°C. Energy consumption ranges from 5,000 to 6,000kWh as evaluated per kilogram of $D_2O$. The long-term operating experience confirmed the plant's reliability and safety.

The plant in Nangal (India) for deuterium extraction from electrolytic hydrogen, with a capacity of 14.11tons of heavy water per year, was designed and supplied by Linde, a West German company, in 1962 [25, 26]. The plant processes hydrogen produced and pre-enriched by water electrolysis to a concentration of 600–800 millionth of HD. Purified to 0.05–0.15 millionth of oxygen and 0.1–0.6 millionth of nitrogen, the feed is concentrated at the first rectification stage up to 4% of HD. Oxygen stripping is performed by catalytic hydrogenation, while other admixtures are removed by freezing. At the second stage a 90% HD is obtained, and at the third stage, after HD catalytic decomposition, a product with a 99.8% deuterium concentration is derived.

Energy consumption in the rectification of enriched hydrogen accounts for 1,800kWh per kilogram of $D_2O$ without regard to power inputs for electrolytic concentration.

Of significant interest is deuterium extraction from hydrogen of ammonia synthesis gas (ammonia–hydrogen mixture, AHM). Unfortunately, the deuterium concentration in hydrogen of nitrogen-hydrogen mixture obtained by the natural gas conversion, may be 10–15% lower compared with average natural concentration of deuterium in water, mainly by hydrogen–water vapor isotope exchange occurring in the CO catalytic conversion.

A concept of an industrial cryogenic rectification facility for deuterium extraction from ammonia synthesis gas was devised, but not implemented, in the USA [27]. Small-scale plants with an annual output of 2tons and 4tons of $D_2O$, respectively, were built in Toulouse (France) and in Hoechst (Germany) [26].

The operating experience of Nangal and Hoechst plants permitted the Linde company to develop an improved process technology for deuterium extraction from ammonia synthesis gas, detailed in the literature [24]. The projected energy consumption in gas compression is 2,400kWh per kilogram of $D_2O$, and the concept is distinguished by several successful engineering solutions, namely:

(1)   the use of process gas of a large ammonia-producing facility (0.5 million tons of $NH_3$ per year) which allows for achieving high output (105tons of $D_2O$ per year);

(2)   the waste flow pressure increased to 0.2MPa which provides substantial energy saving in gas compression for the gas return to ammonia synthesis;

(3)   the use of liquid ammonia extracted from synthesis gas for pre-cooling, which makes it possible to do away with the necessity of building an auxiliary nitrogen cooling cycle unit, and, consequently, to save capital investment;

(4)   the separated gas expansion in turbo-expanders which assures cold generation at a low hydrogen level;

(5)   the use of wet reflux flow with absorber as a simple method of the regenerators' self-purification.


The above-listed solutions bring about an improvement in manufacturer-guaranteed performance characteristics as against the plant for deuterium production from AHM put previously into operation in the Ukraine, U.S.S.R.

The British nuclear research centre at Harwell has devised a concept for the utilization of cryogenic rectification for the deuterium extraction from associated petroleum gases containing 85.4% of hydrogen with the rest consisting of methane and heavier carbohydrates [28]. The main difference between this concept and concepts of processing other hydrogen-containing materials is in the gas preparation stage incorporating several heat exchangers and separators for the carbohydrate condensation and separation. Provision is also made for an auxiliary system of rectification columns to process carbohydrate fractions into end-products suitable for chemical processing. The realization of the products can significantly contribute to the economics of heavy water production from associated petroleum gases. There is no information available on the practical implementation of the concept.

The above concepts of deuterium extraction from hydrogen come under the heading of transit processes: that is, deuterium extraction is performed between hydrogen production and processing, with capacity and location of a heavy water production facility determined by capacity and location of the utilized gas production plant. As distinct from the transit processes, in a circulating process of deuterium extraction from hydrogen, the gas serves only as rectification-suitable working material, with water used as the deuterium source. Here, the deuterium-depleted hydrogen emerging from the rectification system should be sent to a catalytic isotope exchange (CTEX) with water, $H_2 + HDO \leftrightarrow HD + H_2O$, so as to replenish the deuterium content and return it to the rectification system. Unfortunately, the reaction equilibrium constant value is disadvantageous for the hydrogen enrichment. As evidenced by calculations, even at a considerable water vapor rate and at a high temperature of isotope exchange (600°C), the HD concentration in hydrogen fed to the rectification unit will be 1.5 times lower than natural concentration which results in a degradation of performance characteristics.


## 2.4   ISOTOPE EXTRACTION AND CONCENTRATION OF TRITIUM


### 2.4.1   The use of deuterium cryogenic rectification for heavy water purification for nuclear reactor circuit

Hydrogen cryogenic rectification is advantageously employed in heavy water purification from protium and tritium admixtures accumulated during a nuclear reactor operation. Liquid hydrogen cryogenic rectification can be used in the tritium extraction from water flows or heavy water after conversion of extracted isotopes into gaseous state. For this purpose, water electrolysis or catalytic isotope exchange in the water–hydrogen system are employed (see chapter 3).

The vapor pressure HT can be calculated by the GM rule: $P^0_{HT} = \sqrt{P^0_{T_2}P^0_{H_2}}$. Similarly, it is assumed that $P^0_{DT} = \sqrt{P^0_{D_2}P^0_{T_2}}$. The values of the separation factor are given in Table 2.3. Physical properties of all isotopic forms of molecules as well as separation factors for mixtures of isotopes H–T, H–D, and D–T in rectification both of hydrogen and of water are presented in referemce [1].

The contactors' efficiency in the conditions of the liquid hydrogen cryogenic rectification can be estimated on the basis of experimental and trial data on hydrogen enrichment with deuterium by rectification. According to the Los Alamos National Laboratory (U.S.A.), the experimentally determined HETP for a relatively small laboratory-scale packed column with two different commercial packing materials at a vapour rate of up to 89mm/s, was equal to $50 \pm 5$mm [29]. Japanese scientists recorded a slight HETP dependence on vapour rate, which is connected with a near-linear dependence of mass-transfer coefficient on rate. According to data derived from experimental rectification of deuterium-tritium mixture, HETP lies in the range 5–7cm.

In experiments at $T = 22$K, in a column with a diameter of 96mm packed with spiral-prismatic packing material of size 3mm $\times$ 3mm made from caprone thread of 0.2mm in diameter, over a load range of 5,200–6,700m$^3$ stp/(m$^2$·h), was observed a HETP value of 47–50mm, and hydrogen rectification with similar stainless steel wire packing over a load range of 8,500–9,100m$^3$ stp/(m$^2$·h) gave a HETP value of 59–61mm.

An HETP for copper spiral-prismatic packing is equal to 5–6cm [30], and for Dixon rings of 3mm in size a HETP value is equal 4.5–5cm [31].

Practical experience on tritium extraction and concentration by cryogenic rectification has been gained in the Laue-Langevin Institute (Grenoble, France) on the basis of an operable facility for heavy water purification for a high-flux neutron research reactor built with the participation of the Sulzer Company [32]. Put into operation in 1972, the facility began its normal working in 1976 by successfully performing the function of keeping the heavy water concentration at a level of 99.8–99.92% $D_2O$ by continuous heavy water purification from accumulated HDO and DTO admixtures, with an activity at a level of $6.3 \times 10^{10}$–$8.2 \times 10^{10}$Bq/l, owing to a DTO admixture. The facility ensures protium extraction from moderator in an amount of 60l of $H_2O$ per year and tritium extraction in an amount of $8.9 \times 10^{15}$Bq per

**Table 2.3**

Separation Factor of Deuterium–Tritium Mixtures

| $T$, K | $D_2$–DT Mixture | | $D_2$–$T_2$ Mixture | |
|---|---|---|---|---|
| | $\alpha_{ID}$ | $\alpha_{EXP}$ | $\alpha_{ID}$ | $\alpha_{EXP}$ |
| 23 | 1.248 | $1.185 \pm 0.014$ | 1.558 | $1.455 \pm 0.048$ |
| 25 | 1.207 | $1.159 \pm 0.003$ | 1.457 | $1.382 \pm 0.056$ |
| 27 | 1.175 | $1.118 \pm 0.006$ | 1.382 | $1.318 \pm 0.077$ |

year in the form of 98–99% $T_2$, with heavy water loss in the form of waste (60% $D_2O$) accounting for about 130kg of $D_2O$ per year.

Protium and tritium admixture extraction from heavy water is performed by catalytic isotope exchange of water vapors with gaseous deuterium at 473K according to the reactions:

$$DTO + D_2 \leftrightarrow D_2O + DT \text{ (equilibrium constant } K_1 = 0.82), \text{and}$$

$$HDO + D_2 \leftrightarrow D_2O + HD \text{ } (K_2 = 1.77).$$

A portion of heavy water of an amount of up to 20kg/h is supplied from the reactor circuit for three-stage exchange with deuterium. After isotope exchange, the heavy water with a reduced H and T content is recycled to the reactor, and gaseous deuterium is delivered for the rectification to extract HD and DT, and, after that, returned for isotope exchange with water. Hence deuterium gas circulates in a closed loop (at a rate of $45m^3$ stp per hour). To compensate for losses, the loop replenishment with gaseous deuterium is performed at a rate of $2m^3$ stp per week, with the deuterium amount in the facility making up $85m^3$stp.

The extraction of HD and DT admixtures from gaseous deuterium circulating in a closed loop, and their concentration, are performed by cryogenic rectification under pressure of 0.15MPa in a two-stage unit. The unit scheme is shown in Fig. 2.5.

After the heavy water unit and before rectification, gaseous deuterium is delivered by pressure blower to the middle section of the first rectification column 1 for dehydration, purification, and refrigeration through a silica gel adsorber, heat exchanger, active carbon adsorber ($-180°C$), and end heat exchanger at $T = 24K$ (in Figure 2.5, all of this system is conventionally represented by unit 5). Partly purified from HD and HT admixtures, the deuterium flow is withdrawn, through a side bleeder, from the column and, by way of the heat exchangers, returned to the isotope exchange system.

The Grenoble first-stage rectification column with a diameter of 0.25m and a height of 11m is packed with Sulzer regular packing material. The HD mixture enrichment is performed in the column top, and the enrichment with tritium in the DT form in the column bottom. From the top is withdrawn a fraction containing 20% $D_2$ + 80% HD of which the combustion produces heavy water. Withdrawn from the column 1 bottom and enriched with tritium ($D_2$ + DT), deuterium is delivered for further tritium concentration to the second column packed with random packing material of the Dixon ring type. DT is withdrawn from the middle part of the column to the reactor for decomposition in the course of the reaction $2DT \leftrightarrow T_2 + D_2$; after that, in the column bottom this gas mixture is used to extract a 98% tritium ($T_2$) which is periodically withdrawn to special steel reservoirs.

The column 2 top is 20mm in diameter, the bottom 12mm, and the total packing bed height is 9m. The overall efficiency of the rectification unit is estimated as 400TP.

A helium refrigeration cycle is used to compensate for refrigeration loss and to produce wet reflux. Compressed in compressor 3 to a pressure of 1.5MPa, helium is employed as a heat medium in the column 1 still, then expanded in turbo-expander 4 to a pressure of 0.4MPa, supplied to the both rectification columns for retarders cooling, and finally, returned, via heat exchangers 6 to compressor 3, thus closing the loop.

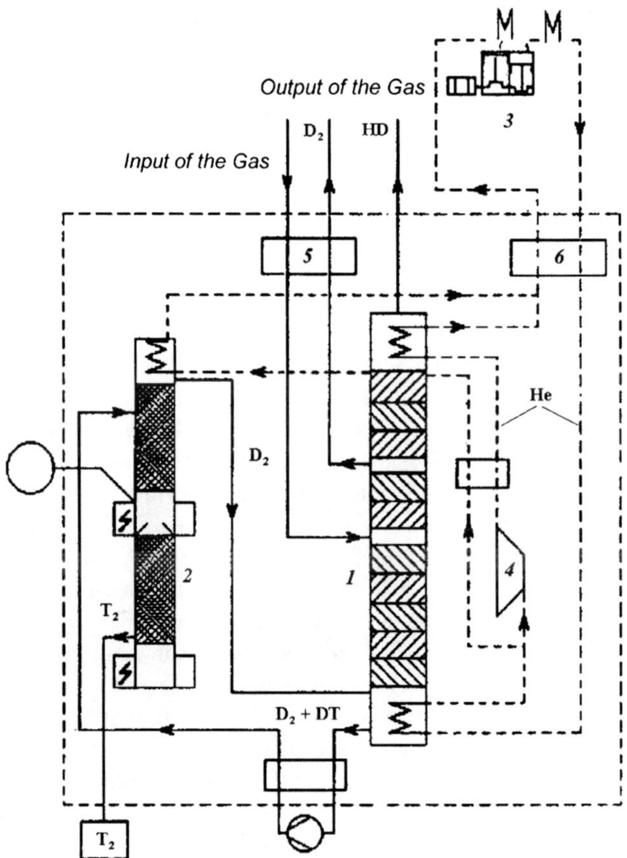


**Figure 2.5**   Scheme of facility for deuterium purification from protium and tritium admixtures by cryogenic rectification in Grenoble, France: 1, first-stage rectification column; 2, second-stage column; 3, helium compressor; 4, turbo-expander; 5, deuterium dehydration, purification and refrigeration system; 6, helium cycle heat exchangers.

Equipment enclosed in a line-dotted box in Figure 2.5 is housed in a single vacuum insulating shell composed of vertical and horizontal parts. The total shell height is 15m, with outer diameters of the vertical and horizontal parts of 800mm and 1200mm, respectively.

The plant described above allows for extracting from the moderator 240,000Ci-Curie of tritium and 100l of $H_2O$ per year, with an electric energy input of 800kWh, and cooling water consumption of 25m³/h.

Based on successful operation of the above plant, Ontario Hydro, a Canadian corporation, built in Darlington a similar, but more powerful, plant called Darlington Tritium Removal Facility (DTRF), for tritium extraction from 350kg/h of heavy water. The plant productivity ranges from 1.7 to 2.5kg (or $630 \times 10^3 - 925 \times 10^3$TBq) of tritium per year. The plant was commissioned in 1987 [33].

Aside from the greater productive capacity, the Canadian plant has several other differences from the Grenoble plant.

Heavy water intended for isotope purification is first of all stripped from non-volatile admixtures and dissolved gases. $D_2O$ purification from tritium is performed as well by catalytic isotope exchange (CTEX) with gaseous deuterium at 200°C, but carried out in eight successive stages. Each stage comprises a water evaporator, an over heater, a catalytic contactor, and a condensing separation system. In each stage, water and gas flows move in one direction, but in passing from one stage to another, the counter flow principle is followed. Stripped from tritium, heavy water at the CTEX last stage is collected into reservoirs from which it returns to the reactor.

Gaseous deuterium containing tritium (in the DT form) transferred/received from water is delivered by compressor from the CTEX system to the cryogenic rectification system for tritium extraction and concentration. For this purpose, gas is at first subjected to dehydratation, cooling by thermal exchange, adsorption refining by silica gel at 77K, and, after further cooling to 30K by thermal exchange, entered into the initial concentration column, or low tritium column (LTC). The column comprises two sections of different diameter and operates at a pressure of about 130kPa.

$^3$He produced by tritium radioactive decay is blown through the condenser and collector, and withdrawn from the system through the catalytic contactor for hydrogen isotope combustion.

Enriched with DT from the LTC bottom, deuterium flow in liquid form is supplied as a reflux to the first section of the rectification column for end concentration, or high tritium column (HTC). The flow delivery from LTC to HTC in liquid phase by the principle of thermosyphon makes it possible to dispense with pumps. Three-sectional HTC with a reactor for DT decomposition is designed to produce highly concentrated (99%) commercial tritium.

With the aim of reducing energy consumption by half and more, compared with a helium cooling cycle, hydrogen cooling of LTC and HTC is adopted. In the thermal insulation system of cryogenic unit a vacuum of up to $10^{-3}$Pa is maintained. The cryogenic unit overall height is 38m.

To prevent losses and assure safety, DTRF is provided with several auxiliary systems including a catalytic reactor for the hydrogen isotope mixture combustion, purging systems, deuterium replenishment systems to compensate for the loss in the gas cycle between CTEX and cryogenic rectification, and tritium fixation and storage units.

An air purification system is connected with all rooms housing tritium-containing equipment. Provision is made for 15 sensors to monitor tritium content. A ventilation system with a capacity of 1700m³/h allows for detritizing the building in a time of no more than five hours since the leak detection.

A pilot demonstration plant for heavy water detritization of AECL reactors in Choke River (Canada) [34] also provides for cryogenic rectification of $D_2$ + DT mixture, which produces sufficiently pure $D_2$ and $T_2$, after the stage of tritium conversion from liquid to gaseous phase by isotope exchange. The plant is shown in Figure 2.6. In this case, isotope exchange is done by a more advanced technique: in a counter-current column with a hydrophobic catalyst (LPCE). Extracted from heavy water, gaseous deuterium containing admix with other hydrogen isotopes is supplied, after dehumidification and purification, to the cryogenic rectification unit consisting of two columns. The first

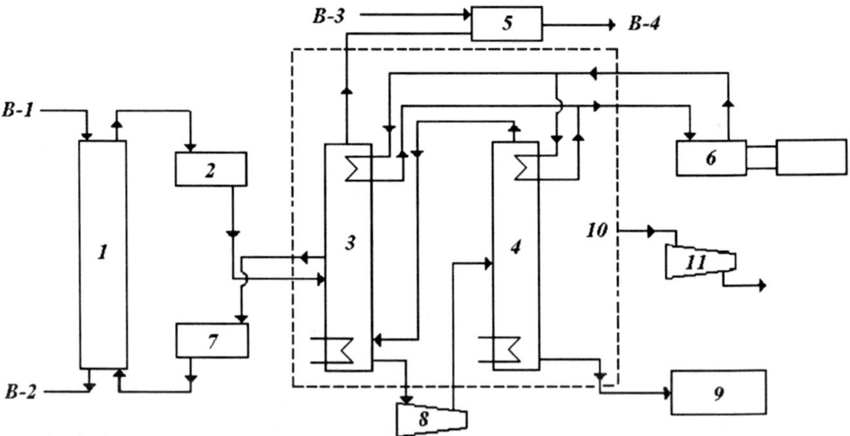

**Figure 2.6**  Scheme of heavy water isotopic purification plant in Choke River, Canada: 1, isotope exchange column with hydrophobic catalyst; 2, main deuterium compressor; 3,4, cryogenic rectification columns of first and second stage, respectively; 5, oxidant; 6, helium refrigerator; 7, expansion tank; 8, compressor; 9, tritide formation; 10, cold unit (enclosed in a line-dotted box); 11, vacuum pump; B-1, heavy water feed flow; B-2, purified heavy water outflow; B-3, oxygen; B-4, outflow of heavy water containing $H_2O$ and HDO.

column 3, with a diameter of 150mm and a height of 15m, filled with Sulzer regular packing material (100TP), is designed to increase tritium concentration by a factor of 25. The second column 4 incorporates an upper section of 50mm in diameter and two lower sections of 12mm in diameter each. A HMEX reactor for DT decomposition into $D_2$ and $T_2$ is accommodated between two lower sections. Tritium from the column bottom is delivered by small portions to the storage unit. It is projected to fix and store tritium in the form of titanium and zirconium tritide. A titanium sponge is capable of absorbing 50g or 18.5TBq of tritium in the form of tritide.

A cryogenic rectification unit of the system of tritium extraction from heavy- water moderator has been modeled in Savannah River, U.S.A. [35]. The scheme shown in Figure 2.7 comprises a catalytic isotope exchange (CTEX) column 1 to convert tritium from liquid to gaseous phase, an initial concentration column 2, three end concentration columns 3–5, and catalytic DT decomposition reactor 6. The scheme is designed to produce tritium with a $T_2$ concentration of over 98%, with tritium concentration of $6.2 \times 10^{-7}$ molar fractions at the inlet of the initial concentration column, and to deplete deuterium return flow to the CTEX column down to a DT content of $3 \times 10^{-4}$ molar fractions.

To detritize and deprotize heavy water, hydrogen cryogenic rectification is adopted as well in a neutron source project at the Oak Ridge National Laboratory (U.S.A.) [36]. To convert H and T into gaseous phase, use is made of catalytic isotope exchange between deuterium and heavy water in counter-current column with hydrophobic catalyst. Enriched with heavy isotopes, water from the column bottom is delivered to an electrolyzer, where it is completely decomposed.

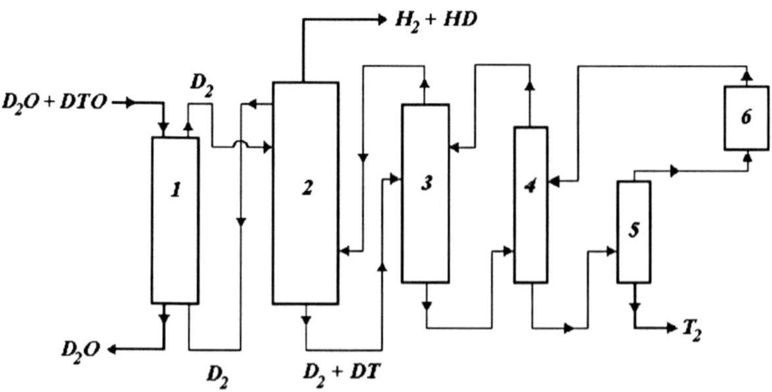


**Figure 2.7**  Scheme of tritium extraction from heavy water in Savannah River, U.S.A.: 1, catalytic isotope exchange column; 2, first-stage cryogenic rectification column (initial concentration); 3, 4, 5, end concentration rectification columns; 6, HMEX reactor.

A portion of the obtained tritium-containing deuterium is returned to the isotope exchange column, and the rest (after purification) is sent to the cryogenic rectification unit (three stages and DT decomposition reactor) to concentrate tritium. From the third-stage column bottom tritium is periodically withdrawn to the storage system (a glove box filled with argon).

As the gas phase moves upwards in the isotope exchange column, it is depleted from heavy isotopes, and in the column head it makes contact with demineralized water washing away D and T residuals, which makes it possible to discharge the gas to the atmosphere.

The cryogenic rectification unit is serviced by a helium refrigerator (the cooling cycle compressor is located outside the unit) and provided with high-vacuum insulation. Withdrawn from the cryogenic rectification unit, the purified deuterium flow is combusted with oxygen from the electrolyzer into the reactor-quality heavy water, which is returned to the reactor.

In India, where nuclear power stations are equipped with heavy-water reactors, a plant has also been developed to detritize heavy-water retarder and to produce concentrated tritium [37]. The trituim transfer from DT-containing heavy-water retarder into the gaseous phase in the form of DT is performed in a counter-current isotope exchange column with hydrophobic catalyst, to which a flow of circulating deuterium is delivered from the bottom. The degree of tritium extraction from the retarder is about 90%. Deuterium, leaving from the top of a column and containing DT impurity, is directed to the block of cryogenic rectification which operates at pressure 0.12–0.15MPa and at a temperature 25–27K.

Issuing from the column head, deuterium admixed with extracted DT is supplied to the second-stage rectification stage consisting of two sections with diameters of 25mm and 20mm. An HMEX reactor for DT decomposition is accommodated between the sections. An Evaporizer and condenser of the first-stage column are heated and cooled by a helium cycle, with the second stage cooled by helium and heated electrically.

In the National Institute of Cryogenic and Isotopic Technologies of Romania, a work package on the development of hydrogen cryogenic rectification process has been implemented, for which the results have been reported [38].

### 2.4.2  Separation of isotopes in the system of deuterium–tritium fuel cycle of thermonuclear power reactor

As is expected, the fuel cycle of a thermonuclear power reactor operating on deuterium–tritium mixture will comprise a fuel regeneration system performing the function of purifying mixture supplied from the plasma chamber, and of separating hydrogen isotopes extracted from the mixture.

At the stage of hydrogen isotope purification, all gaseous chemical compounds should be separated, including helium and regenerated tritium chemically bonded with impurities.

Various methods can be used at the stage of hydrogen isotope separation, but the cryogenic rectification method that has been practically implemented for heavy water production, and for tritium extraction from heavy-water retarder and concentration to the production level, is considered to the most suitable.

In the general case, the degraded D + T fuel mixture represents a mixture of six molecular forms: $H_2$, HD, $D_2$, HT, DT, and $T_2$. At a given nuclear composition of H, D, and T, the relations between molecular forms depend on whether or not the mixture is in thermodynamic isotope equilibrium.

Since at low temperatures isotope equilibrium is not established spontaneously, it is necessary to combine isotope rectification with catalytic equilibration at a sufficiently high temperature, to disrupt intermediate molecular forms of hydrogen for obtaining pure components, as for instance to disrupt HT by the reaction $2HT \leftrightarrow H_2 + T_2$, or DT by the reaction $2DT \leftrightarrow D_2 + T_2$. At Mound, U.S.A., a hydrogen cryogenic rectification system (HISS) has been developed, constructed, and put into operation [39]. The system is designed to process a mixture of all six hydrogen isotopic forms and to obtain a product: tritium with a concentration of no less than 99%, and purified hydrogen (raffinate) with an activity of no more than 770MBq/h.

HISS is composed of three packed columns and three equilibrating reactors (Figure 2.8). The columns operate at 24K, and their height suffices to obtain $H_2$ and $T_2$ at the cascade ends.

Reactors, wherein reactions of HD, HT, and DT decomposition proceed, are filled with metallic catalyst and operated at an indoor temperature (300K). A feed flow of 2.2liters per hour was taken as the project baseline concept, though in the process of modernization it was found that the flow could be several times more intense without lowering the product concentration; however, with an increase in the gas load the tritium content in the raffinate becomes somewhat higher. The rectification columns are filled with Helipack packing material. The design parameters of the rectification columns are presented in Table 2.4.

The evaporators are designed in such a way as to minimize the amount of tritium in the liquid phase. Condensers are cooled by helium refrigerators with an incoming helium temperature of 17K. Copper fins are installed inside the condensers to enhance the surface, with helium passing through the tubes soldered to the condensers outside, which forms a double-layer wall separating helium and tritium.

The columns are enclosed by a shield cooled with liquid nitrogen and provided with multilayer vacuum insulation. The external steel casing is of 0.71m in diameter and 5.8m in height.

Specialists of Ontario Hydro, a Canadian company, have developed a computer program named FLOSHEET to simulate systems of hydrogen isotope separation.

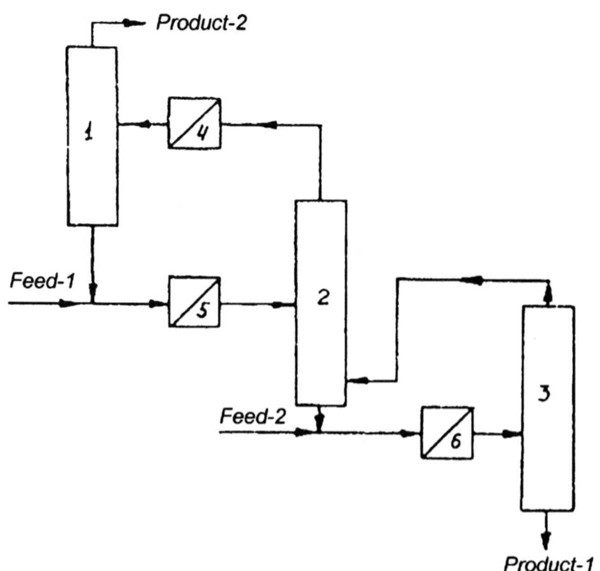


**Figure 2.8**   Scheme of cryogenic rectification of hydrogen isotopes (HISS) at Mound, U.S.A.: 1, 2, 3, rectification columns; 4, 5, 6, HMEX reactors; B-1 and B-2, feed; P-1, enriched product; P-2, depleted product (raffinate).

**Table 2.4**

Design Parameters of Rectification Columns of HISS System at Mound, U.S.A.

| Parameter | Column number | | |
|---|---|---|---|
| | 1 | 2 | 3 |
| Internal diameter, mm | 26.1 | 16.6 | 10.2 |
| Packing bed height, m | 3.66 | 3.66 | 3.66 |
| Vapour velocity, mm/s | 80 | 15 | 54 |
| Reflux ratio | 18.9 | 86.2 | 50 |
| Temperature in condenser, K | 23.9 | 23.7 | 24.2 |
| Pressure in condenser, kPa | 111 | 107 | 107 |
| Condenser cold load, W | 29.8 | 2.3 | 3.2 |
| Temperature in vaporizer, K | 24.3 | 23.8 | 25.2 |
| Vaporizer heat load, W | 30.9 | 2.3 | 3.3 |
| Packing volume, % | 9.5 | 9.5 | 10.7 |
| Tritium holdup, g | 2.73 | — | 7.39 |
| Tritium (radioactivity/concentration) | Raffinate 370MBq/l | | Product 98% |

The program involves modeling of cryogenic rectification cascade as a part of the tritium system test assembly (TSTA) for a four-column system (Figure 2.9). Each column has a concentration and depletion sections. In addition to the four columns, the system incorporates three HMEX reactors to perform the homomolecular isotope exchange, and

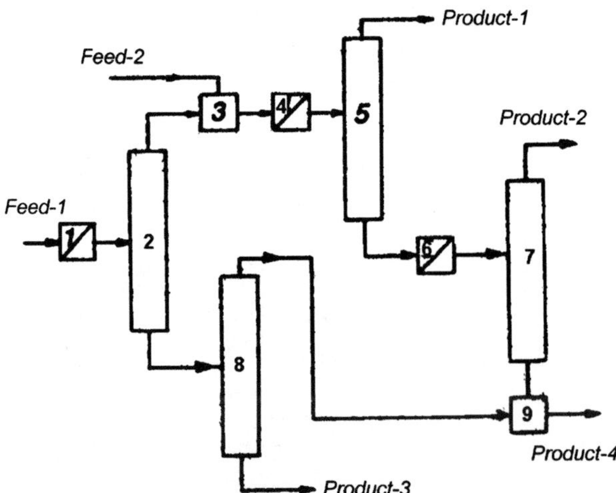


**Figure 2.9**   Four-column scheme of TSTA cryogenic rectification: 1, 4, 6, HMEX reactors; 2, CD-1 rectification column; 5, CD-2 column; 7, CD-4 column; 8, CD-3 column; 3 and 9, mixers; feed flows: F-1, fuel (H, 1%, D and T, 49.5% each); F-2, neutral injecting flow recycle (H, 1.5%, D, 98%, T, 0.5%); product flows: P-1, hydrogen stripped from tritium; P-2, deuterium (> 99%); P-3, tritium (> 99%); P-4, deuterium-tritium mixture (D, 50%, T, 50%).

two flow mixers. A flow charge of 15mol/h of isotope mixture composed of H (1%), and D and T (49.5% each), is supplied for the separation in the HMEX reactor. In addition, 10mol/h of the neutral injecting flow recycle consisting of 1.5% of H, 98% of D and 0.5% of T is introduced into mixer 3.

The characteristics of the rectification columns are presented in Table 2.5. The scheme function is to obtain pure deuterium and tritium flows, and to extract protium-containing forms with a minimum tritium content.

For the calculations, the pressure in the heads of all four columns is taken as 101.3kPa. From the cited calculations of the rectification of six-component mixture consisting of $H_2$, HD, $D_2$, HT, DT, and $T_2$, both with regard to the deviations of liquid–vapor equilibrium from ideality and heat of tritium radioactive decay, and without considering these factors, it follows that taking into account the imperfection and heat generation significantly affects the estimate of impurity content in the separation products.

On the basis of the calculations of Indian scientists (Symposium on Heavy Water Technology, 1989, India), the necessity of taking into account the imperfection of mixture and tritium decay heat is noted, which forces an increase in reflux ratio and cooling of some of the columns [40].

A simpler scheme for the solution to the problem by cryogenic rectification put forward by Kinoshita, a Japanese scientist, was calculated within the FLOSHEET program. The scheme (Figure 2.10) comprises two rectification columns, but each with side withdrawal and recycle, four HMEX reactors, and two mixers. The Kinoshita scheme characteristics are presented in Table 2.6. The pressure in the column is also taken as equal to 101.3kPa.

**Table 2.5**

Characteristics of Rectification Columns of TSTA Isotope Separation Scheme

| Parameter | Column | | | |
|---|---|---|---|---|
| | CD-1 | CD-2 | CD-3 | CD-4 |
| Number of theoretical plates (NTP) | 80 | 85 | 70 | 85 |
| Feed flow, mol/h | 15.0 | 13.75 | 11.25 | 13.1685 |
| Output flow, mmol/h: | | | | |
| Column head | 3.75 | 0.5815 | 7.645 | 12.9 |
| Column bottom | 11.25 | 13.1685 | 3.505 | 0.2685 |
| Reflux ratio | 25.0 | 100 | 10.0 | 12.5 |
| Number of feed injection stage | 50 | 55 | 30 | 45 |
| Holdup, mol: | | | | |
| Condenser | 1.0 | 1.0 | 1.0 | 3.0 |
| Stages | 0.1 | 0.057 | 0.098 | 0.19 |
| Evaporator | 0.75 | 1.3 | 0.75 | 1.0 |

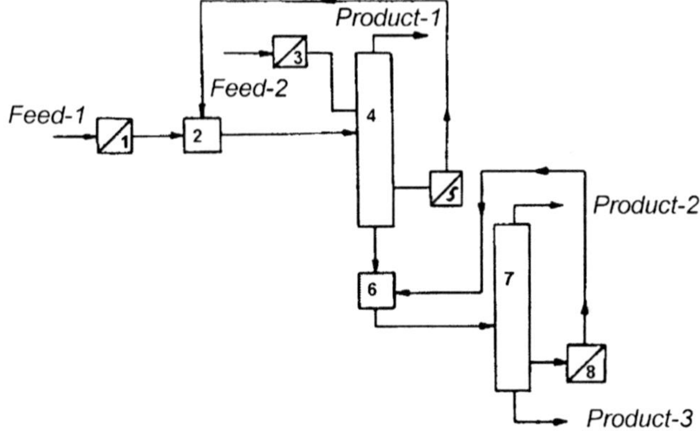


**Figure 2.10**   Two-column cryogenic rectification scheme by Kinoshita: 1, 3, 5, 8, HMEX reactors; 4, rectification column CD-1; 7, rectification column CD-2; 2, 6, mixers; feed flows, B-1 and B-2; product flows: P-1, hydrogen purified from tritium; P-2, deuterium (99%); P-3, tritium (95%).

The cryogenic rectification method, as industrially mastered and offering cost and safety advantages, was adopted for the isotope separation system of the International Thermonuclear Reactor (ITER).

According to the conceptual design of 1990, cryogenic rectification of the ITER reactor isotope separation system (ISS) incorporates four rectification columns CD-1, CD-2, CD-3, and CD-4, five HMEX reactors H-1 to H-5, and appropriate heat-exchange apparatus [41].

The scheme was modeled by the FLOSHEET computer program. To ensure the flows movement through the system without pumping, the columns are operated at three different

**Table 2.6**

Characteristics of Cryogenic Rectification Scheme by Kinoshita

| Parameter | Column CD-1 | Column CD-2 |
|---|---|---|
| Number of theoretical plates (NTP) | 120 | 120 |
| Wet reflux ratio | 300 | 15 |
| Plate number: | | |
| Feed input | 60, 50 | 70 |
| Side withdrawal | 70 | 80 |
| Holdup, mol: | | |
| Condenser | 1.0 | 1.0 |
| Stages | 0.1 | 0.057 |
| Evaporator | 0.75 | 1.3 |
| Recycle from reactor to mixer, mol/h | 50 | 70 |

pressure levels: CD-1 at 155kPa, CD-2 at 130kPa, and CD-3 and CD-4 at 105kPa. Owing to this, only small and tritium-depleted flow should be compressed for the return to the column CD-1. Reactors are operated at ordinary temperatures. The ISS cryogenic rectification unit is shown in Figure 2.11, with the columns' design characteristics presented in Table 2.7.

There are several tritium-containing mixture flows: gas flow B-2 extracted from the aqueous lithium salt blanket (ALSB) via DW-2 and CTEX is shown in Figure 2.12, or solid breeder; gas flow B-2 extracted by the rectification of wastewater in the column DW-1 (Figure 2.12); neutral injecting deuterium flow; pellet injector gas flow; high-temperature isotope exchange (HYTEX) flow; tore outflow.

The hydrogen feeding the cryogenic rectification is taken to be isotopically equilibrated at 300K. In addition, the first column is fed with return flows from the columns CD-2, CD-3, and CD-4.

The column CD-1 function is to extract light hydrogen isotope as completely purified as possible from tritium, and to preliminarily concentrate tritium. Deuterium and tritium are withdrawn from the column bottom for subsequent concentration. The CD-1 wet reflux ratio equal to 3.7 is taken twice as high as the minimum ratio, with the reserve ensuring the reliability of detritization of hydrogen flow withdrawn from the column head.

The column CD-2 function is to extract deuterium with its subsequent purification in the column CD-3. The deuterium high quality is assured by the recycle from the CD-3 head to the CD-1.

The column CD-4 is designed to obtain lower product with a concentration of no less than 80% of T and 20% of D, with a possibility to obtain a 99.9% concentration at a reduced load. Reactor E-3 serves to decompose HT, and reactors E-4 and E-5 in the CD-4 supply line to decompose DT. The reactors are operated at indoor temperature.

Operational stability is maintained by controlling the electric heating of the columns' evaporators. The condensers' cooling is controlled by the maintenance of a constant pressure in the columns. The condensers' capacity is 6.860W.

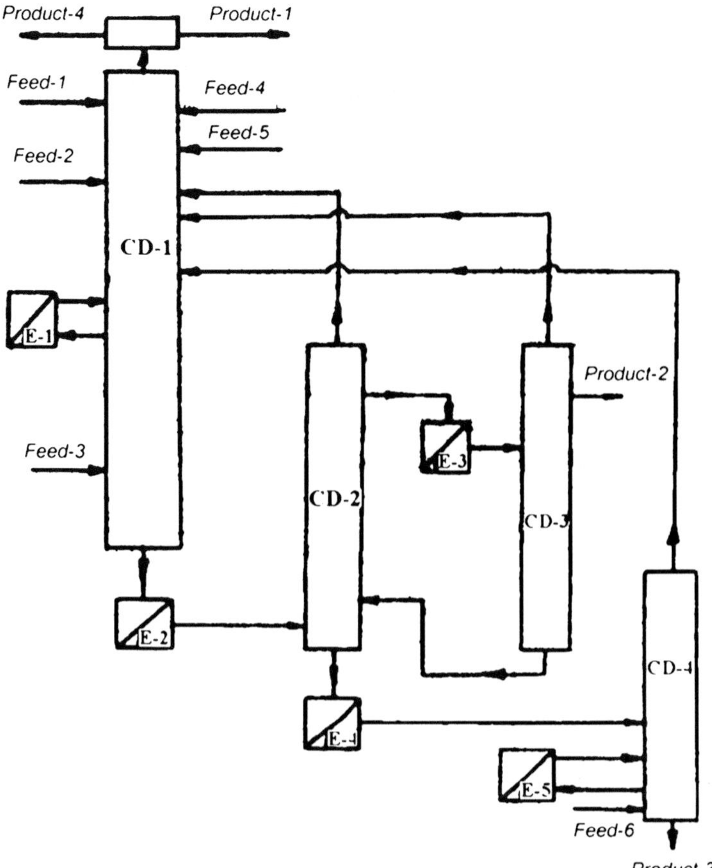


**Figure 2.11**   Scheme of cryogenic rectification of the ITER's isotope separation system: CD-1, CD-2, CD-3, CD-4, rectification columns; E-1, E-2, E-3, E-4, E-5, HMEX reactors; product flows: P-1, protium; P-2, deuterium; P-3, tritium; P-4, return flow to the wastewater treatment system; feed flows: B-1, from the wastewater treatment system; B-2, from the system of gas extraction from the blanket; B-3, neutral injecting deuterium flow; B-4, hydrogen from the high-temperature isotope exchange unit (HYTEX); B-5, pellet transport flow; B-6, tore outflow.

Columns CD-1, CD-2, and CD-3 are housed in a common cold block with a diameter of 1m and a height of 8m. The lower 2m of the block are of 1.5m diameter and accessible for service. Column CD-4 is housed in a separate block with a diameter of 0.5m and a height of 5m.

To store gas in the event of a shutdown, provision is made for three expansion tanks rated at a pressure of 0.16MPa. The overall tritium holdup is 318g, of which 205g fall on column CD-4.

The project in question envisages a unitary system to process all TNR tritium-containing flows (both gas and liquid) allowing for the extraction of tritium product (or, more specifically, of deuterium–tritium mixture) from the flow P-3, and of deuterium from the

**Table 2.7**

Characteristics of the Columns of Cryogenic Rectification Unit of the ITER Reactor's Isotope Separation System

| Parameter | Column | | | |
|---|---|---|---|---|
| | CD-1 | CD-2 | CD-3 | CD-4 |
| Pressure, kPa: | | | | |
| Column head | 150 | 125 | 100 | 100 |
| Column bottom | 155 | 130 | 105 | 105 |
| Temperature, K | | | | |
| Column head | 21.78 | 24.4 | 23.61 | 23.65 |
| Column bottom | 25.28 | 24.77 | 23.79 | 24.84 |
| Number of theoretical plates (NTP) | 130 | 120 | 120 | 65 |
| Diameter, cm: | | | | |
| Column head | 16.7 | 18.20 | 18.0 | 11.1 |
| Column bottom | 15.6 | 10.7 | — | 3.97 |
| Packing bed height, m | 7.15 | 6.60 | 6.60 | 3.60 |
| Wet reflux ratio | 3.7 | 125 | 600 | 10,0 |
| Intermediate vaporizer: | | | | |
| Plate number | — | 100 | — | 42 |
| Power, W | — | 1,537 | — | 500 |
| Overall holdup, mol: | | | | |
| $H_2$ | 366.7 | 0.015 | 0.009 | 0.07 |
| $D_2$ | 283.1 | 661.5 | 720.8 | 72.9 |
| $T_2$ | 9.0 | 8.3 | 0.005 | 35.0 |
| Condenser capacity, W | 1,406 | 2,562 | 2,052 | 700 |

flow P-2, as well as to process hydrogen (protium) waste flows (flow P-1), and water flows D-1 and D-2 with a sanitary-allowable residual concentration of tritium (see Figure 2.12).

To extract tritium from water flows (wastewater W-1, and, when using aqueous lithium salt, water flow from the blanket W-2) two water rectification columns (DW-1 and DW-2, respectively) are utilized. Four cryogenic rectification columns shown in Figure 2.12 as a cryogenic block (CB) serve to extract tritium from the gas phase (hydrogen) with its subsequent concentration.

From the bottoms of the columns DW-1 and DW-2, the tritium-enriched water is delivered to the units of catalytic isotope exchange with hydrogen coming from the cryogenic block (flows P-4 and P-5, respectively). The tritium-enriched hydrogen (flows P-1 and P-2) is returned to the cryogenic block. In addition, the block receives the hydrogen flows B-3 and B-5 associated with pellets injection into the plasma chamber (deuterium–tritium feeding), the hydrogen flow B-4 from high-temperature isotope exchange unit, as well as deuterium–tritium mixture not burnt out in the plasma chamber flow B-6.

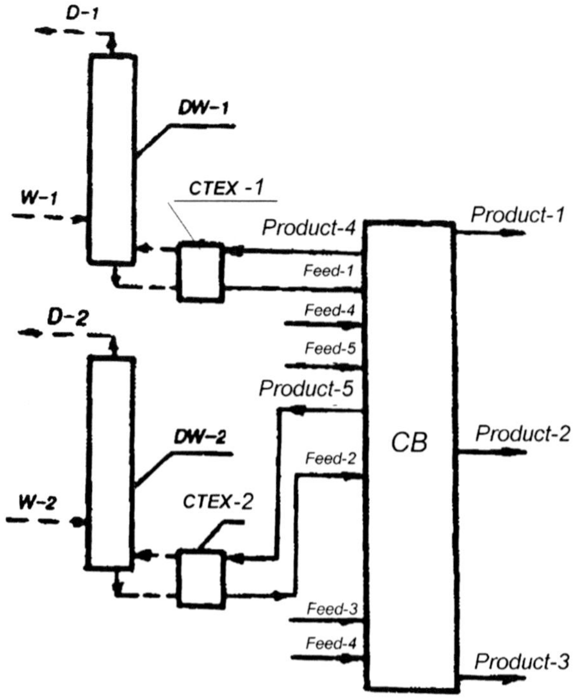


**Figure 2.12**   Unitary system of processing of all TNR tritium-containing flows (both gas and liquid).

As a result of comparison between alternative techniques, the water vacuum rectification was adopted for wastewater purification from tritium in the ISS isotope separation system. The estimated wastewater volume is 200kg/h, with an activity of 3.7GBq/kg, and a residual activity of 370kBq/kg. At first, the water collected from several sources is subjected to the purification from solute salts by ion exchange and distillation. Wastewater purified from tritium is withdrawn from the top of the column depletion section. Enriched with tritium in the bottom of the concentrating section of water rectification column DW-1, the concentrate is delivered for the catalytic vapour-phase isotope exchange with hydrogen CTEX-1, from where tritium-containing gas is supplied to the cryogenic rectification unit to extract tritium. Separated from gas by condensation, tritium-depleted water vapour is returned from the CTEX-1 unit to the corresponding section of the column DW-1. This scheme makes it possible to solve the problem of returning tritium to the fuel cycle.

Water rectification conditions are: a pressure of 11kPa in the head and 26kPa in the bottom, with a temperature of 321K and 329K, respectively. The total design NTP is 555, with depletion section accounting for 480 theoretical plates.

Wet reflux ratio is taken equal to 27. The column diameter is 1.7m, the overall design height of packing bed is 65m, realized, from design considerations, as three series-functioning columns of about 30m in height each, filled with regular packing material.

Despite the considerable dimensions of the water rectification columns, the tritium holdup, owing to a comparatively low concentration, is moderate and accounts for 0.07moles with respect to $T_2O$.

When using an aqueous lithium salt blanket (ALSB), the design isotope separation system incorporates one more water rectification units for the extraction of tritium formed in the blanket (column DW-2). The tritium-containing water is previously freed of solute salts and admixtures.

From the bottom of column DW-2 the tritium-enriched flow ($7 \times 10^{-4}$ fractions of HTO), at a rate of 180kg/h, is delivered, via single-stage catalytic isotope exchange of water vapour with hydrogen CTEX-2, to the cryogenic rectification unit for subsequent concentration. From the column DW-2 head, the tritium-depleted water (degree of tritium extraction from water is taken equal to 60% from 1,043.4 to 418.1GBq/kg) by way of salt separation unit is returned to the blanket. At a design water flow of 1,700kg/h, the column head and bottom diameters are 2.6m and 0.52m, respectively. The column is filled with Sulzer-type regular packing material. The project design is based on the results of practical operation of industrial plants for heavy water end concentration with the use of vacuum rectification method. The rectification conditions are as follows: pressure in the head 16kPa, and in the bottom 23kPa with a temperature of 329K and 336K, respectively. NTP is equal to 175, with a packing bed height of 21m, and the tritium holdup of 1.94mol with respect to $T_2O$.

## REFERENCES


1. B. M. Andreev, Ya. D. Zelvenskii, S. G. Katalnikov *Heavy Isotopes of Hydrogen in Nuclear Technology*, Energoatomizdat, **1987**, 456.
2. W. Jones *J. Chem. Phys.*, **1968**, 48, 207.
3. G. M. Merphy (ed.) *Production of Heavy Water*, **1955**, 394, McGraw-Hill Book Co., New York.
4. D. Bolme, Y. Young, *Trans. Am. Nucl. Soc.*, **1969**, 12, 496.
5. G. M. Keyser, D. B. McConnel, N. Anyas-Weiss, P. Kirkby, In: H. K. Rae (ed.) *Separation of Hydrogen Isotopes*, ACS Sympos. series, **1978**, 126, ACS, Washington.
6. W. P. Bebbington, V. R. Thayer, In: Proc. Second UN Intern. Conf. Peaceful Uses Atomic Energy, **1958**, v.4, P/1065, 527, United Nations, Geneva.
7. R. Zmasek, *Sulzer Technical Review*, Special Number "*Nuclex*", **1972**, 72, 46.
8. M. Huber, W. Meier, *Sulzer Technical Review*, **1975**, 57, 3.
9. "H. K. Sadhukhan, D. Behl, Ramraj, T. S. Iyengar, S. H. Sadarangani, P. K. Vaze, S. D. Soman, National Sympos. on Heavy Water Technology, Bombay (India), **1989**, preprint SA-7, Bhabha Atomic Res. Centr, Trombay.
10. G. Skorobogatov, *Proc. Acad. Sci. USSR, OkhN*, **1961**, 10, 1763.
11. H. Wolf, A. Hopfner, *Ber. Buns. Ger. Phys. Chem.*, **1969**, 73, 480.
12. G. Petersen, M. Benedict, *Nucl. Sci. Eng.*, **1963**, 15, 90.
13. W. Grothw, H. Yhle, A. Murrenholf, *Angew Chem.*, **1956**, 68, 605.
14. A. M. Rozen. *TOKhT*, **1993**, 27, 224.
15. M. Ravoire, K. Grancalat, G. J. Dirian, *Chem. Phys. Phys. Chem. Biol.*, **1963**, 60, 130.
16. V. M. Bakin, Ya. D. Zelvenskii *Atomnaya Energia*, **1971**, 30, 47.
17. I. Kirshenbaum *Heavy Water*. Inostrannaya Literatura, **1953**, 437.
18. Fn. Barr, W. Drew, *Chem. Eng. Progr.*, **1960**, 56, 54.

19. A. M. Rozen. *Khim. Prom.*, **1995**, 207.
20. A. M. Rozen. *Atomnaya Energia*, **1995**, 79, 221.
21. H. Mittelhauses, G. Thodos, *Cryogenics*, **1964**, 4, 368.
22. M. P. Malkov, A. G. Zeldovitch, A. B. Fradkov, I. B. Danilov, In: Proc. 2th UN Intern. Conf. Peaceful Uses Atomic Energy. **1958**, 4, 491, United Nations, Geneva.
23. M. P. Malkov, A. G. Zeldovich, A. B. Fradkov, I. B. Danilov, *Extraction of Deuterium from Hydrogen by Deep Chilling Method*, Gosatomizdat, Moscow, **1961**, 151.
24. Ya. D. Zelvenskii *Isotope Separation by Cryogenic Rectification*, RCTU im. D.I. Mendeleeva, **1998**, 208.
25. D. Gami, A. Ruptal, Third UN Intern. Conf. Peaceful Uses of Atomic Energy, **1964**, Rep 28/P/754, United Nations, Geneva.
26. H. Gutowski *Technica end Economica della Produzione di Acqua Pesant*, Roma, **1971**, 93.
27. G. Banicotes, E. Cimber, M. Sze, *Chem. Eng. Progr. Symp. Ser.*, 39. Nucl En. IX., **1960**, 58, 17.
28. W. Seddon, *Petrol Times*, **1958**, 62, 1588.
29. T. Yamanishi, H. Yoshida, H. Fukui,et al., *Nippon Genshiryoku Kenkyusho*, JAERI-M-88-254, **1988**.
30. I. A. Alekseev, I. A. Baranov, S. D. Bondarenko, S. N. Chernoby, O. A. Fedorchenko, G. A. Sukhorukova, V. D. Trenin, V. V. Uborski, *Fusion Technol*, **1995**, 28 (3), 1579.
31. T. Yamanishi, M. Enoeda, K. Okuno, *J. Nucl. Science and Technol.*, **1994**, 31, 937.
32. Ph. Pautrot, M. Domiani, Separat. Hydrogen Isotopes, *ACS Symp. Series*, **1978**, Washington, 68, 163; Ph. Pautrot, *Fusion Technol.*, **1988**, 14, 480.
33. R. Davidson, *et al. Fusion Technol.*, **1988**, 14, 472.
34. W. J. Holtslander, T. E. Harrison, J. D. Gallagher, *Fusion Technol*, **1988**, 14(9), 484.
35. W. J. Holtslander, T. E. Harrison, J. D. Gallagher, *Fusion Technol*, **1985**, 8, 2473.
36. F. Peretz, ORNL/TM - 12184, 2.
37. H. K. Sadhukhan, T. G. Varadharajan, T. Chandrasekharan, N. K. Nair, N. P. Sethuram, National Sympos. on Heavy Water Technology, Bombay (India), **1989**, prep. PD-5, 10, Bhabha Atomic Res. Centr, Trombay.
38. Progress of Cryogenics and Isotope Separation, **1999**, 3 + 4, 63, ICSI Rm, Valcea.
39. M. Yamaniski, M. Enoeda, K. Okuno, *I. Nucl. Sci. Technol.*, **1994**, 31, 937.
40. S. Mohan, K. Srinivasa, T. G. Varadarajan, H. K. Sadhukhan, National Sympos. on Heavy Water Technology, Bombay (India), **1989**, preprint PD-18, 10, Bhabha Atomic Res. Centr, Trombay.
41. Ph. Pautrot, M. Damiani, In: H. K. Rae (ed.) *Separation of Hydrogen Isotopes*, ACS Sympos. series, **1978**, 163, ACS, Washington.

# – 3 –

# Hydrogen Isotope Separation by Chemical Isotope Exchange Method in Gas–Liquid Systems

## 3.1   TWO-TEMPERATURE METHOD AND ITS MAIN FEATURES

### 3.1.1   Basic two-temperature schemes and cascades of two-temperature plants

The two-temperature method is based on the temperature dependence of the separation factor $\alpha$. Let us consider the simplest scheme of the heavy isotope concentration process (Figure 3.1a). The flow $L_1$ of substance X with a target isotope concentration $x_{F,1}$ is supplied to the upper column 1 as a feed flow. Counter-currently to the substance X flow, the flow $G_2$ of substance Y with a heavy isotope concentration $y_{F,2}$ is delivered to the column 2 bottom. In column 1, a counter-current separation process occurs at a temperature $T_1$ with a separation factor, as in the case of chemical exchange columns with conventional flow conversion. The heavy isotope is concentrated in the flow $L_1$ with the maximum concentration in the column bottom.

In the conventional separation process, the flow $L_1$ from the column 1 bottom should be fed to the flow-conversion system. In the two-temperature separation process, the substance X flow enriched with target isotope is delivered to the second column operated at a temperature $T_2$ with the separation factor $\alpha_2 < \alpha_1$. Consequently, in column 2 the target isotope transfers from the flow $L_2$ to the flow $G_2$, and, with the flow $G_2$, the isotope returns to column 1. Hence the need for the flow-conversion units is obviated. An exchange column operated at a temperature $T_2$ acts as a flow-conversion unit and provides for returning the isotope from the substance X flow to the substance Y flow. The separation plant efficiency is determined by $\alpha_1$, $\alpha_2$, and by the height of the columns. Since in all practically important operated systems employed in two-temperature plants the separation factor decreases with a rise in temperature, then $T_1 < T_2$, and column 1 is referred to as cold, and column 2 as hot.

In the same way as the conventional separation scheme, the two-temperature plant allows for withdrawing the product $B$, and for dumping the flow $L_2 = L_1 - B$, leaving column 2 as waste (Figure 3.1b). The target isotope concentration in the flow $L_2$ will be lower than in the feed flow $L_1$ ($x_{F,2} < x_{F,1}$). This can readily be seen, if both columns are assumed to be infinitely high and thus at the upper end of the cold column and at the bottom end of the hot column, an equilibrium will be established conforming, in the region of low concentrations of heavy isotope, to the equations $\alpha_1 = x_{F,1}/y_{F,1}$, and $\alpha_2 = x_{F,2}/y_{F,2}$. Hence it follows:

$$x_{F,1}/x_{F,2} = \alpha_1/\alpha_2. \tag{3.1}$$

And since $\alpha_1 > \alpha_2$, then $x_{F,1} > x_{F,2}$. In an actual plant with columns of a finite height $x_{F,1}/x_{F,2} < \alpha_1/\alpha_2$, but in all cases, except for non-withdrawal mode, $x_{F,1} > x_{F,2}$. Therefore, isotope exchange reactions at two different temperatures allow for isotope extraction from the feed flow, with its withdrawal at the maximum concentration site – between the columns.

Just like a conventional flow-conversion plant, the two-temperature plant can be a stage of a cascade, and can be operated in the mode of withdrawal of second kind (Figure 3.1c).

To improve the extraction degree, the columns of two-temperature plants must be provided with a depletion section. In this case, the feed flow can be delivered to a specific cross-section of column (between enrichment and depletion units) both of cold and of hot columns. Figure 3.1d represents a scheme with feed flow delivered to the hot column, and with withdrawal of second kind from the enrichment section of a two-temperature plant. The feed flow $F$ forms only a part of the aggregate flow of substance X which also becomes closed.

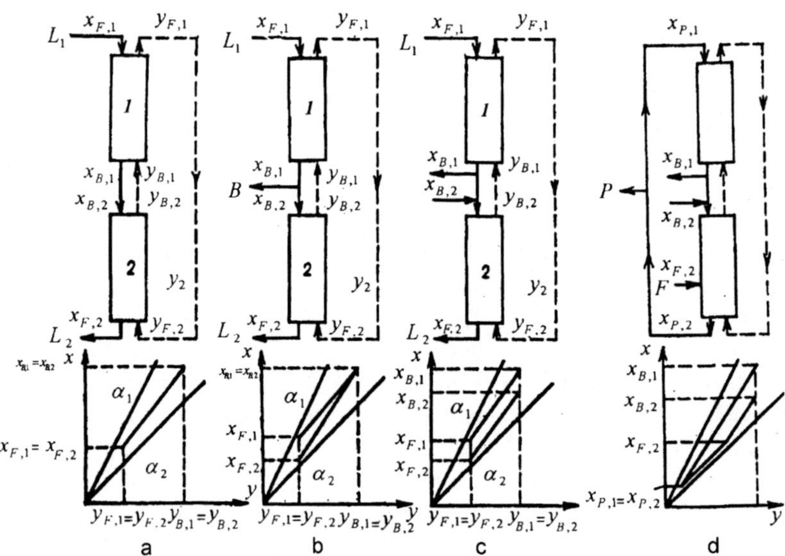

**Figure 3.1**   Schemes and $xy$-diagrams of liquid-fed two-temperature plants with various operating modes: a, without withdrawal; b, with withdrawal of first kind in the enrichment unit; c, with withdrawal of second kind in the concentration unit; d with withdrawal of second kind in the unit with depletion section.

All schemes presented in Figure 3.1 show two-temperature plants fed with substance X flow where, in equilibrium, the heavy isotope is concentrated. This occurs during deuterium extraction from deuterium–protium mixture by the two-temperature method in the water – hydrogen sulphide system.

Figure 3.1 presents as well the *xy*-diagrams for two-temperature plants discussed above, operating in the region of heavy isotope initial concentration.

Similar schemes with corresponding *xy*-diagrams can be readily produced and for cases where the plant is fed by the substance Y flow which, in equilibrium, is depleted of heavy isotope [1]. In this case, the substance X flow becomes closed. Since liquid–gas systems are of interest in isotope separation by the two-temperature method, from here on substance X will be considered as a liquid, and substance Y as a gas.

As opposed to column cascades with flow conversion (see section 1.4), the stages of two-temperature cascades can be interconnected either by a single flow (liquid or gas), or by both flows [1–3]. Figure 3.2 shows a two-stage initial concentration cascade with interstage liquid and gas flows, operating in the mode of withdrawal of second kind. While the cascading reduces capital and operation costs of separation, which are proportional to the columns' volume, the amount of changes in costs proportional to the flows depends on the design of the interstage flows. In the case of interconnection by both flows, the cascade aggregate flows remain identical to one of a single-stage plant, and the costs proportional to the flows do not increase. If the stages are interconnected only by the gas flow, the cascading results only in the aggregate liquid flow increase, and the gas flow remains constant. In the two-temperature method, the greater part of costs proportional to the flows is associated

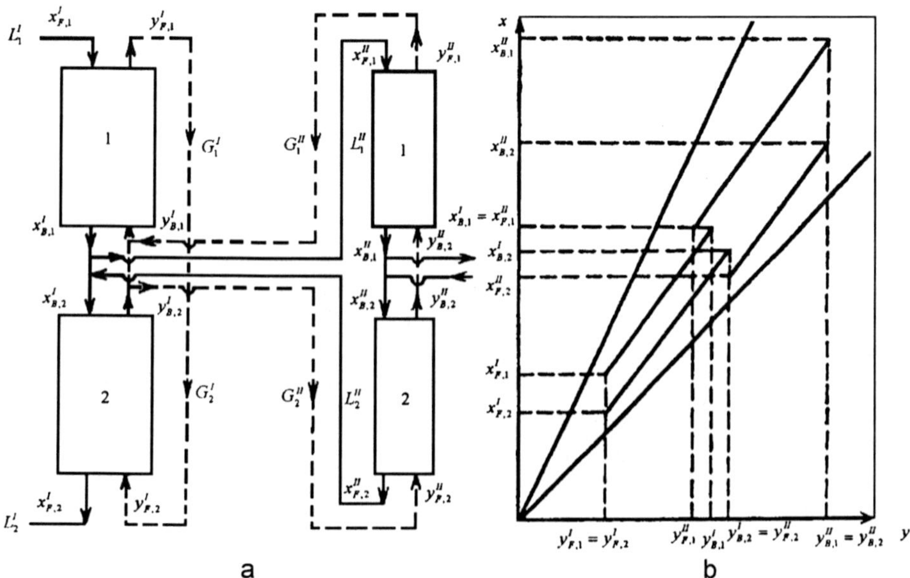

a         b

**Figure 3.2** Scheme (a) and *xy*-diagram (b) of two-stage cascade of two-temperature liquid-fed plants interconnected by both flows with withdrawal of second kind.

with heating of flows fed into the hot column. In the case of the hydrogen sulphide method, both heat consumption and capital investments in heat-exchange apparatus are determined chiefly by the gas flow. Hence the interconnection of cascade stages by the gas flow does not lead to a significant rise in heat consumption, nor to an increase in apparatus dimensions. In this regard, interstage connection by the liquid flow is the most inefficient.

Next, in selecting a scheme of interstage flows, consideration must be given to an additional mass transfer caused by reciprocal solubility of phases. Thus, in a cascade with interstage connection by the liquid flow, a continuous gas transfer from the preceding to the succeeding stage will take place due to the influence of temperature on gas solubility. This effect becomes significant in the cascade of two-temperature hydrogen sulphide plants. When feeding the succeeding stages of the cascade with gas flow, it is essential that the liquid condensate formed in the hot gas cooling be returned to the preceding stage.

In choosing a scheme of interstage flows, no less important are the problems of ensuring on-line control over the cascade operation (and, specifically, of maintaining optimum flow-rate ratio in all cascade columns). That is why the problem of the most rational scheme of flows between two-temperature plants of the cascade remains open, and all three types of interstage connection are currently employed by plants producing heavy-water by the hydrogen sulphide method.

Consider the problem of determining optimum parameters for two-temperature cascaded plants. Not dwelling on the selection of optimum temperatures in hot and cold columns determined for each particular system, it should be noted that in relevant columns of all stages a uniform temperature is maintained. For the initial concentration of heavy-water requiring a high separation degree, progressive-stage cascades with two or three stages are employed. The efficiency of such cascades depends heavily on the correctness of the stage distribution of aggregate separation degree. An analytical solution of the problem, though, has been obtained only for a two-temperature cascade with two stages [4].

The dependence of relative withdrawal $\theta$ corresponding to a minimum volume of exchange columns, on the separation degree for a two-temperature plant with withdrawal of second kind is found to be equal to that for separation columns with conventional flow conversion. Since the flow-dependent components make an important contribution both to operation costs and to capital investments in two-temperature plants, the product unit cost must serve as a criterion in determining $\theta_{opt}$. In this case, $\theta_{opt}$ is a function not only of the separation degree, but also of the quantity $G$ defining the ratio of costs proportional to the column volume and flow (see eq. (1.88)). For industrial two-temperature plants, the value $G = (C_L/C_V)\ln(\alpha_1/\alpha_2)/4$ varies between 3 and 6, and the optimum value $q$ proves to be significantly higher than the value corresponding to the minimum column volume (see Figure 1.7).

Let us consider one more peculiarity differentiating the two-temperature method from the separation processes discussed above. In a two-stage plant or in one stage of a cascade, the maximum concentration in the product is bounded by a point of the operating line intersection with an equilibrium curve, irrespective of the columns height. Up to this point, the operation of two-temperature plants in the region of low content of the target isotope has been considered. But at a high separation degree, the target isotope concentration may prove to be so high that the curvature of equilibrium line at the $xy$-diagram becomes very significant. In this case, the motive force of the isotope exchange process in the cold column bottom decreases. At a considerable content of target isotope, the two-temperature plant efficiency can be improved by varying the flow ratio in some parts of cold and hot columns. The flow ratio $\lambda$, for example, may be decreased by withdrawing a portion of gas flow from the hot column section j-j, and by returning it to a cold column section with the same concentration of $y_j$ in the gas [1,5] (Figure 3.3). From the $xy$-diagram it will be noticed that the enrichment could be much higher than that in a high-temperature plant with infinitely high columns, yet with a flow ratio being equal in all column areas $(x_{B,1} > x^\infty_{B,1})$. By withdrawing a portion of liquid flow from a section of the cold column and returning it to a corresponding section of the hot column, $\lambda$ can be increased in sections of hot and cold columns with a high target isotope content. Apropos, it is worth noting that simultaneous bypassing both by liquid and by gas results in a two-stage cascade with interstage connection by both flows.

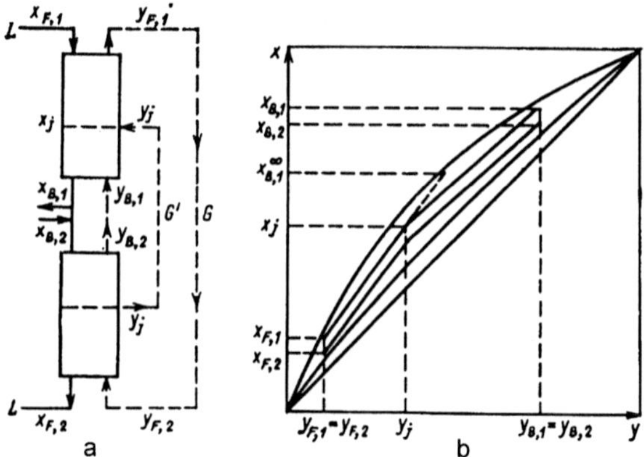

**Figure 3.3**   Scheme (a) and $xy$-diagram (b) of concentrating two-temperature plant with gas bypass.

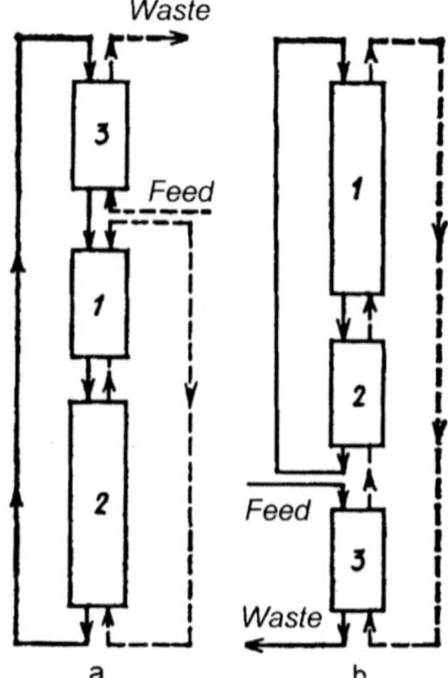


**Figure 3.4**   Schemes of two-temperature plants with feed column fed by gas (a) and by liquid (b).

**The** two-temperature method has the advantage of principal practicability of the separation process by the transit scheme. The scheme capabilities are significantly enhanced by the use of the concept of normalization of the waste flow isotopic composition with the aim of the flow recycling; that is to say, by the use of a so-called feed column. As applied to the two-temperature separation method, two possible variations of the feed column operation exist:

1.   the gas flow is the feed flow (Figure 3.4a). To increase target isotope concentration in the liquid flow feeding the two-temperature plant (in equilibrium, it is typically the liquid phase that is enriched with the target isotope), the separation factor should be at its maximum and the feed column operates at the temperature of a cold column;
2.   the liquid flow is the feed flow (Figure 3.4b). In this case, the highest degree of extraction will be achieved by lowering the separation factor, and consequently, by increasing the temperature in the feed column.

The utilization of feed columns in two-temperature processes is the most efficient, since it permits:

(1)   increase of the target isotope concentration in the flow feeding the two-temperature plant, and enhancement of the efficiency of separation columns;
(2)   loosening of the interdependence of the two-temperature plant and the raw material producer (as the raw material is passed only through the feed column, the two-temperature plant operation is not affected by supply failures, and vice versa);
(3)   avoiding power inputs and extra investments (e.g., for heat-exchange facilities) associated with the feed flow preprocessing, since in the general case the feed flow parameters may not suit to the optimum conditions of separation in two-temperature plants (with expenses on the compression of gas feed flow, such as hydrogen or ammonia–hydrogen mixture (AHM) in heavy-water production often being particularly considerable);
(4)   allowing for possibilities to ensure the optimum composition in closed circuits of gas and liquid, as, for example, a possibility to introduce isotope exchange catalysts improving the mass exchange of additives, as well as corrosion inhibitors for structural materials; a possibility to achieve a higher purity of flows and, consequently, the absence of admixtures poisoning the catalyst and degrading the performance of contactors (e.g. clogging the holes of perforated plates); and a possibility to reduce the section of hot and cold columns by the use of hydrogen for the process gas flow, instead of AHM representing the feed flow.


### 3.1.2   Extraction degree

The extraction degree in separation by the two-temperature method, as well as by other methods, is determined by the ratio between the plant capacity and the amount of isotope arriving with feed flow. Let us define the extraction degree of a plant with withdrawal of first or second kind.

For the withdrawal of second kind (see Figure 3.1c), the extraction degree is:

$$\Gamma = (L_1 x_{F,1} - L_2 x_{F,2})/(L_1 x_{F,1}) = 1 - x_{F,2}/x_{F,1}. \tag{3.2}$$

The maximum extraction degree (extraction degree for the cold and hot columns of infinite height) equals

$$\Gamma_m = 1 - \alpha_2/\alpha_1. \tag{3.3}$$

Thus, the extraction degree $\Gamma_m$ is determined only by the ratio between separation factors in columns 2 and 1.

For the withdrawal of first kind (Figure 3.1b), the extraction degree is:

$$\Gamma_1 = \frac{B x_B}{L_1 x_{F,1}} = \frac{L_1 x_{F,1} - L_2 x_{F,2}}{L_1 x_{F,1}} = 1 - \frac{x_{F,2}}{x_{F,1}} + \frac{B}{L_1} \times \frac{x_{F,2}}{x_{F,1}}. \tag{3.4}$$

By application of eq. (3.1), we obtain the maximum extraction degree

$$\Gamma_{1m} = 1 - \frac{\alpha_2}{\alpha_1} + \frac{B_m}{L_1} \times \frac{\alpha_2}{\alpha_1}, \tag{3.5}$$

where $B_m$ is the maximum withdrawal flow corresponding to the maximum productivity.

From eq. (3.4) it follows that

$$\Gamma_{1m} = K_{x1} B_m / L_1, \tag{3.6}$$

where $K_{x1} = x_B/x_{F,1}$.

By equating the right sides of eqs. (3.5) and (3.6), we obtain

$$B_m/L_1 = (1 - \alpha_2/\alpha_1)/(K_{x1} - \alpha_2/\alpha_1). \tag{3.7}$$

Substituting eq. (3.7) into eq. (3.5) we have the following expression for the maximum extraction degree [1, 5]:

$$\Gamma_{1,m} = \Gamma_m + \frac{\Gamma_m}{(K_{x1}\alpha_1/\alpha_2) - 1} = \frac{\Gamma_m}{1 - (1 - \Gamma_m)/K_{x1}}. \tag{3.8}$$

The maximum extraction degree in a two-temperature plant with withdrawal of first kind is somewhat higher than that with withdrawal of second kind. The maximum extraction degree of an independent two-temperature plant (or the last cascade stage) depends not only on separation factors of the columns but also on the separation degree achievable in the plant: the higher $K$ is, the less the difference between maximum extraction degree values for the withdrawal of first and second kind.

For a plant with columns of a finite height, the practically achievable extraction degree will depend on the relative withdrawal $\theta$. In accordance with eq. (1.63), the actual extraction degree $\Gamma = \theta \cdot \Gamma_m$.

### 3.1.3   Steady state of the two-temperature plant

Let us consider the steady state of a two-temperature plant for the case of low concentration of heavy isotope. This case encompasses hydrogen isotope separation processes occurring in practice. Figure 3.5 shows the general scheme of a plant with columns 1 and 2 with heights $H_1$ and $H_2$, respectively, as well as indicating the agreed notations of isotope concentrations in the flows at the ends of the columns. The steady state of the process can be described by a differential equation system for each column, much as it was described by A. Rozen [6, 8] and K. Bier [9].

$$\begin{cases} L_1 dx_1/dz_1 = K_{0Y,1}(y_1 - x_1/\alpha_1)\alpha_{K1}S_1 \\ G_1 dy_1/dz_1 = K_{0Y,1}(y_1 - x_1/\alpha_1)a_{K1}S_1 \end{cases}, \qquad (3.9)$$

$$\begin{cases} L_2 dx_2/dz_2 = K_{0Y,2}(x_2/\alpha_2 - y_2)\alpha_{K2}S_2 \\ G_2 dy_2/dz_2 = K_{0Y,2}(x_2/\alpha_2 - y_2)a_{K2}S_2 \end{cases}, \qquad (3.10)$$

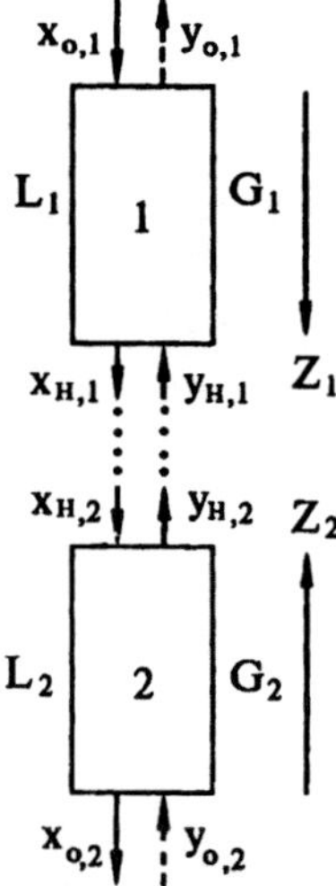


**Figure 3.5**   Scheme of two combined columns.

where $K_{0Y_i}$ is the mass-transfer coefficient; $a_{Ki}$ is the specific surface of phase contact; $S_i$ is the column cross-sectional surface; $z_i$ is the coordinate of the column height; $i = 1, 2$ for the column 1 and 2, respectively.

The solution of the above equations results in the following expressions linking the separation degree with flow ratio and NTU in the cold ($N_{y,1}$) and hot ($N_{y,2}$) columns [10]:

$$K_{x,1} - 1 = \frac{A_1 A_2 (\varphi_x \varphi_y \alpha_1 - \alpha_2)}{A_1 \varphi_y \psi_x (\lambda_2 - \alpha_2) \alpha_1 / \lambda_2 + A_2 (\alpha_1 - \lambda_1) \alpha_2 / \lambda_1}; \qquad (3.11)$$

$$K_{y,2} - 1 = \frac{A_1 A_2 (\varphi_x \varphi_y \alpha_1 - \alpha_2)}{A_1 (\lambda_2 - \alpha_2) + A_2 \varphi_x \psi_y (\alpha_1 - \lambda_1)}, \qquad (3.12)$$

where $K_{x,1} = x_{H,1}/x_{0,1}$; $K_{y,2} = y_{H,2}/y_{0,2}$; $\lambda = G_i/L_i$;

$$A_1 = \exp\left[\frac{\alpha_1 - \lambda_1}{\alpha_1} N_{y,1} - 1\right]; \quad A_2 = \exp\left[\frac{\lambda_2 - \alpha_2}{\alpha_2} N_{y,2} - 1\right].$$

In eqs. (3.11) and (3.12), the type of two-temperature scheme is accounted for by two parameters defined as the ratio of concentrations in liquid and gas flow at $z_i = 0$ ($\varphi_x = x_{0,2}/x_{0,1}$ and $\varphi_y = y_{0,1}/y_{0,2}$), as well as by two parameters depending on the isotope concentration variation in each flow (X-substance and Y-substance) at the boundary between columns 1 and 2,

$$\psi_X = \frac{x_{H,2} - x_{0,2}}{x_{H,1} - x_{0,1}} \quad \text{and} \quad \psi_Y = \frac{y_{H,1} - y_{0,1}}{y_{H,2} - y_{0,2}}.$$

In the non-withdrawal mode, all four parameters are equal to 1 ($\varphi_x = \varphi_y = \psi_x = \psi_y = 1$). In the mode of withdrawal of second kind only one parameter differs from 1 and equals $1 - \theta\Gamma_m$: with a liquid feed flow $\varphi_x = 1 - \theta\Gamma_m$, and with a gas feed flow $\varphi_y = 1 - \theta\Gamma_m$. In the mode of withdrawal of first kind, when $\rho = B/L_1$, there are already two parameters that are different from 1:

$$\varphi_x = 1 - \theta\Gamma_m \text{ and } \psi_x = 1/(1-\rho) - \text{for a liquid feed flow;}$$
$$\varphi_y = 1 - \theta\Gamma_m \text{ and } \psi_y = 1/(1-\rho) - \text{for a gas feed flow.}$$

Eqs. (3.11) and (3.12) are also applicable to the two-temperature plant calculation through NTP. In this case,

$$A_1 = \left(\frac{\alpha_1}{\lambda_1}\right)^{n_1} - 1 \quad \text{and} \quad A_2 = \left(\frac{\alpha_2}{\lambda_2}\right)^{n_2} - 1.$$

It is necessary to note that for different schemes of two-temperature plants, there will also be different dependences of the separation degree on the main parameters of a plant: separation factors ($\alpha_1$ and $\alpha_2$), NTU (or NTP) and flow ratio in the columns $\lambda_i$.

As contrasted with the enrichment in traditional columns, the two-temperature separation method is characterized by a peculiar dependence of the separation degree on the flow ratios with a pronounced optimum at a certain (optimal) flow-rate ratio. For all schemes and modes of operation, the position of the optimum is determined by the values of separation factors $\alpha_1$ and $\alpha_2$, as well as by the ratio of heights of the hot and cold columns. In the calculation by methods of transfer units or NTP, the enrichment dependencies on the flow-rate ratio, resulting from eqs. (3.11) or (3.12), although of a similar nature, lead to some differences between the optimal flow-rate ratio values. In general, the analytic form of the two-temperature plant optimum condition corresponding to the maximum separation degree is based on the equality of so-called separabilities $A_1 + 1$ and $A_2 + 1$.

At the indicated condition, we can derive from eqs. (3.11) and (3.12) the following expression for the minimum NTU (or NTP) required to obtain a desired separation degree at a ratio of NTU (or NTP) in the cold and hot columns equal to $\chi$ (depending on the calculation method, $\chi$ is expressed by $\chi_y = N_{y1}/N_{y2}$, $\chi_x = N_{x1}/N_{x2}$ or $\chi_n = n_1/n_2$ [10]):

$$N_1 = \chi N_2 = a \ln\left[\frac{K-u}{1-u}\right],$$

(3.13)

where by calculating through separation degree $K_{x,1}$

$$1 - u_{x,1} = \frac{\varphi_x \varphi_y \alpha_1 - \alpha_2}{\varphi_y \psi_x (\lambda_2 - \alpha_2)\alpha_1/\lambda_2 + (\alpha_1 - \lambda_1)\alpha_2/\lambda_1},$$

(3.14)

and through separation degree $K_{y,2}$

$$1 - u_{y,2} = \frac{\varphi_x \varphi_y \alpha_1 - \alpha_2}{\lambda_2 - \alpha_2 + \varphi_x \psi_y (\alpha_1 - \lambda_1)}.$$

(3.15)

Table 3.1 [10] shows the optimum of flow-rate ratios $\lambda_1$ and $\lambda_2$ found from the equality of the hot and cold column separabilities with regard to the equation $\lambda_2 = \lambda_1 \psi_x \psi_y$ and parameter $a$.

At equal values of NTU (or NTP) in the cold and hot columns, the minimum overall height of the columns is obtained at $\chi = 1$. This very much simplifies eq. (3.13) for the minimum NTU (or NTP), as well as the expressions for the optimum values of $\lambda_1$ and $\lambda_2$, presented in Table 3.2 for the basic operation modes of two-temperature plants [10].

For the non-withdrawal mode and for the withdrawal of second kind, as is evident from Table 3.2, the optimum flow-rate ratio is equal to the geometric mean (when calculated by NTP), or to the arithmetic mean (when calculated by liquid-phase NTU) of the separation factors $\alpha_1$ and $\alpha_2$, respectively (when calculated by gas-phase NTU, it is the ratio of liquid and gas flow rates that equals the arithmetic mean of the values $1/\alpha_1$ and $1/\alpha_2$, that is, $1/\lambda_0$).

**Table 3.1**

Expressions for calculation of optimum flow-rate ratios ($\lambda_1$ and $\lambda_2$) and parameter $a$ in eq. (3.13) with regard to heights of cold and hot columns

| Parameter | NTU calculation | | NTP calculation |
| --- | --- | --- | --- |
| | Gas phase | Liquid phase | |
| $\lambda_1$ | $\dfrac{1+\chi_y}{\dfrac{\chi_y}{\alpha_1}+\dfrac{\psi_x\psi_y}{\alpha_2}}$ | $\dfrac{\alpha_1\chi_x\psi_x\psi_y+\alpha_2}{(1+\chi_x)\psi_x\psi_y}$ | $\sqrt[1+\chi_n]{\alpha_1^{\chi_n}\alpha_2/(\psi_x\psi_y)}$ |
| $\lambda_2$ | $\dfrac{(1+\chi_y)\psi_x\psi_y}{\dfrac{\chi_y}{\alpha_1}+\dfrac{\psi_x\psi_y}{\alpha_2}}$ | $\dfrac{\alpha_1\chi_x\psi_x\psi_y+\alpha_2}{1+\chi_x}$ | $\sqrt[1+\chi_n]{\alpha_1^{\chi_n}\alpha_2\psi_x\psi_y}$ |
| $a$ | $\dfrac{\alpha_1}{\alpha_1-\lambda_1}=\chi_y\dfrac{\alpha_2}{\lambda_2-\alpha_2}$ | $\dfrac{\lambda_1}{\alpha_1-\lambda_1}=\chi_x\dfrac{\lambda_2}{\lambda_2-\alpha_2}$ | $\dfrac{1}{\ln(\alpha_1/\lambda_1)}=\dfrac{\chi_n}{\ln(\lambda_2/\alpha_2)}$ |

The dependence of the separation degree $K_{x1} = K_{y2}$ on the flow-rate ratio $\lambda_1=\lambda_2=\lambda$ and on $\chi_n$ can be illustrated by a simple example of a two-temperature plant operating in the non-withdrawal mode (the calculations was performed by NTP) for isotope exchange in the hydrogen–hydrosulphide system at $T = 303$ and $403K$ ($\alpha_1 = 2.34$ and $\alpha_2 = 1.84$) [11]. As evident from Figure 3.6, with an increase in the heights of the cold and hot columns, the separation degree dependence on $\lambda$ becomes sharper, which requires us to more accurately maintain a certain level of $\lambda_0$. Figure 3.7 shows the separation degree dependence on $\chi_n$ at the same values of $\alpha_1$, $\alpha_2$ and an optimum flow-rate ratio $\lambda_0$ for each value of $\chi_n$. The same figure presents the dependence $\lambda_0 = f(\chi_n)$ calculated by $\lambda_0 = (\alpha_1^{\chi^n}\alpha_2)^{1/(1+\chi_n)}$ (see Table 3.1). As might be expected, the maximum value of $K$ is obtained at $\chi = 1$, and an increase in $\chi$ will cause the optimum flow-rate ratio value to rise (while a drop in $\chi$ involves an increase in the ratio value). The condition $\alpha_1 > \lambda_0 > \alpha_2$ is fulfilled, however, at any ratio of heights of the hot and cold columns.

Another peculiarity of the two-temperature method is the dependence of the isotope concentration distribution along the column height on the flow-rate ratio. The reason is that the concentration profile depends on the relative positions of operating lines and equilibrium curves for the hot and cold columns determined on the $xy$-diagram, which determines the variation of the motive force of interphase isotope exchange along the column height [1, 5, 12].

Figure 3.8 presents basic patterns of the isotope concentration variations along the column height, in one of the phases in the region of low content of one of the isotopes separated in a non-withdrawal mode. At $\lambda_0$, the process motive force is greatest and steadily increases towards the enriched ends of the cold and hot columns, i.e. in the direction of $z_1$

**Table 3.2**

Expression for calculation for optimum flow-rate ratio $\lambda_1$ and $\lambda_2$ and minimum NTU or NTP at $\chi = 1$

| Operation mode | NTU calculation | | | | | NTP calculation | | |
| --- | --- | --- | --- | --- | --- | --- | --- | --- |
| | Gas phase | | Liquid phase | | NTU | $\lambda_1$ | $\lambda_2$ | NTP $(n_1 = n_2)$ |
| | $\lambda_1$ | $\lambda_2$ | $\lambda_1$ | $\lambda_2$ | $(N_{y,1} = N_{y,2}$ or $N_{x,1} = N_{x,2})$ | | | |
| Non-withdrawal | $\dfrac{2\alpha_1\alpha_2}{\alpha_1+\alpha_2}$ | $\dfrac{2\alpha_1\alpha_2}{\alpha_1+\alpha_2}$ | $\dfrac{\alpha_1+\alpha_2}{2}$ | $\dfrac{\alpha_1+\alpha_2}{2}$ | $\dfrac{\alpha_1+\alpha_2}{\alpha_1-\alpha_2}\ln K_B$ | $\sqrt{\alpha_1\alpha_2}$ | $\sqrt{\alpha_1\alpha_2}$ | $\dfrac{2\ln K_B}{\ln(\alpha_1/\alpha_2)}$ |
| Withdrawal of second kind | $\dfrac{2\alpha_1\alpha_2}{\alpha_1+\alpha_2}$ | $\dfrac{2\alpha_1\alpha_2}{\alpha_1+\alpha_2}$ | $\dfrac{\alpha_1+\alpha_2}{2}$ | $\dfrac{\alpha_1+\alpha_2}{2}$ | $\dfrac{\alpha_1+\alpha_2}{\alpha_1-\alpha_2}\ln\dfrac{K_B-\theta}{1-\theta}$ | $\sqrt{\alpha_1\alpha_2}$ | $\sqrt{\alpha_1\alpha_2}$ | $\dfrac{2\ln[(K_B-\theta)/(1-\theta)]}{\ln(\alpha_1/\alpha_2)}$ |
| Withdrawal of first kind | $\dfrac{2\alpha_1\alpha_2(1-\rho)}{\alpha_1+\alpha_2(1-\rho)}$ | $\dfrac{2\alpha_1\alpha_2}{\alpha_1+\alpha_2(1-\rho)}$ | $\dfrac{\alpha_1+\alpha_2(1-\rho)}{2}$ | $\dfrac{\alpha_1+\alpha_2(1-\rho)}{2(1-\rho)}$ | $\dfrac{\alpha_1+\alpha_2(1-\rho)}{\alpha_1-\alpha_2(1-\rho)}\ln\dfrac{K_B}{1-\theta}$ | $\sqrt{\alpha_1\alpha_2(1-\rho)}$ | $\sqrt{\dfrac{\alpha_1\alpha_2}{1-\rho}}$ | $\dfrac{2\ln[K_B/(1-\theta)]}{\ln[\alpha_1/[\alpha_2(1-\rho)]]}$ |

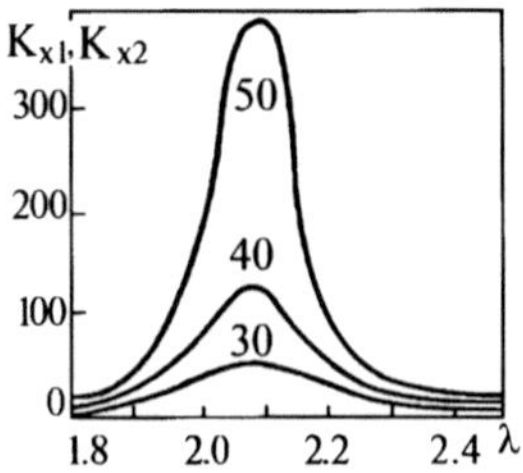

**Figure 3.6**  Dependence of $K_{X,1} = K_{Y,2}$ on $\lambda$ in the non-withdrawal mode at $\chi_n = 1$ (under a curves the values of NTP in one column are specified).

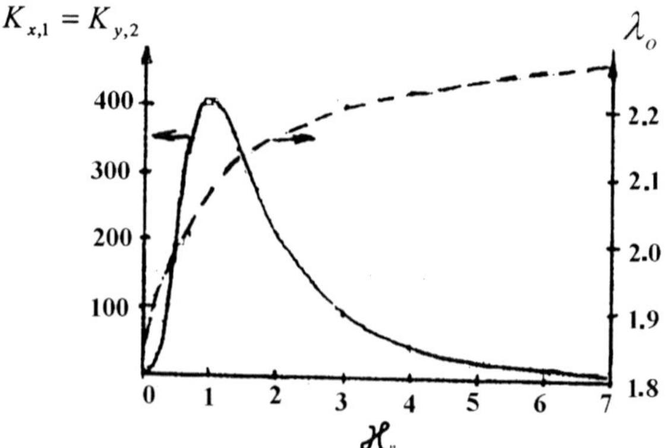

**Figure 3.7**  Dependences of $K_{X,1} = K_{Y,2}$ (continuous line) and $\lambda_0$ (dashed line) on $\chi_n$ in the non-withdrawal mode at $n_1 + n_2 = 100$.

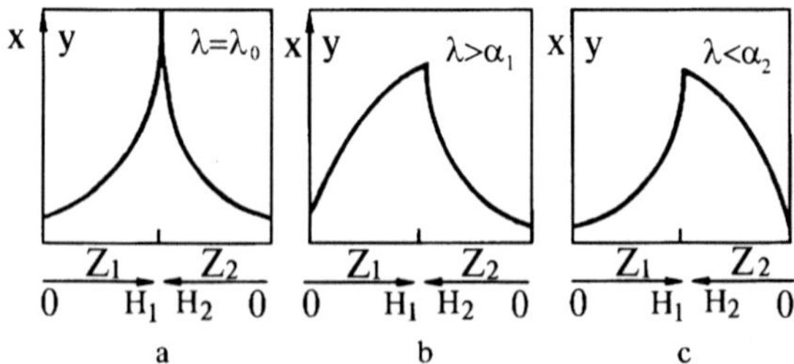

**Figure 3.8**  Basic patterns of the column concentration profiles in the non-withdrawal mode and at various flow-rate ratios $\lambda$.

and $z_2$ coordinates (the concentration profile is represented in Figure 3.8a). At $\lambda > \lambda_0$, the process motive force decreases in the cold column enriched end-area and in the hot column depleted end-area (Figure 3.8b). If $\lambda < \lambda_0$, the column concentration profile will be similar to that presented in Figure 3.8b. In particular cases at $\lambda = \alpha_1$ or $\lambda = \alpha_2$, the motive force through the whole height of the cold or hot column, respectively, is equal, and the concentration–height dependence in this column is linear [13].

For a two-temperature plant operating in the region of medium isotope content, the concentration distribution along the column height depends on the equilibrium conditions determined by the shape of equilibrium curves in the $xy$-diagram operating region. It makes no sense to consider this issue in greater detail since practical application of the two-temperature method is at present confined to the region of low concentration of one of the isotopes in the separated mixture.

If the determination of isotope composition variations in the feed flow, and, hence, the concentration profile in this phase, creates no difficulties, the concentration in the circulating phase cannot be obtained from material balance equations. One of the phase concentrations, and, accordingly, the position operating lines on the $xy$-diagram, can be defined with relative ease by introducing the notion of degree of approximation to the equilibrium conditions at the cold column head ($\eta_1 = x_{01}/(\alpha_1 y_{01})$, or at the hot column bottom ($\eta_2 = \alpha_2 y_{02}/x_{02}$). At a known value of the relative withdrawal $\theta$, it will suffice to define the degree of approximation to the equilibrium conditions in one column, since the following expression is always true:

$$\eta_1 \eta_2 = \alpha_2/(\alpha_1 \varphi_x \varphi_y) = (1 - \Gamma_m)/(1 - \theta \Gamma_m). \tag{3.16}$$

The equations for the degree of approximation to the equilibrium conditions at $n_1 = n_2$ for the withdrawal of second kind were derived with the use of the NTP method [5]:

$$1/\eta_{\text{feed}} = 1/2 + \varphi_x \varphi_y \alpha_1/2\alpha_2; \tag{3.17}$$

$$\eta_{\text{waste}} = 1/2 + \alpha_2/2\alpha_1 \varphi_x \varphi_y, \tag{3.18}$$

where for the liquid feed flow $\eta_{\text{feed}} = \eta_1$ and $\eta_{\text{waste}} = \eta_2$, and for the gas feed flow $\eta_{\text{feed}} = \eta_2$ and $\eta_{\text{waste}} = \eta_1$.

### 3.1.4  Effect of mutual solubility of phases

The above-discussed theory of isotope separation by the two-temperature method means that the isotope exchange occurs between pure substances, i.e. between the liquid phase of X-substance and the gas phase of Y-substance. All practically applied systems, however, including the water – hydrogen sulphide system employed for heavy-water production, are characterized by a considerably high solubility of one phase in the other. This results in generating additional circulating gas and liquid flows of which the rates depend on the plant scheme and operation mode. Let us consider in general terms the most practically interesting case of the effects of mutual solubility of phases for a scheme with the liquid feed flow and with closed gas flow. For this scheme, the additional flows resulting from the temperature influence on the gas solubility and on the pressure of saturated vapours of the liquid phase are shown in Figure 3.9.

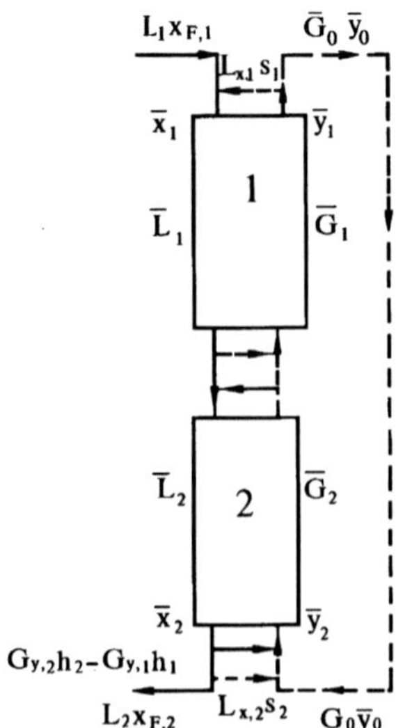



**Figure 3.9** Scheme of flows in two-temperature plants with liquid feed flow to cold column.

Owing to the mutual solubility of the phases, each flow consists of two exchanging substances[10]:

$$\bar{L}_i = L_{xi} + L_{yi} = \bar{L}_i(S_{xi} + S_{yi}) = L_{xi}(1 + s_i);\tag{3.19}$$

$$\bar{G}_i = G_{yi} + G_{xi} = \bar{G}_i(H_{yi} + H_{xi}) = G_{yi}(1 + h_i);\tag{3.20}$$

where $L_{xi}$, $L_{yi}$ are the liquid flow $X$ and dissolved gas flow $Y$, respectively; $\bar{L}_i$ is the flow of the liquid with dissolved gas; $G_{yi}$, $G_{xi}$ are the gas flow of substance $Y$ and vapour flow of substance $X$, respectively; $\bar{G}_i$ is the gas-vapor flow; $S_{xi}$, $S_{yi}$, $H_{xi}$, $H_{yi}$ are mole fractions of substances $X$ and $Y$ in the liquid and gas-vapor flows, respectively: $s_i$ is the solubility of the gas $Y$ in liquid $X$, [mol $X$ /mol $Y$]; $h_i$ is the content of water vapor $X$ in gas $Y$, [mol $X$ /mol $Y$].

Let us notice, if substances $X$ and $Y$ contain different numbers of exchanging hydrogen atoms, $s$ and $h$ must be expressed through the ratio of hydrogen gram-atoms contained in corresponding substances.

The gas–liquid flow-rate ratio can be derived by:

$$\bar{\lambda}_i = \bar{G}_i/\bar{L}_i = G_{yi}(1 + h_i)/L_{xi}(1 + s_i) = \lambda_i(1 + h_i)/(1 + s_i).\tag{3.21}$$

As a consequence of contact with the gas flow leaving the column 1, the feed flow $F$ is saturated by gas in an amount $L_{x1}s_1$. Upon the passage through the cold column, the liquid, heated to the temperature $T_2$, releases gas in an amount $L_{x1}s_1 - L_{x2}s_2$; on leaving the hot column, dissolved gas is separated from the waste flow and refluxed to the column (flow $L_{x2}s_2$). Evidently, the liquid vapor flow which saturates gas in an amount $G_{y2}h_2 - G_{y1}h_1$, is caused by the gas flow heating to the temperature $T_2$. The vapor then condenses as the gas flow cools down, which leads to the additional condensate flow formation. A portion of the liquid vapor will remain in the circulating gas flow. In two-temperature hydrogen sulphide plants operating at $T_1 = 303$ K, for example, the water vapor flow saturating hydrogen sulphide in the cold column is delivered together with gas to the circulation pump, and then to the hot column.

From the material balance equations for substances X and Y at the enriched column end of a two-temperature plant operating in the concentrating mode with the withdrawal of first kind, the following equations are derived:

$$L_{x2} - L_{x1} + B = G_{x2} - G_{x1} = G_{y2}h_2 - G_{y1}h_1, \tag{3.22}$$

$$G_{y1} - G_{y2} = L_{y1} - L_{y2} = L_{x1}s_1 - L_{x2}s_2. \tag{3.23}$$

From eq. (3.22) it follows that

$$L_{x2}(1 - \lambda_2 h_2) = L_{x1}(1 - \rho - \lambda_1 h_1). \tag{3.24}$$

Similarly, from eq. (3.23) we obtain

$$L_{x2}(\lambda_2 - s_2) = L_{x1}(\lambda_1 - s_1). \tag{3.25}$$

And finally

$$L_{x2}/L_{x1} = l = (1 - \rho - \lambda_1 h_1)/(1 - \lambda_2 h_2) = (\lambda_1 - s_1)/(\lambda_2 - s_2). \tag{3.26}$$

This equation allows for relating flow-rate ratios in the cold and hot columns ($\lambda_1$ and $\lambda_2$), of which the differencies are determined by the mutual solubility of phases and withdrawal of the product $B$. Thus, from eq. (3.26) it follows that

$$\lambda_2 = \frac{\lambda_1(1 - s_2 h_1) - s_1 + s_2(1 - \rho)}{\lambda_1(h_2 - h_1) + 1 - \rho - s_1 h_2}. \tag{3.27}$$

By reason of mutual solubility, in the calculation of the flow isotope concentration account must be taken of the content of another phase substance, with its isotope composition. As shown in section 1.1, the effect of mutual solubility of phases on the TDIE value in such a heterogenous system is accounted for by introducing effective concentrations $\bar{x}$ and $\bar{y}$, and effective separation factor $\bar{\alpha}$.

The steady state of the processes occuring in the real conditions of the mutual solubility of phases can be described by a system of differential eqs. (3.9) and (3.10), if such terms as the flows $L_i$ and $G_i$, concentrations $x_i$ and $y_i$, and separation factors $\alpha_i$, are substituted by their corresponding effective values [10]. It is evident that the solution of the set of equations can be obtained in the form of eqs. (3.11) and (3.12), of which all the parameters must be viewed just as their corresponding effective quantities:

$$\bar{K}_{X,1} - 1 = \frac{\bar{A}_1 \bar{A}_2 (\bar{\varphi}_X \bar{\varphi}_Y \bar{\alpha}_1 - \bar{\alpha}_2)}{\bar{A}_1 \bar{\varphi}_Y \bar{\psi}_X (\bar{\lambda}_2 - \bar{\alpha}_2) \bar{\alpha}_1 / \bar{\lambda}_2 + \bar{A}_2 (\bar{\alpha}_1 - \bar{\lambda}_1) \bar{\alpha}_2 / \bar{\lambda}_1}; \tag{3.28}$$

$$\bar{K}_{Y,2} - 1 = \frac{\bar{A}_1 \bar{A}_2 (\bar{\varphi}_X \bar{\varphi}_Y \bar{\alpha}_1 - \bar{\alpha}_2)}{\bar{A}_1 (\bar{\lambda}_2 - \bar{\alpha}_2) + \bar{A}_2 \bar{\varphi}_X \bar{\psi}_Y (\bar{\alpha}_1 - \bar{\lambda}_1)}. \tag{3.29}$$

Apart from the above parameters $\bar{\alpha}_i$ and $\bar{\lambda}_i$ involved in eqs. (3.28) and (3.29), the effective quantities $\bar{\varphi}_X$, $\bar{\varphi}_Y$, $\bar{\psi}_X$, $\bar{\psi}_Y$, $\bar{K}_{X,1}$, and $\bar{K}_{Y,2}$, can also be expressed through effective concentrations.

Here, allowance should be made for the fact that under the mass-transfer conditions in counter-current flow of the phases, not only the concentrations $x_i$ and $y_i$ differ from equilibrium concentrations, but so do those concentrations the difference between which is possibly due to the isotope effects in the evaporation of liquid or dissolution of gas. Next, considering that $\alpha_{PH}$ and $\alpha_S$ approach unity, it can be assumed that, for the calculation of the effective concentrations in the column, $x_{PH} = x$ and $y_S = y$. Parameters $\bar{\varphi}_X$, $\bar{\varphi}_Y$, $\bar{\psi}_X$, and $\bar{\psi}_Y$ will then be expressed in the following manner [10]:

$$\bar{\varphi}_X = \frac{\bar{x}_{0,2}}{\bar{x}_{0,1}} = \varphi_X \frac{m + s_2 / (\varphi_X \varphi_Y)}{m + s_1} \times \frac{1 + s_1}{1 + s_2}; \tag{3.30}$$

$$\bar{\varphi}_Y = \frac{\bar{y}_{0,1}}{\bar{y}_{0,2}} = \varphi_Y \frac{1 + h_1 m}{1 + \varphi_X \varphi_Y h_2 m} \times \frac{1 + h_2}{1 + h_1}; \tag{3.31}$$

$$\bar{\psi}_X = \frac{\bar{x}_{H,2} - \bar{x}_{0,2}}{\bar{x}_{H,1} - \bar{x}_{0,1}} = \psi_X \frac{1 + s_2 / \lambda_2}{1 + s_1 / \lambda_1} \times \frac{1 + s_1}{1 + s_2}; \tag{3.32}$$

$$\bar{\psi}_Y = \frac{\bar{y}_{H,1} - \bar{y}_{0,1}}{\bar{y}_{H,2} - \bar{y}_{0,2}} = \psi_Y \frac{1 + h_1 \lambda_1}{1 + h_2 \lambda_2} \times \frac{1 + h_2}{1 + h_1}. \tag{3.33}$$

Parameters $\varphi_X$, $\varphi_Y$, $\psi_X$, and $\psi_Y$, have been evaluated above, and $m = x_{01}/y_{01} = \alpha_1 \eta_1$ is dependent on the plant operation mode and on $\theta$. For the concentrating plants with withdrawal of first and second kind operating at $\theta$ close to unity, it can be assumed that $m \approx \alpha_1$, and for non-withdrawal mode, at $n_1 = n_2$ and $\lambda = \lambda_0$, that $m \approx \lambda_0$.

It is convenient to characterize the plant performance by the separation degree determined from isotope concentrations in the "pure" phase related to the effective separation degree by the following equation:

$$\bar{K}_{X,1} = \bar{x}_{H,1}/\bar{x}_{0,1} = (K_{X,1} + K_{Y,1}s_1/m)/(1+s_1/m). \tag{3.34}$$

The separation degree $K_{Y,1}$, involved in the equation, can be derived from $(K_{x,1}-1)/(K_{y,1}-1) = \lambda_1/m$. If $K_{X,1}$ and $K_{Y,1} \gg 1$, then $K_{Y,1}=K_{X,1}m/\lambda_1$, and eq. (3.34) will take the form

$$\bar{K}_{X,1} = K_{X,1}(1+s_1/\lambda_1)/(1+s_1/m). \tag{3.35}$$

Similarly, we can obtain

$$\bar{K}_{Y,2} = \bar{y}_{H,2}/\bar{y}_{0,2} = (K_{Y,2} + K_{X,2}\varphi_X\varphi_Y h_2 m)/(1+\varphi_X\varphi_Y h_2), \tag{3.36}$$

which at $K_{Y,2}$ and $K_{X,2} \gg 1$ will become

$$\bar{K}_{Y,2} = K_{Y,2}(1+h_2\lambda_2)/(1+h_2 m). \tag{3.37}$$

We now need to focus upon the determination of the optimal flow-rate ratio, which, under conditions of the phases' mutual solubility, may differ from the value obtained without regard to the additional circulating flows (i.e by the equations given in Tables 3.1 and 3.2). Since the optimum conforming to the maximum separation degree is characterized under these conditions by the equality of separabilities $\bar{A}_1 + 1$ and $\bar{A}_2 + 1$ derived from the values $A_1$ and $A_2$ by the substitution of $\alpha_i$, $\lambda_i$ and $N_i$ for the corresponding effective quantities, the equations for the optimum values $\bar{\lambda}_1$ and $\bar{\lambda}_2$ can be taken from Table 3.1 by substituting all quantities for their effective values. To find the parameters $\bar{\psi}_X$ and $\bar{\psi}_Y$ appearing in the equations, it is necessary to know the values $\lambda_1$ and $\lambda_2$ that correspond to the optumum values $\bar{\lambda}_1$ and $\bar{\lambda}_2$.

If eqs. (3.28)–(3.37) relating the separation degree with two-temperature plant parameters are applicable to all schemes and operation modes, the additional circulation flows are determined for a specific scheme of a plant. That is why the optimum flow ratios for similar plants with liquid or gas feeding may differ from each other.

For a two-temperature plant with closed gas flow (see Figure 3.9) operating with withdrawal of first kind, for example, substituting $\bar{\lambda}_1$ and $\bar{\lambda}_2$, with regard to eqs. (3.21) and (3.27), into the expression $\bar{\alpha}_1/\bar{\lambda}_1 = \bar{\alpha}_2/\bar{\lambda}_2$ (the optimum mode condition at $\bar{n}_1 = \bar{n}_2$ in NTP calculation) results in a second-degree equation in terms of the optimum flow-rate ratio [3]:

$$\lambda_1^2(1-s_2 h_1)+\lambda_1[(h_2 - h_1)C - s_1 + s_2(1-\rho)] - C(1-\rho-s_1 h_2)=0, \tag{3.38}$$

where $C = \bar{\alpha}_1\bar{\alpha}_2 (1+s_1)(1+s_2)/[(1+h_1)(1+h_2)]$.

The equations for the withdrawal of second kind or for the non-withdrawal mode are simplified, since $\rho = B/F = 0$.

The optimum flow-rate ratios obtained, the determination of the minimum $\overline{N}_i$, required to achieve a desired separation degree, becomes a simple problem. At $= \overline{A}_1 + 1 = \overline{A}_2 + 1$, the following equation is derived from eqs. (3.28) and (3.29):

$$\overline{N}_1 = \overline{\chi} N_2 = \overline{a} \ln\left[(\overline{K} - \overline{u})/(1 - \overline{u})\right],  \tag{3.39}$$

where, when calculated through the separation degree $\overline{K}_{X,1}$,

$$1 - \overline{u}_{x1} = \frac{\overline{\varphi}_x \overline{\varphi}_y \overline{\alpha}_1 - \overline{\alpha}_2}{\overline{\varphi}_y \overline{\psi}_x (\overline{\lambda}_2 - \overline{\alpha}_2)\alpha_1/\overline{\lambda}_2 + (\overline{\alpha}_1 - \overline{\lambda}_1)\overline{\alpha}_2/\overline{\lambda}_1},  \tag{3.40}$$

and when calculated through the separation degree $\overline{K}_{Y,2}$,

$$1 - \overline{u}_{y2} = \frac{\overline{\varphi}_x \overline{\varphi}_y \overline{\alpha}_1 - \overline{\alpha}_2}{\overline{\lambda}_2 - \overline{\alpha}_2 + \overline{\varphi}_x \overline{\varphi}_y (\overline{\alpha}_1 - \overline{\lambda}_1)}  \tag{3.41}$$

next, in the calculation through NTU for liquid and gas flows,

$$\overline{a} = \frac{\overline{\alpha}_1}{\overline{\alpha}_1 - \overline{\lambda}_1} = \overline{\chi}_y \frac{\overline{\alpha}_2}{\overline{\lambda}_2 - \overline{\alpha}_2} \quad \text{and} \quad \overline{a} = \frac{\overline{\lambda}_1}{\overline{\alpha}_1 - \overline{\lambda}_1} = \overline{\chi}_x \frac{\overline{\lambda}_2}{\overline{\lambda}_1 - \overline{\alpha}_2};  \tag{3.42}$$

and finally, when calculated through NTP

$$\overline{a} = \frac{1}{\ln\left(\overline{\alpha}_1 \middle/ \overline{\lambda}_1\right)} = \frac{\overline{\chi}_n}{\ln(\overline{\lambda}_2/\overline{\alpha}_2)}.  \tag{3.43}$$

### 3.1.5   Unsteady state of two-temperature plant

An unsteady-state theory taking into account all processes occuring in two-temperature plants remains to be elaborated; there exists only an approximate solution of the unsteady-state problem.

With regard to the holdup at the column enriched end, the unsteady state of a two-temperature plant with $n_1 + n_2$ theoretical plates in two columns is described by a set of $n_1 + n_2 + 1$ differential equations of material balance. Since the solution of such a set of exponential equations has been found only for the non-withdrawal mode [14], let us consider an approximate calculation of the two-temperature plant equilibrium time, based on assumptions of the column transfer equivalence to the external withdrawal, and of a similarity of concentration profiles at all theoretical plates of separation (see section 1.4).

It might be well to point out that at lower plates (with liquid feeding) and at upper plates (with gas feeding), the deviations from a concentration profile similarity, which is characteristic for other theoretical plates of separation, are unavoidable [11]. Because of this, for the columns with small NTP (NTU) values, the last assumption is hardly applicable, which must be taken into account in testing laboratory-scale and pilot plants.

Compared with the unsteady state of columns with flow conversion discussed in section 1.4, the calculation of the two-temperature plants is distinguished by the following features:

1.  The equilibrium accumulation is calculated in conformity with the enrichment equation typical for the two-temperature method of enrichment. The equation is taken into account for the cold and hot columns, and for the enriched end of the plant (accumulation $M_K$ in the liquid reservoir with a volume $\Delta H_K$).
2.  The initial transfer is determined using the two-temperature plant maximum extraction degree.

For the most typical case of the start-up period (non-withdrawal mode, $\lambda = \lambda_0$, $n_1 = n_2 = n$), the column concentration profiles in NTP calculation are determined by the equations $x = x_F (\alpha_1/\alpha_2)^{n/2}$ and $y = x/\lambda_0$ and. The equilibrium accumulation in a single column is obtained by the integration of

$$M_\rho = \Delta H_x \int_0^n (x - x_F)dn + \Delta H_y \int_0^n (y - x_F)dn,$$

or

$$M_\rho = \left(\Delta H_x + \frac{\Delta H_y}{\lambda_0}\right) x_F \int_0^n \left(\frac{\alpha_1}{\alpha_2}\right)^{n/2} dn - (\Delta H_x + \Delta H_y)x_F n, \tag{3.44}$$

where $\Delta H_x$ and $\Delta H_y$ are holdups of one theoretical plate (TP) of the column for substances $x$ and $y$, respectively:

$$M_\rho = 2\left(\Delta H_x + \frac{\Delta H_y}{\lambda_0}\right) x_F \left[\left(\frac{\alpha_1}{\alpha_2}\right)^{n/2} - 1\right] \Big/ \ln\frac{\alpha_1}{\alpha_2} - (\Delta H_x + \Delta H_y)x_F n. \tag{3.45}$$

With feeding by the liquid flow $L_1$, the relaxation time, with regard to the initial transfer $j_0 = Lx_F(1 - \alpha_2/\alpha_1)$ and equation $\tau_0 = M_\Sigma/j_0 (M_\Sigma = 2M_\rho + M_K)$, is determined by the expression [1, 11]:

$$\tau_0 = \frac{\dfrac{4(\Delta H_x + \Delta H_y)/\sqrt{\alpha_1\alpha_2}}{\ln(\alpha_1/\alpha_2)}\left[\left(\dfrac{\alpha_1}{\alpha_2}\right)^{n/2} - 1\right] - 2n(\Delta H_x + \Delta H_y) + \Delta H_K \left[\left(\dfrac{\alpha_1}{\alpha_2}\right)^{n/2} - 1\right]}{L_1\left(1 - \alpha_2/\alpha_1\right)}.$$

$$\tag{3.46}$$

For the gas–liquid systems, in the isotope exchange at pressures of up to 10MPa, the gas-phase holdup, as opposed to the liquid-phase holdup, can be disregarded. Then, in conformity with the equation $\tau = \tau_0 \ln[1/(1-\varphi)]$, the accumulation time required for the achievement of a degree $\varphi$ of approximation to the steady state in the plant with liquid feeding equals

$$\tau = \frac{1}{L_1}\frac{\alpha_1}{\alpha_1-\alpha_2}\left\{\left[\frac{4\Delta H_x}{\ln(\alpha_1/\alpha_2)}+\Delta H_K\right]\left[\left(\frac{\alpha_1}{\alpha_2}\right)^{n/2}-1\right]-2n\Delta H_x\right\}\ln\frac{1}{1-\varphi},\qquad(3.47)$$

and for the plant with gas feeding $G_2$ we obtain

$$\tau = \frac{1}{G_2}\frac{\alpha_1}{\alpha_1-\alpha_2}\left\{\left[\frac{4\Delta H_x\sqrt{\alpha_1\alpha_2}}{\ln(\alpha_1/\alpha_2)}+\Delta H_K\sqrt{\alpha_1\alpha_2}\right]\left[\left(\frac{\alpha_1}{\alpha_2}\right)^{n/2}-1\right]-2n\Delta H_x\right\}\ln\frac{1}{1-\varphi}.\qquad(3.48)$$

In the same manner one can obtain the equation for the accumulation time calculation through NTU. For the liquid feeding, for example,

$$\tau = \frac{1}{L_1}\frac{\alpha_1}{\alpha_1-\alpha_2}\left\{\left(2\Delta H\frac{\alpha_1+\alpha_2}{\alpha_1-\alpha_2}+\Delta H_K\right)\left[\exp\left(\frac{\alpha_1-\alpha_2}{\alpha_1+\alpha_2}N\right)-1\right]-2N\Delta H\right\}\ln\frac{1}{1-\varphi},\qquad(3.49)$$

where $\Delta H$ is a holdup of one transfer unit; and $N$ is NTU in one column ($N_{Y,1} = N_{Y,2}$ or $N_{x,1} = N_{x,2}$).

The experimental data on the accumulation curve of the two-temperature hydrogen sulphide plant for the initial concentrating of heavy-water operating in non-withdrawal mode, are in close agreement with the results of calculation by eq. (3.47) [15].

## 3.2 TWO-TEMPERATURE HYDROGEN SULPHIDE METHOD

### 3.2.1 Phase equilibrium and isotope equilibrium

The two-temperature hydrogen sulphide method (or *GS*-process) is at present the most economical-to-operate industrial method of heavy-water production. The main advantages of the two-temperature hydrogen sulphide method are due to:

- virtually unlimited primary resourses (water);
- a high rate of the water – hydrogen sulphide isotope exchange reaction;
- a high degree of heat recovery since the heat generation is the principal expenditure item;

- a possibility of assuring the required thermal conditions in the cold column through the use of cooling water, and in the hot column through the use of heat medium with relatively low parameters.

In the region of a low content of heavy isotopes (deuterium or tritium), the following isotope exchange reactions can occur in this system:

$$H_2O_{(lq)} + HDS_{(gas)} = HDO_{(lq)} + H_2S_{(gas)}, \tag{3.50}$$

$$H_2O_{(lq)} + HTS_{(gas)} = HTO_{(lq)} + H_2S_{(gas)}, \tag{3.51}$$

$$D_2O_{(lq)} + DTS_{(gas)} = DTO_{(lq)} + D_2S_{(gas)}. \tag{3.52}$$

Both for these reactions, and for the similar reactions occuring in the gas phase, the isotope equilibrium has been much studied [15–22]. The equilibrium constants have been either determined experimentally, or calculated from the spectrum data (for the gas-phase reactions). The temperature dependencies of the separation factors of mixtures H–T, H–D, and D–T (or, generally, of isotopes A and B) are expressed by

$$\alpha_{gas-lq} = a_{gas-lq} \exp(b_{gas-lq}/T); \tag{3.53}$$

$$\alpha_{gas} = a_{gas} \exp(b_{gas}/T). \tag{3.54}$$

The temperature dependence of binary separation factors in the region of low heavy isotope content, i.e. for the isotope exchange reactions (3.50), (3.51), (3.52) written as

$$A_2O_{(lq)} + ABS_{(gas)} = ABO_{(lq)} + A_2S(gas), \tag{3.55}$$

is less sharp than that in the region of high heavy isotope content (see Table 3.3), where the following isotope exchange reactions occur:

$$ABO_{(lq)} + B_2S_{(gas)} = B_2O_{(lq)} + ABS_{(gas)}. \tag{3.56}$$

**Table 3.3**

Constants of equation describing temperature dependence of separation factors for binary isotope mixtures in water – hydrogen sulphide system

| Exchanging isotopes | High isotope content | $a_{gas-lq}$ | $b_{gas-lq}$ | $a_{gas}$ | $b_{gas}$ |
|---|---|---|---|---|---|
| Protium–deuterium | H | 0.855 | 305 | 1.002 | 237 |
|  | D | 0.862 | 308 | 0.988 | 245 |
| Protium–tritium | H | 0.819 | 426 | 1.006 | 336 |
|  | T | 0.812 | 433 | 0.962 | 355 |
| Deuterium–tritium | D | 0.951 | 122 | 0.994 | 103 |
|  | T | 0.952 | 122 | 0.991 | 105 |

For the reaction (3.50) utilized in heavy-water production, it is the Bigeleisen equation that has enjoyed the widest application

$$\alpha_{HD,gas} = 1.051\exp(218/T),\tag{3.57}$$

obtained by calculations from spectrum data, as well as the equation

$$\alpha_{HD,gas} = 1.010\exp(233/T),\tag{3.58}$$

which, in combination with the Kirschenbaum equation [23]

$$1/\alpha_{HD\,PH} = P^0_{HDO}/P^0_{H_2O} = 1.1596\exp(-65.43/T)\tag{3.59}$$

is widely employed in the industrial plant design calculation. From eqs. (3.58) and (3.59) one can find that

$$\alpha_{HD\,gas-lq} = 0.871\exp(298/T).\tag{3.60}$$

The efficiency of the two-temperature hydrogen isotope separation method depends on the operation conditions, primarily on pressure and temperature, in the cold and hot columns, which should be selected in terms of the phase diagram of the hydrogen sulphide – water system (see Figure 3.10). The diagram represents the phase-transition curves: 1 is the hydrogen sulphide liquefaction, which represents the temperature dependence of the hydrogen sulphide saturated vapor pressure; 2 is the crystalline hydrate formation from gaseous hydrogen sulphide and water; 3 is the crystalline hydrate formation from liquid hydrogen sulphide and water. The curves intersect at the quadrupole point $A$ ($T = 302.6K$;

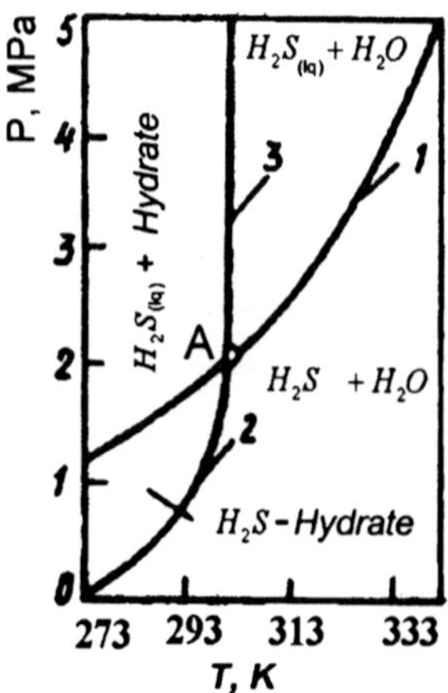


**Figure 3.10**   Phase diagram of H₂O–H₂S system.

$P = 2.26$MPa), where four phases coexist: gaseous and liquid hydrogen sulphide, water, and solid hydrogen sulphide crystalline hydrate of composition $8H_2S{\cdot}46H_2O$ similar in appearance to a snow or loose ice. It is obvious that the operation conditions (temperature and pressure) in a column with counter-current flows of gaseous hydrogen sulphide and water are bounded by the phase diagram area below curves 1 and 2.

Various experimental data [24–26] resulting from the research on the conditions of the hydrogen sulphide crystalline hydrate formation from gas and water (see curve 1 in Figure 3.10), are adequately described by a linear dependence between the logarithm of crystalline hydrate formation pressure and inverse temperature in accordance with

$$\lg P = 12.325 - 3{,}625/T, \qquad (3.61)$$

where $P$ is the pressure (MPa), and $T$ is the temperature (K).

The following expression describing the curve 2 can also be employed [19]:

$$T = 294.73 + 9.3987 \ln P. \qquad (3.62)$$

The pressure dependence of the hydrogen sulphide boiling (liquefaction) temperature obtained through the experimental data generalization [3.18] is characterized by

$$T = 213.19 + 20.12\ln P + 2.328(\ln P)^2 - 0.04895(\ln P)^3 + 0.06622(\ln P)^4. \qquad (3.63)$$

In a heterogenous system comprising hydrogen sulphide saturated with water vapor and water with dissolved gas, the hydrogen sulphide liquefaction temperature varies only slightly, and its dependence on the hydrogen sulphide pressure (see curve 2 in Figure 3.10) is represented by the following equation obtained from the experimental data [24]:

$$T = 157.854 + 112.66\ln P - 57.59(\ln P)^2 + 16.76(\ln P)^3 - 1.64(\ln P)^4. \qquad (3.64)$$

An increase in the temperature difference between the cold and hot columns improves the two-temperature plant separation efficiency, enhancing both the extraction degree and the separation degree. A decrease in the cold column temperature $T_1$, though, demands pressure reduction in accordance with curve 1 or 2 bounding the region of admissible performance characteristics of a two-temperature plant in the phase diagram. With the temperature and pressure values exceeding those corresponding to the quadrupole point $A$ coordinates, upon lowering $T_1$, the positive effect of the separation factor $\alpha_1$ enhancement in the cold column overrides the negative effect of the columns' throughput rate decrease due to the pressure drop (temperature $T_1$ and pressure $P_1$ vary in accordance with curve $I$ in the phase diagram). When changing $T_1$ and $P_1$ below the quadrupole point, account must be taken of curve 2 characterized by a steeper slope. In this case, even a minor decrease in $T_1$ results in a significant pressure drop and deterioration of the columns' throughput rate, which is no longer compensated for by $\alpha_1$ enhancement. That is why the optimum operation conditions in the cold column are determined by the quadrupole point coordinates, with $T_1 = 303$–$308$K and $P_1 = 2.0$–$2.2$MPa.

With the temperature increase in the hot column, the heat consumption required for the heating of flows rises not only through a larger temperature difference in the cold and hot columns, but also because of a larger amount of water vapor needed to saturate hydrogen sulphide. The high water vapor content in the hot column gas flow demands of enlargement column cross-section. Besides, proper allowance should be made for the negative influence on the separation efficiency of additional circulating flows brought about by mutual phase solubility. When evaluating the hot column temperature, it is necessary to apply, as the optimum criterion, an economic indicator allowing for separation energy costs (chiefly associated with heat consumption) and for the plant size (cost). A series of calculations (computer calculations included) demonstrated that at a pressure of $2.0 - 2.2$MPa, the optimum temperature in the hot column $T_2$ ranges from 393 to 403K.

The composition of liquid and gas phases in equilibrium in the water – hydrogen sulphide system at various temperatures and pressures was studied [19]. The equation to calculate water vapor content in hydrogen sulphide was obtained through the computer processing of experimental data:

$$\begin{aligned}
\ln H = &-28.4116 - 5.1971\ln P + 0.6270(\ln P)^2 - 0.02829(\ln P)^3 \\
&+ 0.1179T - 1.2238\times10^{-4}T^2 + 0.3653\times10^{-7}T^3 + 0.023154T\ln P \\
&- 7.62\times10^{-6}T^2(\ln P)^2 - 0.4046\times10^{-4}T^2\ln P + 9.5097\times10^{-10}T^3(\ln P)^3 \\
&+ 0.75736\times10^{-8}T^3(\ln P)^2 + 0.2011\times10^{-7}T^3\ln P,
\end{aligned} \qquad (3.65)$$

where $H$ is the water vapor mole fraction in the gas-and-vapour phase; $P$ is the pressure, atm ($1\,\text{atm} = 1.013 \times 10^5$Pa).

The hydrogen sulphide water solubility can be found from the following equations:

(a)  in the region from the quadrupole point to 373K:

$$\begin{aligned}
\ln S = &\,5.0375 + 0.1604P - 4.0602\times10^{-3}P^2 + 0.4484\times10^{-4}P^3 \\
&- 0.44833T + 5.7530\times10^{-6}T^2 + 8.0270\times10^{-8}T^3;
\end{aligned} \qquad (3.66)$$

(b)  from 373 to 444K :

$$\begin{aligned}
\ln S = &-0.066875 + 0.2114\times10^{-2}P - 0.1237\times10^{-4}P^2 + 3.4556\times10^{-4}T \\
&- 4.5473\times10^{-7}T^2 - 0.3825\times10^{-5}PT + 0.67872\times10^{-10}P^2T^2,
\end{aligned} \qquad (3.67)$$

where $S$ is the mole fraction of water-dissolved hydrogen sulphide.

Table 3.4 represents the values of hydrogen sulphide humidity $h = H/(1 - H)$ and hydrogen sulphide water solubility $s = S/(1 - S)$ found from the above equations, as well as separation factors $\alpha_{HD,gas}$, $\alpha_{HD,PH}$ and $\alpha_{HD,gas-lq}$ calculated by eqs. (3.58)–(3.60). The hydrogen sulphide

**Table 3.4**

Mutual solubility of phases $h$ and $s$ and effective separation factor $\overline{\alpha}$ of isotope mixture H–D

| T, K | Separation factor | | | | $P = 2.0\,MPa$ | | | $P = 2.1\,MPa$ | | | $P = 2.2\,MPa$ | | |
|---|---|---|---|---|---|---|---|---|---|---|---|---|---|
| | $\alpha_{gas}$ | $\alpha_{gas-lq}$ | $\alpha_{PH}$ | $\alpha_{S}$ | $h$ | $s$ | $\overline{\alpha}$ | $h$ | $s$ | $\overline{\alpha}$ | $h$ | $s$ | $\overline{\alpha}$ |
| 293 | 2.2361 | 2.4070 | 1.0780 | 0.9940 | 0.0019 | 0.0349 | 2.3538 | 0.0019 | 0.0368 | 2.3513 | 0.018 | 0.0386 | 2.3493 |
| 303 | 2.1782 | 2.3277 | 1.0701 | 0.9931 | 0.0032 | 0.0286 | 2.2822 | 0.0031 | 0.0301 | 2.2804 | 0.0030 | 0.0317 | 2.2787 |
| 313 | 2.1254 | 2.2537 | 1.0627 | 0.9923 | 0.0052 | 0.0239 | 2.2133 | 0.0051 | 0.0252 | 2.2115 | 0.0050 | 0.0264 | 2.2108 |
| 323 | 2.0771 | 2.1903 | 1.0558 | 0.9915 | 0.0094 | 0.0203 | 2.1449 | 0.0081 | 0.0214 | 2.1474 | 0.0079 | 0.0224 | 2.1459 |
| 333 | 2.0326 | 2.1305 | 1.0495 | 0.9907 | 0.0130 | 0.0175 | 2.0833 | 0.0126 | 0.0184 | 2.0831 | 0.0122 | 0.0194 | 2.0829 |
| 343 | 1.9916 | 2.0757 | 1.0435 | 0.9900 | 0.0198 | 0.0154 | 2.0203 | 0.0191 | 0.0162 | 2.0208 | 0.0185 | 0.0170 | 2.0211 |
| 353 | 1.9535 | 2.0253 | 1.0379 | 0.9894 | 0.0296 | 0.0138 | 1.9575 | 0.0285 | 0.0145 | 1.9587 | 0.0275 | 0.0153 | 1.9597 |
| 363 | 1.9185 | 1.9787 | 1.0326 | 0.9887 | 0.0433 | 0.0126 | 1.8941 | 0.0416 | 0.0133 | 1.8961 | 0.0400 | 0.0139 | 1.8981 |
| 373 | 1.8858 | 1.9357 | 1.0276 | 0.9881 | 0.0624 | 0.0117 | 1.8296 | 0.0598 | 0.0123 | 1.8326 | 0.0575 | 0.0129 | 1.8366 |
| 383 | 1.8553 | 1.8958 | 1.0229 | 0.9876 | 0.0886 | 0.0111 | 1.7631 | 0.0848 | 0.0117 | 1.7671 | 0.0813 | 0.0123 | 1.7709 |
| 393 | 1.8268 | 1.8568 | 1.0185 | 0.9871 | 0.1247 | 0.0106 | 1.6842 | 0.1189 | 0.0110 | 1.6998 | 0.1138 | 0.0116 | 1.7046 |
| 403 | 1.8002 | 1.8240 | 1.0143 | 0.9866 | 0.1743 | 0.0097 | 1.6232 | 0.1656 | 0.0103 | 1.6301 | 0.1580 | 0.0108 | 1.6364 |
| 413 | 1.7752 | 1.7917 | 1.0103 | 0.9861 | 0.2428 | 0.0089 | 1.5499 | 0.2299 | 0.0094 | 1.5584 | 0.2077 | 0.0100 | 1.5739 |
| 423 | 1.7517 | 1.7614 | 1.0065 | 0.9857 | 0.3393 | 0.0080 | 1.4744 | 0.3179 | 0.0085 | 1.4845 | 0.2691 | 0.0090 | 1.5133 |
| 433 | 1.7295 | 1.7330 | 1.0030 | 0.9852 | 0.4792 | 0.0070 | 1.3975 | 0.4487 | 0.0075 | 1.4091 | 0.4227 | 0.0080 | 1.4194 |
| 443 | 1.7087 | 1.7063 | 0.9996 | 0.9848 | 0.6919 | 0.0060 | 1.3196 | 0.6422 | 0.0064 | 1.3324 | 0.6005 | 0.0069 | 1.3440 |

dissolving in water produces only a slight isotope effect which can be determined from the values of the pressure of saturated vapors of hydrogen sulphide and sulfur deuteride [28]:

$$1/\alpha_{HD,S} = 1.043 \exp(-8.037/T).$$ (3.68)

The mutual solubility of phases affects the separation factor $\alpha_{HD,gas-lq}$ accounted for by introducing the effective separation factor $\bar{\alpha}$ (commonly designated as $\beta$) defined through the ratio of effective concentrations which take into account the mutual solubility of phases [18, 29, 30]. The effective separation factor $\bar{\alpha}$ presented in Table 3.4 was calculated by the equation for a low content of heavy isotope. As can be seen from Table 3.4, the mutual solubility of phases in the water – hydrogen sulphide system leads to the separation factor variations, which is particularly true for the hot column.

To calculate $\bar{\alpha}$ at the heavy isotope mole concentration exceeding 1–5%, it is requisite to use the full, and not abridged, separation factor computational equation. In so doing, the concentration dependencies of separation factors $\alpha_{gas}$ and $\alpha_S$ must be taken into account. The isotope composition influence on the quantities $s$ and $h$ is generally disregarded, and the factor $\alpha_S$ is taken equal to unity.

The effective separation factor calculations for the isotope exchange reaction (3.50) within the temperature range 303–403K, and pressure range 1.76–2.28MPa, with deuterium concentration varying from 0.010 to 99% were executed [31]. At a low deuterium content and at a pressure of 1.76MPa, for example, the temperature dependence of the effective separation factor will be determined by eq. (3.69):

$$\bar{\alpha}_{HD,gas-lq} = -0.00736\,T + 4.534.$$ (3.69)

### 3.2.2 Kinetics of isotope exchange: Packing materials

The high rate of heterogenous isotope exchange between water and hydrogen sulphide is a controlling consideration which determines the wide industrial application of the hydrogen sulphide method for heavy-water production. To describe the method, an inter-phase isotope transfer model taking into consideration the following stages [32] is suggested:

(1)  diffusion in water towards interphase boundary;
(2)  isotope exchange reaction between water and dissolved hydrogen sulphide near liquid surface;
(3)  phase isotope exchange reaction between dissolved and gaseous hydrogen sulphide;
(4)  diffusion in hydrogen sulphide from interphase boundary.

It was assumed that it is the second stage that limits the isotope-exchange rate. The ionic mechanism of the water-dissolved hydrogen sulphide isotope exchange is considered in references [33–35]. The study of mass exchange in the liquid phase [35] demonstrates that with increasing pressure, the chemical component contribution into overall resistance to mass transfer diminishes, and at a pressure of over 2.0MPa gets insignificant, and it is the diffusion resistance that becomes rate-limiting.

Since the hydrogen sulphide dissolution in water pertains to fast processes, the third stage of isotope transfer can be disregarded. In hydrogen isotope separation both in packed and in tray columns, the contribution of diffusion resistance in hydrogen sulphide (fourth stage) is insignificant and the rate-controlling function passes on to the processes occuring in the liquid phase: isotope-exchange reaction and reagent diffusion. The transfer of a heavy isotope (e.g. deuterium) from the gas into liquid phase can be represented in the following manner:

$$H_2O_{(lq)} + HDS_{(S)} \underset{k_R}{\overset{k_F}{\longleftrightarrow}} HDO_{(lq)} + H_2S_{(S)}, \tag{3.70}$$

where $k_F$ and $k_R$ are the rate constants of the direct and reverse isotope-exchange reactions.

With regard to the near-constancy of the values $[H_2O]$ and $[H_2S]$ at low heavy isotope concentrations, the second-order isotope-exchange reaction can be viewed as a quasi-first-order reaction, of which the rate is expressed by:

$$-d[HDS_{(s)}]/d\tau = d[HDO_{(lq)}]/d\tau = \bar{k}_F[HDS_{(s)}] - \bar{k}_R[HDO_{(lq)}], \tag{3.71}$$

where $\bar{k}_F = k_F [H_2O_{(lq)}]$ and $\bar{k}_R = k_R [H_2S_{(s)}]$ are the rate constants of the direct and reverse quasi-first-order reactions.

In the liquid phase, the isotope exchange and diffusion of reagents proceed simultaneously. With a steady-state isotope concentration profile in the column, the isotope transfer in the liquid phase from the interphase boundary is performed by a steady diffusion of reagents (dissolved gas and liquid). Under steady conditions, from the material balance equations it follows that the rates of reagents diffusion and of isotope-exchange reaction are equal to:

$$\left. \begin{array}{l} D_{H_2S}d^2C_y/dl^2 = \bar{k}_F C_y - \bar{k}_R C_x \\ -D_{H_2O}d^2C_x/dl^2 = \bar{k}_F C_y - \bar{k}_R C_x \end{array} \right\}, \tag{3.72}$$

where $D_{H_2S}$ and $D_{H_2O}$ are the coefficients of molecular diffusion of hydrogen sulphide dissolved in water, and of water self-diffusion; $l$ is the coordinate normal to the interphase boundary; $C_y = [HDS_{(s)}]$ and $C_X = [HDO_{(lq)}]$ expressed in kmol/m$^3$.

The set of eqs. (3.72) was solved by the operational method using the assumption that $\bar{k}_F \gg \bar{k}_R$ ($\bar{k}_F / \bar{k}_R = K = \alpha_{lq}[H_2O_{(lq)}]/[H_2S_{(s)}] \gg 1$ since $[H_2O_{(lq)}] \gg [H_2S_{(s)}]$, and the dissolved gas–liquid isotope separation factor $\alpha_{lq} > 1$), under the following boundary conditions [35]: at $l = 0$ $C_Y = C_{0Y}$, $dC_Y/dl = C'_{0Y}$; $C_X = C_{0X}$, $dC_X/dl = C'_{0X}$.

The set of eqs. (3.72) is true for the diffusion boundary layer where the molecular mass-transfer process dominates over the turbulent transfer. The solution of the equations set leads to the following expression for the chemical component of the mass-transfer coefficient:

$$\beta_{IE} = \sqrt{\overline{k}_F D_{H_2S}}\,[H_2S_{(s)}]. \qquad (3.73)$$

From the above equation it follows that as the isotope exchange rate grows, the mass-transfer coefficient $\beta_{IE}$ increases without limit. But it by no means implies a limitless acceleration of the mass-transfer rate. The gradient of isotope concentration $C_{0,Y}$ depends both on the isotope exchange kinetics and on the motive force at the interphase boundary which decreases with the reaction rate increase. If this decrease in motive force is taken into account, the phase resistance additivity equation, as applied to the system under consideration, will be written as follows:

$$1/K_{0,Y} = 1/\beta_y + 1/(\alpha\beta_x) + 1/(\alpha_S\beta_{IE}); \qquad (3.74)$$

$$1/K_{0,X} = 1/\beta_x + \alpha/\beta_y + \alpha_{lq}/\beta_{IE}, \qquad (3.75)$$

where $\alpha_{lq} = \alpha/\alpha_S$ is the isotope separation factor of the reaction proceeding in the liquid phase (i.e. between water and dissolved hydrogen sulphide).

The isotope exchange between water and dissolved hydrogen sulphide may proceed by two mechanisms:

Mechanism I

$$H_2O + HDS_{(s)} \underset{k_{-1}}{\overset{k_1}{\longleftrightarrow}} H_2DO^+ + HS^- \underset{k_2}{\overset{k_{-2}}{\longleftrightarrow}} HDO + H_2S_{(s)}. \qquad (3.76)$$

The rate constants for dissociation of HDS and $H_2S$ molecules are designated as $k_1$ and $k_2$, and rate constants for the reverse reactions – as $k_{-1}$ and $k_{-2}$. The ionic reaction rate constants are vitrually independent of the isotopic composition. From the equality of the rates of HDO formation and HDS consumption expressed by

$$d[HDO_{(lq)}]/d\tau = k_{-2}[H_2DO^+][HS^-] - k_2[HDO_{(lq)}][H_2S_{(s)}]; \qquad (3.77)$$

$$-d[HDS_{(s)}]/d\tau = k_1[HDS_{(s)}][H_2O_{(lq)}] - k_{-1}[H_2DO^+][HS^-], \qquad (3.78)$$

and with $k_{-1} = k_{-2}$, we obtain

$$d[HDO_{(lq)}]/d\tau = (1/2)k_1[H_2O_{(lq)}][HDS_{(s)}] - (1/2)k_2[H_2S_{(s)}][HDO_{(lq)}]. \qquad (3.79)$$

From the correlation between the above equation and eqs. (3.71) and (3.73) it follows that

$$\overline{k}_F = (1/2)k_1[H_2O_{(lq)}] \quad \text{and} \quad \beta_{IE}^l = \sqrt{(1/2)D_{H_2S}k_1[H_2O_{(lq)}]}\,[H_2S_{(S)}]. \qquad (3.80)$$

Mechanism II

$$HDS_{(s)} + OH^- \underset{k_{-3}}{\overset{k_3}{\leftrightarrow}} HDO + HS^- ; \tag{3.81}$$

$$HS^- + H_2O \underset{k4}{\overset{k_{-4}}{\leftrightarrow}} H_2S_{(s)} + OH^- . \tag{3.82}$$

Since the rate constants of these reactions are almost identical, and $[HDS] \ll [H_2S]$, the first reaction is the slowest. Hence

$$d[HDO_{(lq)}]/d\tau = k_3[OH^-][HDS_{(s)}] - k_{-3}[HS^-][HDO_{(lq)}]. \tag{3.83}$$

From the correlation between the above equation and eq. (3.71) it follows that

$$\bar{k}_F = k_3[OH^-]. \tag{3.84}$$

With the concentration $[OH^-]$ expressed through the equilibrium constant $K_4$ of the reaction (3.82), and considering that the equilibrium constants of reactions (3.81) and (3.82) are interrelated as $K_3/K_4 = \alpha_{lq}$, eq. (3.84) can be rewritten as

$$\bar{k}_F = k_{-3}\alpha_{lq}[H_2O_{(lq)}][HS^-]/[H_2S_{(s)}];$$
$$\beta_{IE}^{II} = \sqrt{D_{H_2S}a_1 k_{-3}[H_2O_{(lq)}][HS^-][H_2S_{(s)}]}. \tag{3.85}$$

The above equations allow for the mass-exchange efficiency calculation. It can be illustrated by calculating the chemical components $\beta_{IE}^{I}$ and $\beta_{IE}^{II}$ associated with isotope exchange reaction proceeding by mechanism I and mechanism II, respectively, at $T = 303K$ and $P = 0.1MPa$. Under these conditions, the rate constants for the hydrogen sulphide hydration and proton transfer are equal to $k_1[H_2O] = 4.3 \times 10^3$ s$^{-1}$ [36] and $k_3 \approx k_4 = 5 \times 10^{10}$ m$^3$/(kmol·s) [37], and $D_{H_2S} = 1.6 \times 10^{-9}$ m$^2$/s [38]. From Henry's law constant for hydrogen sulphide dissolving in water ($K_H = 55.2MPa$ [39]), the concentrations $[H_2S_{(S)}] = [H_2O]P/K_H$ and $[HS^-] = \sqrt{K_{diss}}[H_2S_{(s)}]$ are determined (hydrogen sulphide dissociation constant $K_{diss} = 10^{-7}$ [40]).

The calculation results are as follows: at atmospheric pressure $\beta_{IE}^{I} = 0.66$ kmol/(m$^2$·h) and $\beta_{IE}^{II} = 0.047$ kmol/(m$^2$·h) ($\beta_{IE}^{I}/\beta_{IE}^{II} = 14$).

With growing pressure, the contribution of the exchange by mechanism II must further reduce since, in accordance with eqs. (3.80) and (3.85), the ratio $\beta_{IE}^{I}/\beta_{IE}^{II} \sim \sqrt[4]{P}$. As a consequence, at a pressure of 2.2MPa, which is the admissible limit at the temperature 303K in the H$_2$O–H$_2$S system, $\beta_{IE}^{I}/\beta_{IE}^{II} = 30$.

The isotope exchange reaction, hence, proceeds chiefly by the dissociative mechanism II. In conformity with eq. (3.80), the isotope exchange rate increases in proportion to the hydrogen sulphide pressure. As the pressure grows, the chemical component contribution into the total resistance to mass transfer decreases, and, at a pressure of over 2MPa, becomes apparently insignificant, and the diffusuon resistance  takes the rate-limiting function.

A high reaction rate complicates experimental investigation into the kinetics of CHEX reaction between water and hydrogen sulphide. That is why presented below are the experimental data on the efficiency of heterogenous process performed in counter-current flows of water and hydrogen sulphide. The experimental data on the dependence of HTU ($h_{0x}$) on the flow rate obtained for the column filled with Levin's spiral-prismatic packing material (unit size 3.5mm $\times$ 3.5mm $\times$ 0.2mm), at a temperature of 300K within a pressure range of 0.1–0.2MPa are presented in Figure 3.11.

As is evident from Figure 3.11, the isotope exchange rate rises steeply with increased pressure, and the flow-rate dependence of HTU becomes weaker which agrees well with dependences discussed above (see Figure 1.1). The experimental data represented in Figure 3.11 agree with a quasi-linear dependence of $\beta_{IE}$ on the hydrogen sulphide pressure.

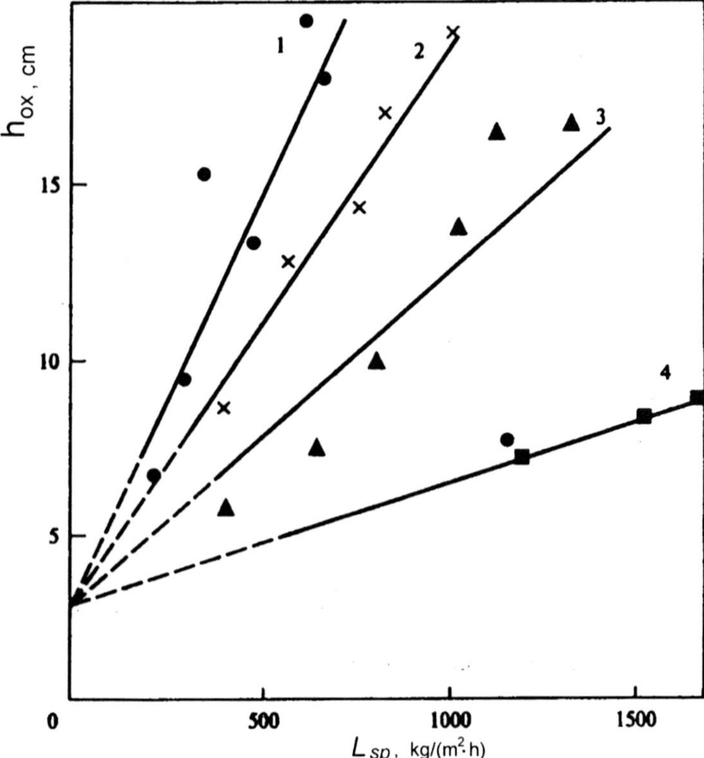


**Figure 3.11**   Pressure influence on HTU for spiral-prismatic packing (3.5mm $\times$3.5mm $\times$0.2mm) for H–D isotope exchange in $H_2O$–$H_2S$ system at 300K: 1, 2, 3, 4, at 0.1, 0.3, 0.82, and 2.0MPa, respectively.

As pressure grows from 0.1 to 2.0MPa, the $\beta_{IE}$ value increases by a factor of about seven, which is indicative of a significant decline in the dissolved hydrogen sulphide diffusion coefficient $D_{H_2S}$ with a rise in pressure. If the coefficient remained constant then, in conformity with eq. (3.80), with such a rise in pressure, the $\beta_{IE}^l$ value should have increased by a factor of 20 [41].

The temperature influence on the chemical component $\beta_{IE}^l$ will depend on the activation energy of hydrogen sulphide hydration accounting for about 25kJ/mol. In accordance with eq. (3.80), the temperature influence on the diffusion coefficient of dissolved hydrogen sulphide, and on the hydrogen sulphide solubility, must also be taken into account. Even though both the diffusion coefficient and the hydrogen sulphide hydration rate constant grow with temperature, the mass-transfer efficiency depends only weakly on the temperature owing to a significant reduction of $H_2S$ solubility [42].

The chemical component influence can be reduced and even eliminated not only by pressure increase but also by introducing water-soluble substances (salts), of which interaction with hydrogen sulphide leads to a rise in $OH^-$ and $HS^-$ ion concentration in the liquid phase. When, for example, anions of an acid weaker than hydrogen sulphide are added, the hydrolysis rate of the anions will be higher than that of $HS^-$ ions. With a rise in $OH^-$ and $HS^-$ ion concentration, the contribution of mechanism 2 to the total exchange rate becomes predominant. Figure 3.12 represents experimental and calculated NTU dependences on the $HS^-$ ion concentration [35]. Experimental data were obtained with the use of spiral-prismatic packing with unit size of 1.5mm $\times$ 1.5mm $\times$ 0.2mm and specific surface $a = 3880m^2/m^3$ at $T = 303K$, $P = 0.1MPa$ and $G/G^* = 0.9$; the calculation was performed by eqs. (3.85) and (1.37). In the calculation, the quantity $h_Y$ was disregarded, and the quantity $h_X$ determined by the diffusion resistance in water, was found from

$$Nu_x = 5\times10^3 d_E^{1.5} Re_X^{0.8} Pr_X^{0.5},$$

(3.86)

obtained from experimental data on the $H_2O$–$HTO$ mixture rectification [43]. As Figure 3.12 suggests, the calculated values of NTU agree closely with the experimental ones.

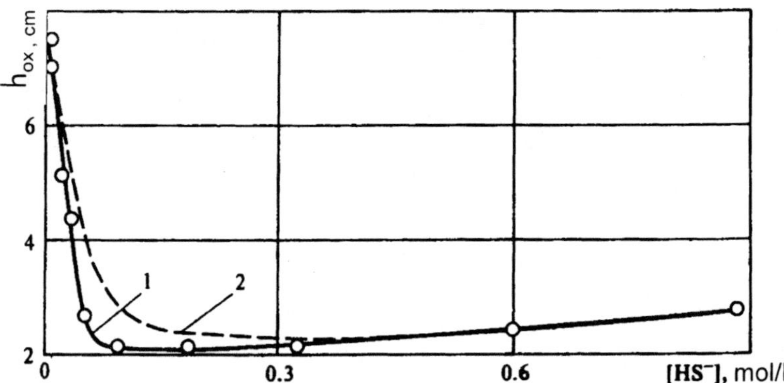


**Figure 3.12**   Experimental (1) and calculated (2) dependences of $h_{0X}$ in an $H_2O$–$H_2S$ system on concentration of $HS^-$ ions.

The horizontal part of the curve represents the establishment of an instantaneous isotope exchange regime, when the reaction rate does not affect the mass transfer process, and the diffusion of water molecules becomes the rate-limiting stage of the process.

Figure 3.13 shows the experimental HTU dependences on flow rate, obtained as well with the use of spiral-prismatic packing with unit size of 1.5mm $\times$ 1.5mm $\times$ 0.2mm in a system without additives and with additives at an optimum concentration of $K_3PO_4$ salt which is the most efficient among a relatively great number of water-soluble isotope exchange accelerating agents [41, 44]. As is evident from Figure 3.13, the instantaneous isotope exchange regime is established in the presence of $K_3PO_4$. The identity of $h_{OX}$ values for the systems H–D and H–T demonstrates the diffusion resistance in the liquid phase being, in this case, the rate-limiting stage ($\beta_Y > \beta_{IE} > \beta_X$), and the mass-transfer efficiency will be virtually uniform in the isotope exchange both of deuterim and of tritium which is due to a weak dependence of the self-diffusion coefficient of water molecules on its isotopic composition. By subtracting component $h_x$ determined by the horizontal straight line in Figure 3.13 from the quantity $h_{OX}$, we obtain $h_{IE}$, and then the $\beta_{IE}a$ values are calculated by eq. (1.37) from the $h_{IE}$ values, in accordance with straight lines 1 and 2: for the H–T isotope exchange $\beta_{IEa} = 2000$ kmol/(m$^3$·h), and for the H–D isotope exchange $\beta_{IEa} = 2300$kmol/(m$^3$·h).

Based on the packing specific surface equal to 3880m$^2$/m$^3$, and allowing for the fact that the active surface of phase contact accounts for 80–90% of the geometric surface area, one will find that for the H–T isotope exchange $\beta_{IE} = 0.61$kmol/(m$^2$·h), and for the H–D isotope exchange $\beta_{IE} = 0.70$kmol/(m$^2$·h). The derived values agree well with the calculated value $\beta_{IE}^l = 0.66$kmol/(m$^2$·h) given above.

The drawback of small irregular packing materials consists of an increased role of the scale-up factor for large-diameter columns. In this case, preference is given to regular packing materials of which the structured packing was tested in the water – hydrogen

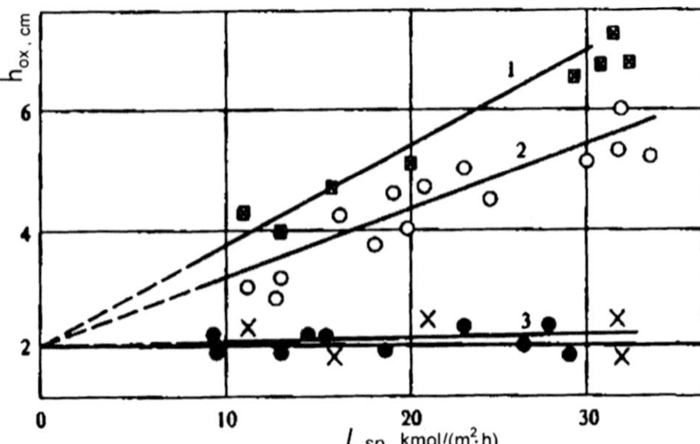

**Figure 3.13**  The experimental HTU dependence on liqued flow rate for spiral-prismatic packing (1.5mm ×1.5mm ×0.2mm) in $H_2O$–$H_2S$ system at 300K and $P = 0.11$MPa for isotope exchange: H–T (■, ×) and H–D (○, ●) without additives (■, ○) and with 4 weight % $K_3PO_4$ additive (×, ●).

sulphide system. Figure 3.14 shows the test results for a structure packing material made of corrugated stainless-steel gauze with equilateral triangle-shaped ($h = 3.5$ mm) corrugation at an angle of 45° to the packing axis. Figure 3.14 presents as well the flow-rate dependences of NTU obtained both in the absence of additives and in the presence of $K_3PO_4$. The experiments were performed in a triple isotopic mixture H–D–T which made it possible to simultaneously (under uniform hydrodynamic conditions) obtain data relating to the H–T and H–D mixtures. As in the case of small irregular packing, the $K_3PO_4$ introduction causes the liquid-phase diffusion to become the rate-limiting stage.

The $\beta_{IE}a$ values calculated from the flow-rate dependence of NTU were equal to 500 and 530 kmol/(m³·h) for the isotope exchange of H–D and H–T, respectively. With regard to the packing specific surface accounting for 1400 m²/m³, the $\beta_{IE}$ values equalled 0.36 and

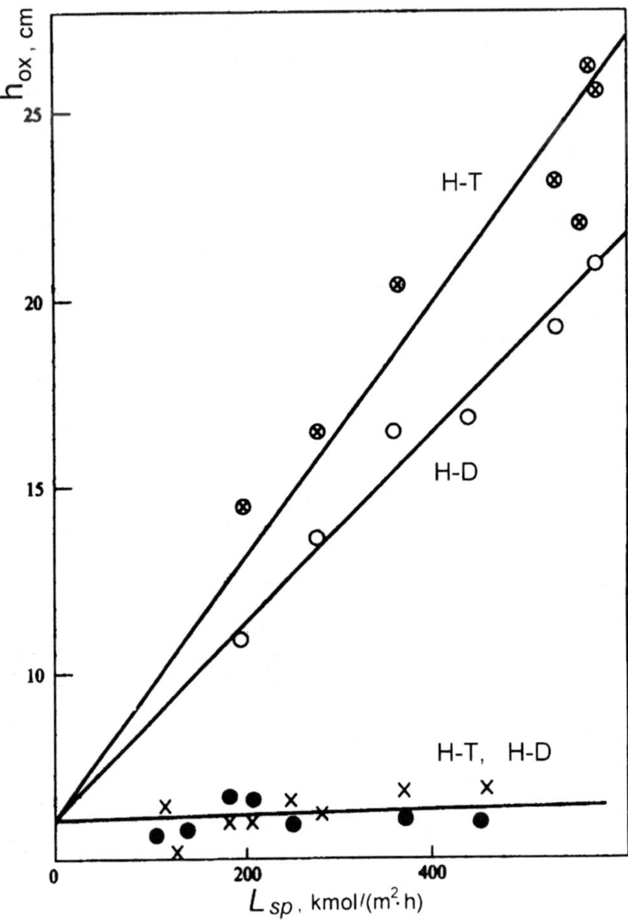


**Figure 3.14**  Liquified flow rate influence on HTU for gauze structured packing in $H_2O$–$H_2S$ system at 300K and $P = 0.1$ MPa for isotope exchange: H–T ($\otimes$, ×) and H–D ($\bigcirc$, ●) in the system without additives ($\otimes$, $\bigcirc$) and  with 4 weight % $K_3PO_4$ additive (×, ●).

0.38kmol/(m$^2$·h), respectively. The equality of $\beta_{IE}$ values for the H–D and H–T isotope exchange derived from experimental data both on random and on regular packing demonstrates the absence of a noticeable kinetic isotope effect during the hydrogen isotope exchange in the H$_2$O–H$_2$S system.

The industrial-scale separation columns for heavy-water production are provided with tray contactors. In the first industrial plant (Aleksin, USSR), the columns of 2.8m in diameter were fitted with Kittel perforated plates which were subsequently replaced with more efficient sieve trays [45]. The sieve trays with 3mm holes (free cross-section content of 9.6%) and overflow weirs of 30mm height, were spaced 300 and 400mm in the cold and hot columns, respectively. Owing to the chemical component contribution to the total mass-transfer resistance (temperature and pressure values were $T_1$ = 298K, $P_1$ = 0.88MPa and $T_2$ = 373K, $P_2$ = 0.98MPa), and to the mixing effect in the liquid radial movement around the tray, the tray performance factors in cold and hot columns accounted for only 35 and 50%, respectively [45, 46].

The results of experimental testing of sieve trays and random (dump) packing materials in the column of 600mm in diameter points as well to the chemical component influence on mass-transfer efficiency in the cold column at $P_1$ = 0.8MPa. Figure 3.15 shows the dependence of the performance factors of sieve trays differing in the free cross-section content (10% and 14%) on the gas linear velocity related to the total column cross-section, at $\lambda$ = 2. At 300K, irrespective of the free cross-section and intertray space (300 and 400mm), the performance factor is characterized by curve 1. Owing to the chemical kinetics influence, the tray performance factor lowers with an increase in flow rate, while with the introduction of exchange-accelerating additive K$_3$PO$_4$, it enhances markedly and no longer depends on the flow rate (straight line 2).

At 373K ($P_2$ = 0.92MPa), regardless of the flow rate, free cross-section, intertray space, and presence of additives (including those inhibiting the crystalline hydrate formation), the tray performance factor increases to about 60% (curve 3).

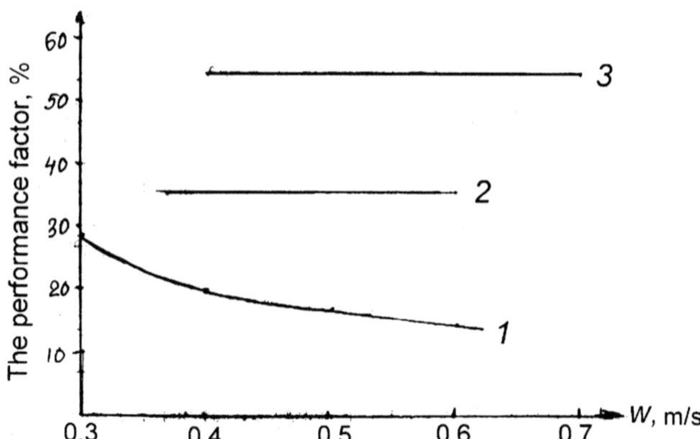


**Figure 3.15**   Sieve tray performance factor in test column of 600mm in diameter, at a temperature of 300 K (1, "pure" system; 2, with K$_3$PO$_4$ additive), and at a temperature of 373 K (straight line 3).

The absence of the flow-rate influence on the mass-transfer efficiency in the hot column, as wel as in the cold column with $K_3PO_4$ introduced in the liquid phase, points to a minor role of the isotope exchange reaction rate in these conditions.

No less important results were obtained for random packing materials characterized by a higher limit flow rate, and considered to be efficient contactors for mass-transfer columns. Figure 3.16 shows the dependences of HTU ($h_{0X}$) on $w$ for the Intalox ceramic packing with unit size 25mm, at $T_1 = 300K$ and $P_1 = 0.85MPa$. The $K_3PO_4$ introduction doesn't affect the throughput capacity and hydraulic resistance of the packing bed layer but brings about the HTU reduction by a factor of 1.4–2 (straight line 2). Since the tests were performed in intermediate and turbulent operation modes, the obtained HTU dependence on the flow rate was defined by a significant enhancement of the active phase contact surface.

Similar dependences were obtained at $T_2 = 290–300K$ and $P_2 = 0.72MPa$ with the use of metal ring packing with unit size 25mm. With the introduction of isotope exchange accelerating additives, HTU decreases by half, and even in the presence of a considerable amount of an additive inhibiting the formation of $8H_2S\cdot46H_2O$, accounts for 0.4m at $w = 0.8m/s$. It follows from the results of experimental testing, that the flow rate in the packed column can be doubled as against tray column, without any increase of the column hydraulic resistance.

Single-flow bubble cup trays were employed at major plants for heavy-water production by hydrogen sulphide method in Dana and Savannah River (USA) [47]. But in the course

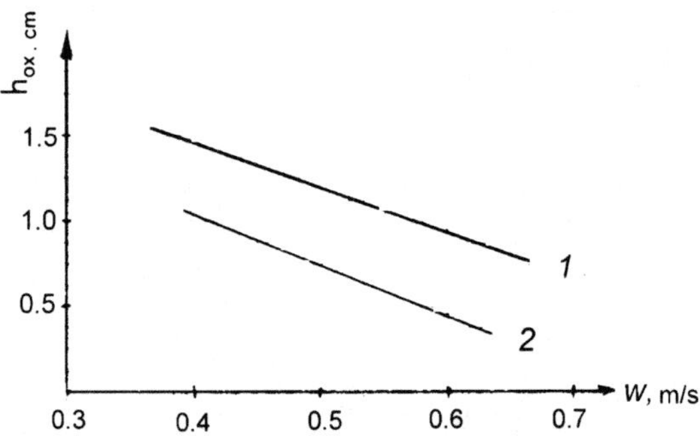


**Figure 3.16**    Flow rate influence on HTU for Intalox packing at isotope exchange in "pure" system (straight line 1) and with additive (straight line 2).

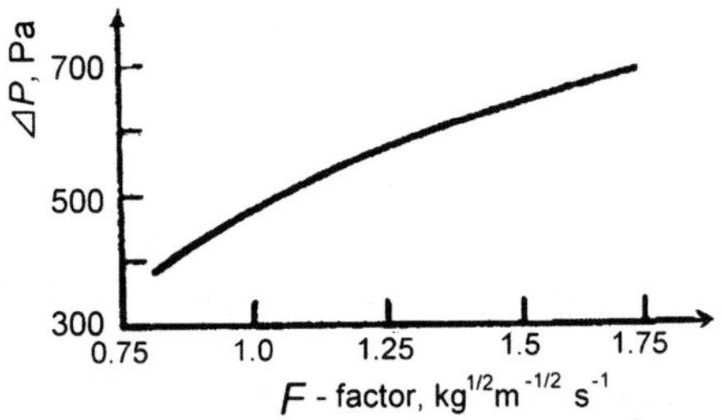


**Figure 3.17**    Dependence of hydraulic resistance of sieve tray on $F$-factor.

of operation, by reason of the trays corroding, they were replaced with novel sieve trays. An industrial plant with a cold and hot columns of 3.36 and 3.66m in diameter, respectively, and with inter-tray space of 457mm, was complete with trays characterized by the parameters given below. Bubble cup trays: 298 cups in the cold column and 300 cups in the hot column of 101.6mm in diameter, each cup with 50 slits of 32mm in height and 3mm in width. Sieve trays: holes of 6.35mm in diameter, with a free cross-section content of 9.1%, and overflow weirs of 30mm in height [48].

Table 3.5 presents the operating characteristics of the above trays, and Figure 3.17 shows the dependence of the sieve tray hydraulic resistance on gas flow rate expressed in terms of the $F$-factor ($F = w_0 \rho^{1/2}$, $w_0$ is the gas velocity, m/s; $\rho$ is the gas density, kg/m$^3$). Replacing the bubble cup trays with sieve trays made it possible to increase the trays performance factor by 19%, to improve the productivity by 12%, and to lower the columns hydraulic resistance by 30%. What is more, the sieve tray is about 1.6–1.8 cheaper than the bubble cup tray. The columns of the world's largest two-temperature plants constructed in Canada during the ensuing years are complete with sieve trays.

The sieve tray technological calculation for the major Canadian heavy-water plants was initially based on the data for other systems (air–water, petroleum chemical), and on experimental data from the operation of pilot plants in the USA [48, 49]. Shortly after the Canadian plants were commissioned, it became clear that the trays' actual efficiency was lower than designed one, which resulted in a substantial loss in productivity. H. Neuburg and K. Chuang [50, 51] have devised a technique for calculation of efficiency of large industrial sieve trays of up to 8.1m in diameter, which takes into account the liquid flow distribution, mutual capture of phases, and gas and liquid mixing in central and peripheral overflow units. The calculation results agree with tray efficiency measured at the Canadian heavy-water plants.

D. Spagnolo and K. Chuang [53] investigated hydrodynamic and mass-transfer properties of sieve trays provided with a layer of 25mm cellular packing made of type 316L stainless steel (type 421 Yorkmesh packing), for heavy-water production by the hydrogen sulphide method at a temperature of 305K and pressure of 2.17MPa. The packing mounting on the conventional sieve trays brings about an increase in the tray resistance, a reduction of carryover, and a 3–20 % gain in the performance factor. The advantages of the packed

### Table 3.5

Comparison between efficiencies of bubble cup trays and sieve trays in first-stage columns of Savannah River Plant [48]

| Tray type | $T$,K | $P$, MPa | Flow rate | | Tray performance factor, % |
| | | | Gas flow, kmol H$_2$S/h | $F$-factor | |
|---|---|---|---|---|---|
| Bubble cup | 303 | 1.90 | 7,358.3 | 1.2 | 60 |
| | 413 | 1.95 | 7,358.3 | 1.4 | 69 |
| Sieve | 303 | | | | |
| | 413 | | 8,255.5 | 1.4 | 69 |
| | | | 8,255.5 | 1.6 | 75 |

trays manifest themselves in a wide range of gas flow rates (at a $F$-factor of 1.2–3.2), but the peak efficiency is achieved in the presence of a gas bubble flows regime.

The same authors have compared hydrodynamic, mass-transfer, and heat-exchange properties of a regular packing with these of sieve trays in a column of 311 mm in diameter. The comparison was performed with the use of Sulzer's Mellapack 250Y regular packing [54], offering a lower hydraulic resistance and higher throughput capacity in the hydrogen sulphide processes compared with sieve trays. The packing as well has advantages for the foaming problem solution. In the heat exchange under conditions of direct phase contact, the packing assures a higher heat-transfer efficiency. Apart from losses due to a high cost, a major obstacle to the packing application in large-diameter columns is the difficulties associated with assurance of the uniform liquid distribution.

In the literature [55] the pressure drop calculation for sieve trays has been described. The importance of this parameter lies in the fact that it determines the gas circulation energy input, which, in its turn, is the major component of electric power consumption in hydrogen sulphide plants. In addition, the pressure difference data are necessary for the estimate of the exchange column operation conditions (e.g. thermal profile of the cold column). Such complications as a gradual plugging of the sieve tray holes can also be correctly assessed by controling the variations in tray pressure difference with time. The paper analyzes the existing methods of calculation and offers recommendations on the selection of equations for the calculation of the sieve tray pressure drop in the hydrogen sulphide process conditions.

It is advised as well to note the pressure difference at various velocities of gas and liquid flows when the plant is put into operation, when the trays are good and clean. The data should be used to derive correlation coefficients for the calculation equations, as well as to judge the condition of the trays in the functioning plant.

An important problem in heavy-water production by the hydrogen sulphide method is the prevention of foam formation, which is typical for this method owing to the fact that in the cold column, for example, the process is performed over the pressure and temperature range close to that of the $H_2S$ liquefaction. The trace impurities in the feed water flow also intensify the foaming. The necessity of suppressing the foam formation with the use of antifoaming additives was encountered as early as in the operation of the first heavy-water plants in the USA, where uncontrollable foaming resulted in unstable functioning of the plant and a substantial impairment in heavy-water productivity. This problem has been considered by Canadian [56, 57] and Argentinian [58] researchers.

The antifoaming chemical agents (water-insoluble silicone oil with low surface tension, surface-active emulsifiers, etc.), used by the Canadian heavy-water plants cost several millions of dollars per year. Besides, use of the agents gives rise to technological problems which manifest themselves within several years after commissioning. Since 1972, a program of searching for novel and more workable antifoaming agents for heavy-water production by the hydrogen sulphide method has been implemented in Canada. The requirements for these agents are: antifoaming efficiency, low volatility, chemical stability, non-toxicity and biodegradability, easy handling, low prices, and improved mass-transfer efficiency.

In laboratory-scale and pilot plants, more than 30 non-ionic surface-active agents have been tested, four of which were utilized in Canadian industrial plants. At two of three Canadian heavy-water plants, since 1978, one of these nonionic surface-active agents, namely polysiloxane glycol, has been subjected to prolonged testing which evidenced a gain in productivity owing to a partial alleviation of technological problems associated with foam formation, and a fourfold reduction in the antifoaming agent costs. In spite of the above, the tests of newer and cheaper antifoaming agents (i.e. non-silicone and non-ionic surface-active agents) have been continued with the aim of further improving the mass-transfer properties.

### 3.2.3  Heat recovery

Owing to the absence of traditional flow conversion and the possibility of the recovery of heat making up the principal item of the separation energy expenses, the hydrogen sulphide method is rendered to be a most efficient heavy-water production technique. The basic heat recovery schemes for the flows leaving the hot column are shown in Figure 3.18 by the example of a single-stage two-temperature plant.

In the simplest scheme, the heat of a hot gas or liquid flow is transferred to a corresponding flow entering the hot column (Figure 3.18a). Because of the heat under-recovery in the liquid heat exchanger LL, there is a need to preheat the liquid in the preheater H, and, owing to the heat underrecovery in the gas–gas heat exchanger GG, the hydrogen sulphide flow entering the cold column should be further chilled in a cooler C, and the gas supplied to the hot column should be further heated in the heater H. The drawback to this scheme is a low heat-transfer coefficient in the gas–gas surface heat exchanger. Because of this, a high heat-recovery degree of the gas flow leaving the hot column can be ensured only by a large heat-exchange surface. A large size of the gas recuperative heat exchanger

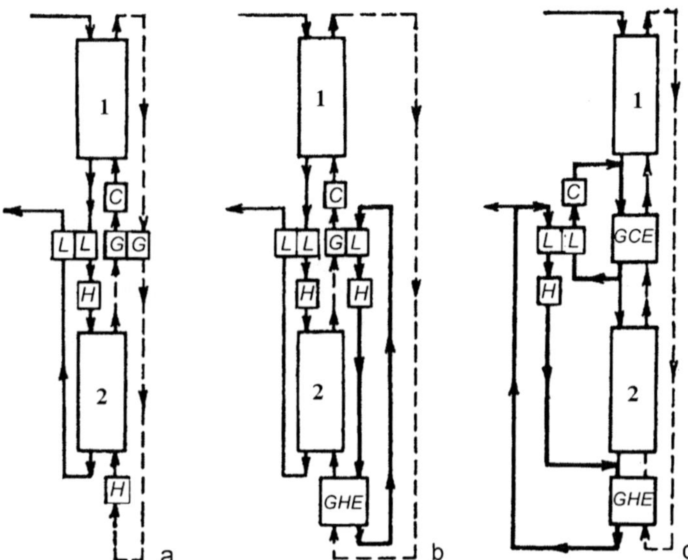


**Figure 3.18**    Schemes of heat recovery in single-stage two-temperature plant.

increases the capital costs and augments power consumption for the gas circulation (owing to a substantial hydraulic resistance in the heat exchanger).

The scheme shown in Figure 3.18b, does not include the gas recycling heat exchanger. The gas issuing from the hot column is cooled in the surface gas–liquid heat exchanger GL, giving up its heat to the liquid of the auxiliary flow circuit. The cold gas heating is performed by the direct contact of hydrogen sulphide with the auxiliary flow circuit water in the gas heating heat exchanger GHE. For the gas heating to the hot column temperature, it is necessary to overheat the liquid entering GHE in the preheater H, or to feed some vapor into GHE, with the vapor amount determined by the under-recuperation in the heat exchanger GL. The scheme with the vapor feeding into GHE was implemented at the Dana and Savannah River plants.

The direct contact ensures a high efficiency of the gas–water heat exchange and can be performed in the column bottom with the same contactors as those used for the isotope exchange. The most efficient is the heat-exchange scheme where the gas cooling is also performed in the direct contact of the phases in the gas cooling heat exchanger GCE [59–62]. This heat-exchange scheme is shown in Figure 3.18c. The auxiliary lower trays of the cold column can be employed as the GCE. The heat of the gas cooled in the liquid heat-recovery heat exchanger LL by the liquid circulating as well through GCE and heat exchanger LL, is transferred via the liquid of the second auxiliary flow circuit to the gas heated in the GHE. Heat loss due to the underrecuperation in the heat exchanger LL can be compensated for by the heat medium supplied to the preheater H or by the vapor supplied directly to the GHE. The latter scheme provides a substantial curtailment of the heat-exchange apparatus dimensions, and is applied at Canadian heavy-water plants.

The heat transfer in the heat exchangers with direct phase contact is performed so that the issuing gas is fed to the cold or hot isotope exchange column with corresponding temperature, with the aim of eliminating the need for gas reheating or aftercooling.

The calculation of the GCE and GHE heat exchangers consists of determining the circulating flows and their temperatures, as well as the required number of trays (or packing bed height). From the GCE heat balance it follows that

$$I_1 - I_O = (i_O - i_1)/\lambda_{GCE}, \tag{3.87}$$

where $I_1$ and $I_O$ are the gas flow enthalpy at the GCE inlet and outlet, respectively, J/mol $H_2S$; $i_1$ and $i_O$ are the liquid flow enthalpy at the GCE inlet and outlet, respectively, J/mol $H_2S$; and $\lambda_{GCE}$ is the ratio of hydrogen sulphide and water flow rates in the GCE.

The wet hydrogen sulphide enthalpy is defined as the sum of enthalpies of dry hydrogen sulphide and water vapor:

$$I = I_Y + hI_X = I_Y + h(i_X + i_h), \tag{3.88}$$

where $I_Y$ is the dry hydrogen sulphide enthalpy, J/mol $H_2S$; $I_X$ is the saturated water vapor enthalpy, J/mol $H_2O$; $i_X$ is the liquid water enthalpy, J/mol $H_2O$; and $i_h$ is the water evaporation heat, J/mol $H_2O$.

**Table 3.6**

Enthalpy of wet hydrogen sulphide and enthalpy of water saturated with hydrogen sulphide, J/mol

| $T$, K | $i_h$ | $i_x$ | $i_S$ | $P = 2.0$MPa | | | $P = 2.1$MPa | | | $P = 2.2$MPa | | |
|---|---|---|---|---|---|---|---|---|---|---|---|---|
| | | | | $I_y$ | $I$ | $i$ | $I_y$ | $I$ | $i$ | $I_y$ | $I$ | $i$ |
| 293 | 44,088 | 1,537 | 177,827 | Precipitation of hydrogen sulphide crystalline hydrate | | | | | | | | |
| 303 | 43,697 | 2,258 | 15,958 | −438 | −351 | 1,787 | −538 | −441 | 1,759 | −667 | −529 | 1,731 |
| 313 | 43,299 | 2,989 | 14,166 | −25 | 216 | 2,651 | −51 | 185 | 2,631 | 127 | 103 | 2,611 |
| 323 | 42,892 | 3,728 | 12,463 | 518 | 910 | 3,486 | 450 | 828 | 3,472 | 381 | 749 | 3,457 |
| 333 | 42,473 | 4,476 | 10,888 | 987 | 1,597 | 4,302 | 926 | 1,518 | 4,292 | 864 | 1,437 | 4,282 |
| 343 | 42,040 | 5,230 | 9,466 | 1,434 | 2,370 | 5,106 | 1,379 | 2,282 | 5,099 | 1,324 | 2,138 | 5,031 |
| 354 | 41,591 | 5,991 | 8,204 | 1,865 | 3,273 | 5,903 | 1,816 | 3,172 | 5,898 | 1,765 | 3,074 | 5,892 |
| 363 | 41,126 | 6,757 | 7,104 | 2,282 | 4,355 | 6,996 | 2,237 | 4,229 | 6,692 | 2,191 | 4,106 | 6,688 |
| 373 | 49,643 | 7,527 | 6,157 | 2,688 | 5,694 | 7,486 | 2,646 | 5,527 | 7,483 | 2,604 | 5,374 | 7,481 |
| 383 | 40,142 | 8,301 | 5,253 | 3,084 | 7,376 | 8,275 | 3,046 | 7,154 | 8,274 | 3,007 | 6,945 | 8,272 |
| 393 | 39,624 | 9,077 | 4,677 | 3,474 | 9,547 | 9,064 | 3,437 | 9,228 | 9,062 | 3,401 | 8,943 | 9,061 |
| 403 | 39,089 | 9,855 | 4,185 | 3,857 | 12,388 | 9,852 | 3,822 | 11,927 | 9,851 | 3,787 | 11,520 | 9,850 |
| 413 | 38,539 | 10,634 | 3,653 | 4,235 | 16,174 | 10,638 | 4,201 | 15,506 | 10,637 | 4,168 | 14,381 | 10,636 |
| 423 | 37,976 | 11,412 | 3,279 | 4,609 | 21,366 | 11,423 | 4,557 | 20,366 | 11,423 | 4,543 | 19,532 | 11,423 |
| 433 | 37,401 | 12,190 | 2,981 | 4,981 | 28,745 | 12,203 | 4,948 | 27,199 | 12,204 | 4,916 | 25,878 | 12,204 |
| 453 | 36,817 | 12,966 | 2,752 | 5,305 | 39,780 | 12,981 | 5,317 | 37,288 | 12,981 | 5,285 | 35,180 | 12,982 |

The enthalpy of water saturated with hydrogen sulphide is determined in the same manner:

$$i = i_X + s(I_Y + i_S),\qquad(3.89)$$

where $i_S$ is the heat of hydrogen sulphide dissolution in water, J/mol $H_2S$.

The values $h$ and $s$ appearing in eqs. (3.88) and (3.89) are calculated by eqs. (3.65)–(3.67) (see also Table 3.4).

Shown in Table 3.6 are the values of $I$ and $i$ obtained by generalization of both experimental and calculation data on the water and hydrogen sulphide enthalpy.

Owing to the water vapour condensation, the gas enthalpy changes more than the liquid one. As a consequence, $\lambda_{GCE} < 1$, while in the isotope exchange columns $\lambda > 1$. Hence, an additional liquid flow $L_{cl} = L_{GCE} - L_l$ is required to ensure the desired flow-rate ratio in the GCE. The GCE heat exchange can be represented in graphical form by the coordinates $I = f(T)$. As Figure 3.19 suggests, the gas enthalpy rises steeply as temperature increases (since the gas humidity augments), while the temperature dependence of liquid enthalpy is practically linear (since the water heat capacity has only a weak dependence on the temperature, and dissolved hydrogen sulphide affects $i$ only at a low temperature).

Similar to the mass-transfer processes, the height of a counter-current heat exchanger with direct phase contact can be determined by the graphical plotting of theoretical heat-exchange plates (THEXP).

As can be seen from Figure 3.19, in theory the case is possible when the liquid and gas temperatures are equal both in the upper and in the lower GCE cross-sections. The heat-exchange process driving force, however, will be equal to zero, and an infinitely high GSE is required to realize the process.

To create the useful temperature difference at the heat exchanger cold end, it is necessary to overcool the liquid flow, while some underheating of liquid will be observed at the hot end.

From the condition $\Delta I = \Delta i/\lambda_{GCE} \approx C_{pH_2O}\Delta T/\lambda_{GCE}$ it follows that the greater the liquid temperature difference at the heat exchanger inlet and outlet, the higher $\lambda_{GCE}$, and consequently,

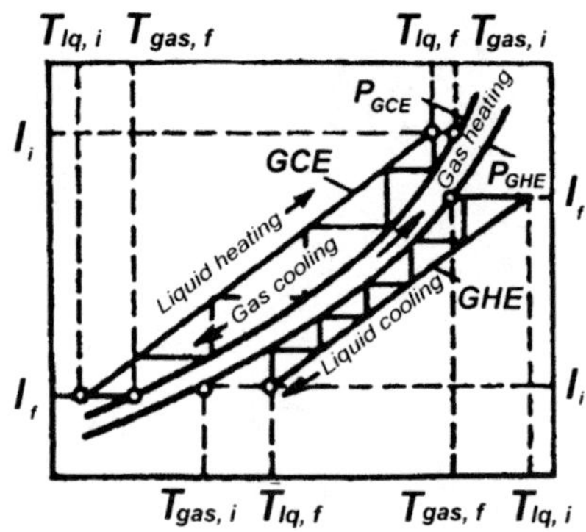


**Figure 3.19**   Calculation of counter-current heat exchangers with direct phase contact.

the smaller $L_{c1}$. The lowest temperature at the GCE inlet will be determined by the flow $L_{c1}$ since the flow temperature at the cooler C outlet cannot be lower than the temperature of hydrogen sulphide crystalline hydrate precipitation. From these considerations, the temperature of liquid at the GCE inlet is determined, and the heated flow temperature at the GCE outlet is so selected that on the one hand the liquid underheating does not exceed 2–3 degrees, and on the other the THEP number is not too high (no more than 5–10). On this basis, $\lambda_{GCE}$, as well as the flow $L_{c1}$ and its temperature, are determined.

In the same manner, the gas heating heat exchanger calculation is done. In this case, however, even with an infinite GHE height, it is impossible to achieve equality of the gas and liquid temperatures either at the hot or at the cold ends of the GHE (Figure 3.19). As the gas humidity, and consequently its enthalpy, are pressure dependent, for the GHE calculation in Figure 3,19 is shown the dependence $I = f(T)$ corresponding to the GHE pressure which differs from the GCE pressure by the hot column hydraulic resistance. In the GHE calculation, proper allowance must be made for the fact that owing to the gas heating on compession in the circulation pump, the gas temperature at the heat exchanger inlet will be higher than that at the cold column outlet.

In actual practice, the liquid arriving at the GHE is superheated, and its temperature at the outlet is also higher than that in the cold column.

In the GHE calculation, the THEXP number (no more than 10) and the rate of an additional flow circulating through GHE $L_{c2} = L_{c1}$ are set since the optimum condition of the recovery heat exchanger LL operation is the equality of heat exchanging. On this basis, the liquid temperature at the GHE inlet and the additional circulating flow temperature are determined. A high heat-transfer coefficient and a large phase contact surface allow for small dimensions of heat exchangers with direct phase contact. Owing to the curvature of the line determining the temperature dependence of the wet gas enthalpy (Figure 3.19), however, it is difficult to attain a heat recovery degree exceeding 50–60%. The heat recovery degree can be significantly augmented when the heat recovery process is performed in several heat-exchange stages.

Figure 3.20 shows a two-temperature plant with two heat-exchange stages, which allows for reducing the consumption of an expensive (compared with cooling water) refrigerant. The necessity of utilizing the refrigerant may arise in the absence of cooling water with required parameters or in the case of the cold column operation at a lower temperature (with the use of a hydrogen–ammonia system, for example, the cold column temperature is 230–240K).

A peculiarity of this scheme is that most of the heat is withdrawn at a high enough level, and only the rest with the refrigerant. The most efficient method is to use the thermal energy withdrawn at a high heat level for the cold generation (i.e. with absorption refrigerating units operated by hot water). In the two-stage heat exchange scheme, both GHE and GCE should be provided with two circulating flows of heat-transfer medium (i.e. water) in each heat exchanger with direct phase contact. Leaving the heat exchanger 8, the heat-transfer medium flow passes through the main heat-recovery heat exchanger, where it gives up the heat to the main heat-transfer medium curculating flow which heats the gas in the stage I of GHE 5′, and reheats the second circulating flow passing through the both stages of GHE. Following the main heat-recovery heat exchanger, most of the heat-transfer medium is supplied to the refrigerating machine (RM) 10, and then it returns to the GCE first stage.

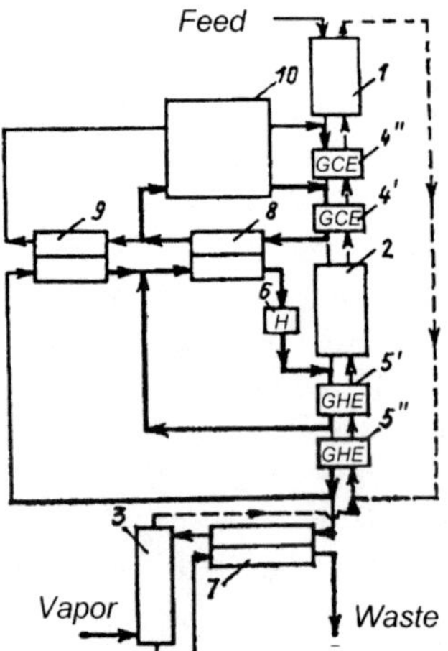



**Figure 3.20**   Scheme of two-temperature hydrogen sulphide plant with two heat-exchange stages in the heat-recovery system: 1, cold column; 2, hot column; 3, hydrogen sulphide stripping column; 4′, 4″, 5′, 5″, stages I and II of GCE and GHE, respectively; 6, preheater; 7, 8, 9, heat-recovery heat exchangers; 10, refrigerating machine.

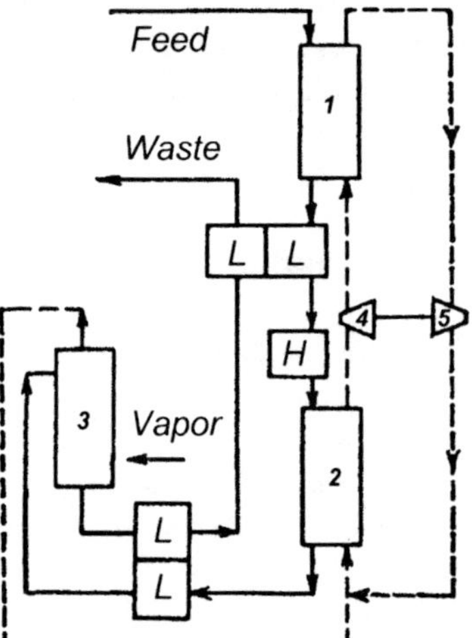



**Figure 3.21**   Scheme of two-temperature hydrogen sulphide plant with cold and hot columns operated at different pressure values: 1, cold column; 2, hot column; 3, hydrogen sulphide stripping column; 4, turbine; 5, compressor.

The rest of heat-transfer medium passes through the auxiliary heat-recovery heat exchanger 9, then it is cooled in the RM, and utilized in the GCE stage II for the required cooling of gas on its way from the hot to the cold column. The above heat-exchange scheme implementation does not involve additional (external) heat-transfer medium input (except for cooling water).

Regarding the reduction of the heating vapor and cooling water discharge, of interest is a scheme for the two-temperature hydrogen sulphide plant, of which the cold and hot columns are operated at different pressures (Figure 3.21). The pressure in hot column 2 is several times higher than that in cold column 1. Leaving the hot column at a temperature $T_2$, the wet gas is cooled due to adiabatic expansion to a pressure $P_1$ in turbine 4, and is supplied to the cold column in which the temperature is $T_1$. External work performed by the turbine is used to compress gas in compressor 5 on the way from the cold to the hot column.

The turbine–compressor system can ensure a higher energy recovery than that in a scheme with heat exchangers with direct phase contact. Indeed, the total external adiabatic efficiency of large gas turbines is about 0.85–0.92 while the mechanical efficiency accounts for 0.95. The total turbine efficiency, therefore, makes up 0.85. For the same efficiency of the compressor where an inverse process takes place, we obtain the energy recovery degree equal to 0.7.

If the selected operation parameters are those at which the gas compression in the compressor ensures the gas temperature increases from $T_1$ to $T_2$, the heat energy consumption will be chiefly associated with under-recuperation in the liquid–liquid heat exchangers, and cooling water is unnecessary. In addition, the scheme under consideration makes it possible to increase the extraction degree and separation efficiency by extending the temperature operating range in the separation columns.

### 3.2.4  Schemes of industrial plants

Heavy-water industrial production by the two-temperature hydrogen sulphide method was first realized in the U.S.S.R. [8, 45]. In 1946–1949, an industrial plant with an output of four tons of $D_2O$ per year was constructed in Aleksin, Tula region. Insufficient knowledge of the process industrial applicability, the lack of the process engineering theory, and the absence of experience in developing such plants did not allow the most efficient parameters of the process to be achieved. An eightfold initial concentration of deuterium was performed in a single-stage two-temperature plant with a cold and a hot column of 2.8m diameter and 30m height each. Owing to the plant's low pressure (0.8–0.9MPa), the hot column temperature ($T_2$) did not exceed 373K, and with the cold column temperature ($T_1$) equal to 298K, $\Delta T$ was only 75°, which resulted in a rather poor, by present standards, separation efficiency. With the water feed flow of 34.5tons per hour, the extraction degree accounted for only 13.1%. Nevertheless, the plant was in operation for about 30 years.

In 1952–1953 two facilities were commissioned in Dana and Savannah River, USA [47]. At the Dana facility, initial concentration of deuterium up to 15–20% was performed in a five-stage cascade with 700 trays in total, while in Savannah River a two-stage cascade with 450 trays in total was used to attain the same concentration.

After commissioning, the facility was operated at full power for a relatively short time. The Dana facility produced 450–500tons, and the Savannah River facility 540tons of $D_2O$ per year. Being more complex (six five-stage plants with four concurrently operating cold and hot columns at each stage) the Dana facility required more time for inspection and maintenance than the Savannah River facility, comprising 24 independently operating plants. In early 1957, the Dana facility was shut down, and by the end of the year, 16 plants of the

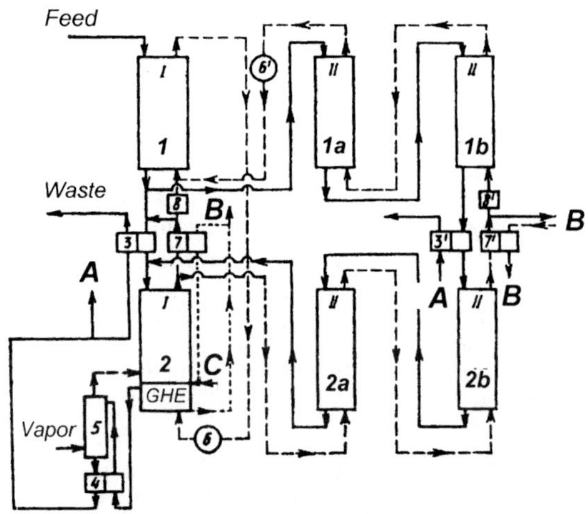


**Figure 3.22**    Scheme of two-temperature hydrogen sulphide plant at Savannah River site (U.S.A.).

**Table 3.7**

Comparison of dimensions of columns of two-temperature hydrogen sulphide plants at Savannah River and Bruce facilities

| Parameter | Savannah River facility, plant with an output of 22.4 tons of D$_2$O per year | | Bruce facility, plant with an output of 400 tons of D$_2$O per year | | |
|---|---|---|---|---|---|
| | Stage I | Stage II | Stage I | Stage II | Stage III |
| Separation degree | 4–6 | 250–190 | 4 | 6 | 120 |
| Number of concurrently operating columns | 1 | 1 | 3 | 1 | 1 |
| Diameter, m: | | | | | |
|    cold column | 3.35 | 1.98 | | 7.0 | 3.0 |
|    hot column | 3.66 | 1.98 | 9.2 | 3.0 | |
| Height, m: | | | | | |
|    cold column | 35.4 | 40 × 2 | | | |
|    hot column | 35.4 | 40 × 2 | 92 | 82 | 64 |
| Number of trays: | | | | | |
|    cold column | 70 | 85 × 2 | | | 110 |
|    hot column | 70 [a] | 85 × 2 | 170 [b] | 150 [b] | 110 |

[a] including ten trays for gas heating.

[b] including 15–20 trays for gas cooling (in the column of stage I, an additional 15–20 trays serve for gas heating).

Savannah River facility were mothballed, and until 1981 the facility operated at one third of full capacity. The flow diagrams, processing equipment, and performance characteristics of the above facilities are adequately covered in the literature [1, 11, 63–65]. A Savannah River plant is shown in Figure 3.22, with the main dimensions of the separation columns presented in Table 3.7. The source water fed to the upper part of the cold column 1 of stage I passes down the column and via heat exchanger 3 is delivered for hot column 2 refluxing. The adopted heat recovery scheme corresponds to Figure 3.18b (the ten bottom trays are used for the gas heating). The wastewater is withdrawn from the eleventh tray of the hot column and via liquid heat exchanger 4 supplied to column 5 for hydrogen sulphide stripping.

Gas circulation is performed by centrifugal blower 6 from the column 1 head through the 10 bottom trays of column 2, where the gas is heated by water, and then through cold column 1 of stage I. About a third of the gas and liquid flows are supplied to stage II, comprising two series-connected cold columns (1a and 1b) and two hot columns (2a and 2b). The liquid arriving at the hot column is heated in the heat exchanger 3' by the wastewater heat. The gas emerging from hot column 2b is cooled in heat-recovery heat exchanger 7', losing heat to the liquid of the additional circuit, and then in cooler 8'. Centrifugal blower 6 is utilized for the gas circulation through stage II.

All heat is delivered to the plant as live steam, and fed to hydrogen sulphide stripping column 5, maintained at 493K. The product is withdrawn from the column as a condensate formed in heat exchangers 7'and 8'. The condensate merges with liquid flow from the cold column 1, heats up in heat exchanger 3 by the wastewater heat and arrives at column 2 as a reflux.

Owing to the development of nuclear-power engineering based on CANDU-type heavy-water reactors, an extensive program for heavy-water plant construction was implemented in Canada (see Table 3.8).

At these plants, initial deuterium enrichment of up to 20–30% is performed by the hydrogen sulphide method, and the end concentration by water vacuum rectification. The main dimensions of the Bruce plant are presented in Table 3.7, and its scheme is shown in Figure 3.23. The design of all Canadian plants, except for the Glace Bay facility,

**Table 3.8**

Plants for heavy-water production by two-temperature hydrogen sulphide method

| Plant location | Design capacity, tons per year | Year of commissioning | Note |
| --- | --- | --- | --- |
| Canada | | | |
|   Port Hawkesbury | 400 | 1970–1972 | Shut down in 1985 |
|   Glace Bay | 450 | 1976 | Shut down in 1985 |
|   Bruce, Block A | 800 | 1973 | Shut down in 1984 |
|   Bruce, Block B | 800 | 1978 | Shut down in 1997 |
| India | | | |
|   Kota | 100 | 1984 | |
|   Manuguru | 185 | 1988–1991 | |
| Argentina | | | |
|   Arroyito | 250 | 1985 | |
| Romania | | | |
|   Drobeta-Turnu | 270 | 1987 | |
| China | N/A | | |

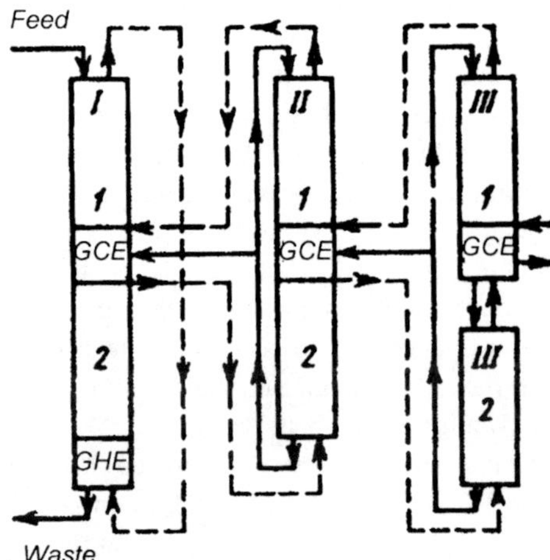


**Figure 3.23**   Scheme of flows for two-temperature facility at Bruce Plant (Canada).

follows the same scheme. In the first two stages, cold column 1 and hot column 2 are arranged one above the other in a single casing. The main performance characteristics of the Canadian plants are similar to those of the Savannah River facility ($T_1$ = 303–308K, $T_2$ = 403–413K; $P$ = 1.9–2.1MPa), with the difference that, as is seen from Figure 3.23, a three-stage cascade is applied here, and the stages are coupled by the gas flow. Since the successive stages are fed with the gas saturated with water vapor, it is necessary that not only the gas leaving the cold column, but also a quantity of water equal to the amount of condensate formed in the process of gas cooling in GCE, should be returned to the preceding stage. At the Glace Bay plant, the two-temperature cascade stages are coupled only by liquid flows.

In the 1980s, two plants for industrial heavy-water production by the two-temperature hydrogen sulphide method were built in India: in Kota with an output of 100tons, and in Manuguru with an output of 185tons of $D_2O$ [66–68]. Both of these plants employ three-stage cascades. A concentration of up to 15% is obtained in the hydrogen sulphide plants, with further concentration to 99.8% performed by the water vacuum rectification method. The cost of exchangers amounts to 45% of the overall plant cost. In the Manuguru plant construction, in addition to the experience gained in the construction and operation of the Canadian plants, the on-site experience of the Kota plant construction was used. Initially, the Manuguru plant construction schedule time from the project zero point to the full-scale mechanical start-up was set at 4.5 years. But in reality the schedule time was exceeded by 17 months, and, instead of July 1987, the plant was put into operation in December 1988. Because of heavy capital investments, project optimization aimed to minimize the plant's volume and maximize heat recovery. In the first stage, provision is made for concurrently operating columns owing to the high rates of liquid flows processed at this stage, and to limitations of manufacturing large-diameter columns. The dimensions and processing characteristics of the Kota plant's exchange columns are presented in Table 3.9 [57].

**Table 3.9**

Dimensions and processing characteristics of exchange columns of Kota heavy-water plant

| Stage | Number of columns | Height, $m$ | Diameter, $m$ | Number of trays | $P$, MPa | $T$, K |
|---|---|---|---|---|---|---|
| 1st, cold | 3 | 49 | 4.25 | 246 | 2.1 | 305 |
| 1st, hot | 3 | 50 | 4.50 | 275 | 2.2 | 408 |
| 2nd, cold | 1 | 51 | 4.25 | 252 | 2.1 | 305 |
| 2nd, hot | 1 | 38 | 4.50 | 211 | 2.2 | 406 |
| 3rd, cold | 1 | 54 | 2.30 | 108 | 2.1 | 305 |
| 3rd, hot | 1 | 54.5 | 2.30 | 116 | 2.2 | 404 |

## 3.2.5  Industrial safety and environmental protection operational safety

As mentioned above, an advantage of the two-temperature hydrogen sulphide method lies in the fact that deuterium is extracted from the most abundant natural resource: water. But the source water must first be freed of dissolved and suspended matter, and the hydrogen sulphide must be removed from wastewater flow. That is why the water is filtered and acidified to decompose carbonates. Filters with ion-exchange materials, such as sulfonated coal and ion-exchange resins, are used for water softening. Dissolved gases (oxygen, carbon dioxide, noble gases) are removed by deaeration without heating (or vacuum deaeration), or at an elevated temperature (365–370K). The presence of dissolved gases is undesirable since as the water temperature increases (in the hot column) the gases escape into the hydrogen sulphide which results in the accumulation of $CO_2$ and noble gases in the cycle. This ballast in the hydrogen sulphide flow will reduce the two-temperature plant output. In addition, dissolved oxygen readily oxidizes hydrogen sulphide by

$$2H_2S + O_2 = 2H_2O + S. \tag{3.90}$$

The released sulfur clogs the column contactors (trays, packing) and heat exchangers, which augments the resistance and impairs throughput rate. Irrespective of whether the wastewater is discharged or utilized in other production processes, it is freed of dissolved hydrogen sulphide in the so-called stripping column. The stripping column is a counter-current mass-transfer unit (generally packed) where withdrawal (stripping) of hydrogen sulphide from the arriving reflux water is performed with the use of live steam supplied from the bottom. The counter-current flow makes it possible to achieve a high degree of water purification from hydrogen sulphide.

Residual hydrogen sulphide content in the wastewater of industrial plants generally does not exceed 20–50% of the maximum allowable concentration. The stripping column can operate both at atmospheric pressure (in this case the hydrogen sulphide extracted from water is returned to the two-temperature unit with the use of a special-purpose compressor) and at an elevated pressure (when the combined hydrogen sulphide and water vapor pressure exceeds the gas pressure in the two-temperature unit, and there is no need for the compressor).

Figures 3.20–3.22 show the schemes of the two-temperature plant with a stripping column operating at 493K and ensuring the hydrogen sulphide return to the gas flow supplied to the hot column (at a working pressure of the two-temperature plant $P = 2.1\,MPa$).

The stripping column can be calculated with the use of $xy$-diagram. The equilibrium line is determined by Henry's law constant ($K_H$) for hydrogen sulphide and by the stripping column pressure $P$; the operating line position depends on the hydrogen sulphide concentration in the column inlet and outlet flows. Since the equilibrium line is straight, NTP is readily found by the grapho-analytical method:

$$n = \ln \frac{c_l'' K_H/(P - c_g'')}{c_l' K_H/(P - c_g')} / \ln(\lambda K_H/P), \tag{3.91}$$

where $c_l''$ and $c_l'$ are the dissolved hydrogen sulphide concentrations in the liquid phase in the upper and lower cross-sections of the stripping column, mole fractions; $c_g''$ and $c_g'$ are the dissolved hydrogen sulphide concentrations in the gas-vapor phase in the upper and lower cross-sections of the stripping column, mole fractions; $\lambda$ is the ratio of gas and liquid mole flow rates ($\lambda = (c_l'' - c_l')/(c_g'' - c_g')$).

The concentration $c_l''$ is determined from the data on phase equilibrium in the hydrogen sulphide – water system, and the $c_l'$ values are assigned from the allowable concentration of water-dissolved hydrogen sulphide (i.e. from the sanitary standard equal to $10^{-4}\%$). It is evident that $c_g' = 0$, and the concentration $c_g''$ depends on $\lambda$.

Next, with due regard to HETP and to the contactor's throughput capacity, the stripping column height and diameter are determined.

The operating experience of large-scale heavy-water industrial plants suggests that a relatively small stripping column is capable of ensuring efficient purification of wastewater from dissolved hydrogen sulphide (the overall losses of hydrogen sulphide including the carry-over by the wastewater, do not exceed 50–100 grams per one kilogram of heavy water).

For the initial fillup of separation columns and loss compensation, a hydrogen sulphide production unit is required. Traditionally, the hydrogen sulphide is obtained from the intraction of NaHS and $H_2SO_4$, and yet the method of hydrogen sulphide direct synthesis from elements, which allows for excluding liquid-sulfur-containing waste, is more efficient. Hydrogen sulphide synthesis is done with/over a sulphide catalyst at 600–620K.

The issues of environmental security and industrial safety are of prime importance in the operation of a two-temperature hydrogen sulphide plant, the characteristic property of which is the abundance of hydrogen sulphide. The Indian Kota plant, with an output of 100tons of $D_2O$ per year, requires some 200tons of hydrogen sulphide as the processing cycle feeding and, in addition, 50tons of stored liquified $H_2S$ to compensate for the hydrogen sulphide losses in the cycle [69]. What is more, account must be taken of the fact that the hydrogen sulphide is not only toxic (the maximum allowable concentration in the industrial plant air accounts is $10\,mg/m^3$), inflammable, and explosive (explosivity limits in the air are 4.3–46%), but also highly corrosive, in the presence of water, to carbon steel.

An example of how all industrial safety aspects at design, construction, installation, and operation stages are taken into account is the plant of heavy-water production by the two-temperature hydrogen sulphide method in Kota (India) [69]. The plant is sited at a distance of about 700m from the Rajasthan nuclear power plant. In deciding on the plant location, the following criteria were used: in addition to availability of water, vapor, and convenient access roads, the area must be practically unpopulated. The adopted concept provides for a restricted close-control area of 1.6km in radius round the plant, and a limited population access area of 6km in radius. Safety issues associated with equipment placement are: the exchange columns and $H_2S$ production and storage unit are positioned so that any leakage is swept away by wind without forming local concentrations. To ensure the emergency evacuation of personnel, provision is made for  safe passage between the equipment, and evacuation routes are properly marked. All main pipelines are raised at a height of about 6m from the ground. The central control board is located beyond the risk area, and provided with a protective shelter for operating personnel.

At the stage of design, selection and manufacture of various equipment, including control and electrical power systems, particular attention was also given to personnel safety issues. Because of a very high corrosivity of hydrogen sulphide, investigation of various types of corrosion with processing environment modeling, and selection of suitable engineering materials, have been performed. All employed materials were subjected to intense scrutiny comprising:

- full-scale radiographic inspection of all welds;
- ultrasonic testing;
- dye penetration testing or magnaflux examination;
- hydraulic tests.

Stringent control techniques were applied at all stages of equipment and pipeline manufacturing.

The high explosivity of hydrogen sulphide was taken into account in the electrical equipment selection, and all working zones were categorized according to the hazard level.

For the operation stage, an accident prevention package was worked out. The plant is so engineered that various working areas and individual equipment can be remotely isolated. All main equipment is connected to two flare systems. The prevention of emissions and release of water contaminated by hydrogen sulphide is ensured by valve and pump isolation systems. The plant is provided with a closed drainage system to which all the equipment is connected. Appropriately diluted waste water is discharged at a distance of 2.8km from the plant. Provision is made for the emergency air reserve, $N_2$ storage, emergency power plant, and storage of breathing air. Occurrence of $H_2S$ in the plant's environmental air is controlled by a system comprising lead acetate test paper, tube analyzers, $H_2S$ concentration monitors (41 monitors in various plant areas), and up to 10 portable $H_2S$ monitors. Eight external monitors are designed to control the hydrogen sulphide concentration level in various directions outside the plant.

Personnel protection measures comprise: gas masks in various protection shelters for the personnel emergency evacuation, isolating breathers for all personnel bound to work in the plant's hazardous zones, local breathing air units for prolonged work in various hazardous zones, and supply of breathing air to the central control board (CCB). In the event of a hydrogen sulphide contamination hazard, the safest place for the plant personnel is the CCB room, which is provided with a breathing air system for 50 people for two hours. Serving as an auxiliary facility, the local weather station monitors weather conditions (wind speed and direction), which helps to determine the direction of personnel evacuation in case of $H_2S$ release.

The protective measures generally comprise: (a) a system of protective shelters; (b) control over the personnel and transport vehicles arriving and departing in all working areas; (c) medical aid system; and (d) evacuation.

The impact of $H_2S$ on personnel health and safety, and on the environment is considered in the reports [70, 71]. From 1970 to 1980, there were 15 events of exposure of personnel to $H_2S$ without fatalities but involving intensive care measures, recorded at the Canadian heavy-water plant. During 1982 to 1989, the Indian Kota plant accounted for one fatal accident, one event of acute exposure, one event of medium severity and three to four events with weak health effects associated with the influence of $H_2S$. Although the toxicity of hydrogen sulphide has been known since the early seventeenth century, the full mechanism of the effect of $H_2S$ on humans remains unclear. The organismal response to various concentrations of $H_2S$ for different exposure times is considerd in reference [70]. The other reference [71] presents an overview of recent investigations in hydrogen sulphide toxicity, and examines the therapy modes, including role of various antidotes utilized in $H_2S$ poisoning treatment.

In the chemical and petrochemical industries, flare systems are used for harmful gas emissions combustion and high-altitude venting into the atmosphere. At heavy-water hydrogen sulphide plants with large amounts of hydrogen sulphide (generally in the region of 150–200tons of $H_2S$ per 100tons of $D_2O$ annual output), all scheduled and emergency $H_2S$ and $H_2$ emissions are effected through flare systems. As a rule, the flares burn continuously which prevents the risk of potential explosions. Many explosions in gas flare systems have been reported in the literature [72–74]. Reference [75] describes experience in designing and operating flare systems for the Kota heavy-water plant. The paper examines the causes of explosions initiated by flashbacks in flare systems, and provides recommendations for flare design and technological parameters.

An experimental investigation of different methods of sulphide oxidation in acid liquid wastes in the Kota plant has been performed [76]. Dissolved hydrogen sulphide oxidation by chlorine gas, hypochlorite solution, and air was tested. It was desired to lower the sulphide ion content to no more than 2 ppm. In all cases, the end products of the oxidation are sulfur and sulfate ions, and the controlling factors in their formation are pH for oxidation by chlorine, and temperature and pH for oxidation by hypochlorite. As is shown, sulphide ion oxidation with air is a slow process, and it can be substantially catalyzed by trace manganous salts ($MnSO_4$, $MnCl_2$). In any case, the necessity of oxidizing the sulphides in wastes leads to an increase in production costs. For the Managuru plant, use of the chlorination method was recommended.

Risk assessment techniques in heavy-water production by two-temperature hydrogen sulphide method are discussed in some sources [77–80]. The studies take account of the process technology data, properties of hydrogen sulphide, technical parameters of all equipment types, and the character of the plant-scale safety measures. Proper allowance is also made for available records of in-plant accidents, and for the influence of human factors on the plant's operation. The papers describe and analyze the causes of process control faults.

To determine a possible frequency of emergency situations at hydrogen sulphide heavy-water plants, mathematical models have been developed.

Reference [81] emphasizes a need for permanent training to raise the skill levels of personnel and reduce the adverse effect of human factors on the process. The paper presents a program comprising different forms of personnel training as applied to various types of heavy-water production plants, including hydrogen sulphide plants.

One of the personnel protection measures is the warning of gas contamination of particular plant or system zones. A deciding factor is the promptness in determining $H_2S$ concentration in the air. For on-line monitoring of $H_2S$ concentration in the air, an analyzer based upon the chemiluminescent reaction between $H_2S$ and $O_3$ was developed [82]. The analyzer was field tested with determination of daily variations of the hydrogen sulphide concentration level within the Kota plant territory. The $H_2S$ concentration in the air was demonstrated to be substantially lower than the admissible concentration limit.

Carbon steel is the main structural material for the hydrogen plant equipment both in the USA and Canada, and, subsequently, in other countries. Under the influence of hydrogen sulphide, a protective ferrous sulphide film is formed which inhibits further corrosion. But a high liquid-flow velocity causes the film to erode, and the corrosion rate increases dramatically. That is why the development of a technique for stable sulphide film formation during the start-up period seems to be of profound importance [83–85].

Reference [83] defines the types of ferrous sulphide formed under hydrogen sulphide process operation conditions: mackinwite $FeS_{0.83}$, troilite $FeS$, pyrrhotite $Fe_{0.8}S$, marcasite $FeS_2$, and pyrite $FeS_2$. It is demonstrated that the corrosion rate is very high at the initial stage, but decreases drastically with time. This stage of a low corrosion rate corresponds with a sharp increase of the pyrite–pyrrotite content in the sulphide film. The corrosion rate becomes rather low when the pyrite–pyrrotite content in the sulphide film accounts for 90% which is achieved within 700–1,000hours. The paper studies the conditions of the stable film formation as related to the time of exposure to hydrogen sulphide environment, temperature admixtures to source water, hydrogen sulphide concentration, and pH variation due to the addition of NaOH dilute solutions. As a result it was found that for the formation of a stable sulphide film with good adhesion properties on carbon steel in the $H_2S$–$H_2O$ system, the sulphide film should consist of pyrite–pyrrotite to the extent of 90%. The formation of such a film was achieved within 500hours. A higher pH value (about 11) and a higher temperature (up to 120°C) are preferable, with a high hydrogen sulphide concentration being an additional favourable factor. Though the best results are obtained in the liquid phase, an adequate film formation can also be achieved at elevated temperatures and higher values of the source water pH. A newly devised technique of carbon steel surface immunization was tested at the Kota plant and produced good results.

The problems of carbon steel protection at Argentinian heavy-water plants are discussed in references [84, 85]. Tests were performed to immunize the surface of carbon steel utilized in heavy-water production by hydrogen sulphide method. The steel was treated with saturated hydrogen sulphide solution at a pressure of 2.3MPa and a temperature of 125°C in the NaOH solution with a concentration of $5 \times 10^{-3}$M. The paper examines properties, composition, and adhesiveness of the resultant film and studies the corrosion rate data.

Reference [86] is concerned with the problem of depositions in the stripping columns of hydrogen sulphide heavy-water plants with the aim of identifying the nature and qualitative properties of the depositions. To reduce the depositions to a minimum, it was suggested that the deionized water be used as a source flow, which was implemented.

The elaborated measures, and the employment of specially trained personnel, provide a sufficient degree of safety for heavy-water production by the two-temperature hydrogen sulphide method.


### 3.2.6  Production control

The maximum efficient use of large-scale hydrogen sulphide plants can be achieved only with reliable on-line process control to ensure optimum parameters in the columns (temperature, pressure, and flow-rate ratio $\lambda$). The most complicated problem is the maintenance of the optimum flow-rate ratio $\lambda_0$ ($\lambda_0$ deviations exceeding 1% are not permissible, and a 15% deviation leads to the complete loss of efficiency).

Since industrial-class flow meters do not provide the required accuracy of flow measurements, control over the flow-rate ratio of the hydrogen sulphide plants is based on the dependence of the column concentration profile on the $\lambda$ value (see section. 3.1.3). At an optimum flow-rate ratio, the deuterium concentration in water (hydrogen sulphide) on the middle tray of the hot column will be equal to that on the middle tray of the cold column. With departure from $\lambda_0$ in the direction of increasing gas flow rate (or decreasing liquid flow rate), the concentration on the middle tray of the cold column will be higher than the concentration on the middle tray of the hot column ($x_1 > x_2$). Conversely, at $\lambda < \lambda_0$, $x_1 < x_2$. Using the results of intermittent (every 4–8 hours) analytical monitoring of the deuterium concentration (by using mass spectrometry) in the sample water from the middle cross-sections of the cold and hot columns, it is possible to maintain the optimum flow-rate ratio, by manual correction, within an accuracy of $\pm 0.5\%$. The correction is done by adjusting the liquid flow, with an unchanged hydrogen sulphide flow rate that is generally set to the maximum allowable level. In determining the optimum flow-rate ratio in two-temperature plants at stages II and III of the cascade, account must be taken of the curvature of equilibrium lines, which at $\lambda = \lambda_0$ results in the ratio between the deuterium concentration in water at the middle tray of the cold column and the corresponding concentration in the hot column being greater than unity. The optimum value of the ratio, which depends on the concentration range in which the cascade stage operates, augments with increase in the stage number.

At present, more accurate ultrasonic flow meters insulated against operating environment and, consequently, not prone to hydrogen sulphide corrosion, as well as flow infrared analyzers of the water isotopic composition, are available. At the Bruce plant, the trays' operation was controlled with the use of gamma radiography. The foam formation at the sieve trays, clogging of holes, as well as damage to trays and structural elements, are controlled by the radiation source ($^{60}$Co), and a detector mounted on the opposite side of the column. It should be noted that at early stages of the Canadian industrial facility's development, the process breakdowns associated with the instability of the trays' hydraulic regime represented a chief cause of productivity decline, and of an elevated hydrogen sulphide concentration in the wastewater. Since the foam formation at the trays is associated either with admixtures entering the plant with incoming hydrogen sulphide (as a result of oil leakages through the compressor seals), or with intentionally introduced, and gradually accumulated, additives (for water treatment), the control over the process flow impurity content (in a low concentration range from $10^{-4}$ to $10^{-7}\%$) called for the development of novel analysis techniques.

It is important to monitor the impurity content of the circulating hydrogen sulphide. Apart from the above-mentioned admixtures introduced with source water ($CO_2$ and noble gases), the hydrogen sulphide can accumulate hydrogen produced by equipment corrosion; that is, by the interaction of hydrogen sulphide with iron. The purity of hydrogen sulphide is generally maintained at a level of 99.5%, which is accomplished by the scavenging of hydrogen sulphide circulating through a special-purpose purification system using either hydrogen sulphide liquefaction, or its preferential absorption by source water.

A mathematical model, describing the steady state of a separation cascade of two-temperature plants and rectification columns for the final concentration of heavy water, was developed to ensure the efficient operation of the Savannah River plant. Similarly, simulation software for the separation process, as applied to other plants for heavy-water production by the hydrogen sulphide method, was developed.

### 3.2.7  Performance characteristics and ways of improvement

The largest two-temperature hydrogen sulphide plants were engineered in Canada (see Table 3.8). The operation experience of American plants and the available recommendations according to which the units were enlarged in size (by a factor of 18 compared with Savannah River plants), the cold and hot columns, as well as heat exchangers of the direct phase contact, were combined into a single unit (see Table 3.7). Owing to these advances, the steel intensity of separation units was reduced by a factor of about 2.3.

Average performance characteristics of the Canadian heavy-water plants are presented in Table 3.10.

**Table 3.10**

Average performance characteristics of hydrogen sulphide plant

| Parameter | For a plant with a productivity of 400 tons per year (48.3kg per hour) | Consumption factor per 1 kg of $D_2O$ |
| --- | --- | --- |
| Source water flow | 470–650kg/s | 35t |
| Cooling water flow | 2,300kg/s | 170t |
| Hydrogen sulphide consumption | 4kg/h | 0.08kg |
| Heat consumption | 300–320MW | 6.4MW·h |
| Power consumption | 34–40MW | 0.7MW·h |
| Personnel strength | 180–220 | — |
| Capital investment | USD150 million [a] | — |
| Cost price | — | USD100 [a] |

[a] based on 1974 prices.

From the cost price composition of heavy water produced at Canadian plants (60% depreciation, 20% heat energy, 7% electric power, 13% maintenance) it can be seen that the major part of expenditure is determined by heavy capital investment. Since some 20% of the total cost of the two-temperature plant falls on the pipelines, further capital cost reduction was suggested by shortening the length of pipeline communications with the replacement of heat-exchange units by coils mounted inside the columns.

But the most important reduction of separation costs can be achieved by decreasing the cost of separation columns. An example of such an improvement is the Savannah River plant, where the bubble cup trays were replaced with more efficient and inexpensive sieve trays, which also resulted in an increase in plant productivity by a factor of 12% (Table 3.5). At the Canadian plants, the electric power charge was also reduced because of the sieve trays and direct contact heat exchangers having lower hydraulic resistance compared with casing-pipe heat exchangers.

It seems likely that further reduction of the cost of the separation columns can be obtained by intensifying the hydrogen sulphide – water isotope exchange with the use of accelerating agents (see section 3.2.2) or more efficient novel contactors.

The second largest contribution to the cost of heavy water falls on the heat energy. Since the heat recovery level in the liquid–liquid heat exchangers is sufficiently high, the major heat consumption is associated with heating and saturating with vapor the gas arriving at the hot column. That is why all improvements to the heat recovery schemes discussed in section 3.2.3 were generally associated with gas heating and cooling issues. The operation of two-temperature hydrogen sulphide plants involves high energy consumption, so power supply reliability is an important factor in assuring process efficiency. Although the Canadian plants were supplied with steam from nuclear power plants, the Bruce plant was provided with a steam generator operated by petroleum. To compensate for under-recuperation and heat loss to the environment, live steam supply should not be used (as it was in the two-temperature hydrogen sulphide plants), but instead by heating circulating liquid flows in a gas furnace. This will not only avoid the need to use relatively costly steam, but it will also give a greater independence to the production process.

An ingenious heat recovery scheme is based on the use of an intermediate water-immiscible heat medium. This scheme offers such apparent advantages as the minimum number of heat-exchange apparatuses and the absence of gas–gas heat exchangers. As heat mediums, we suggest using lighter (compared with water) paraffin hydrocarbons with a chain length of 9–12 carbon atoms (the density is $750-800 kg/m^3$), or heavier organic matters (silicone oil, fluorinated hydrocarbons, amines). Unfortunately, apart from an increase in the liquid flow passing through the hot column and an additional operation of water – organic mixture separation, the drawback to this scheme is a relatively low level of heat energy recovery, since a great deal of heat is withdrawn in the water cooler and with the wastewater flow. It seems that the schemes discissed in section 3.2.3 (Figures 3.20 and 3.21) hold more promise.

In addition to the suggested improvements, consisting generally of the enlargement of separation units, and in the use of more efficient contactors and heat-recovery schemes, the separation efficiency can be increased by the implementation of novel schemes which make it possible not only to increase the heat recovery level (owing to a two-stage heat recovery system) but also to widen the operating temperature range (by the difference in pressure between the cold and hot columns) [11].

Furthermore, a careful study of physico-chemical properties of the water – hydrogen sulphide system, as well as of the kinetics of isotope exchange at various contactors [11], has revealed the following main avenues of process intensification [87]:

1.  with the obtained additives [26, 44] it becomes worthwhile modifying the process operation parameters (lowering the pressure and temperature in the columns) to enhance the separation degree and improve separation efficiency;
2.  it is possible to intensify isotope exchange with the use of activating additives and more efficient contactors;
3.  the utilization of novel separation schemes intensifies isotope exchange because of an increase in the process driving force and the possibility of operating with a closed cycle of liquid of optimum composition [11].

In the context of a continuing demand for large-scale heavy-water production facilities and with regard to the development of Russian gas fields containing considerable amounts of hydrogen sulphide, of particular interest is the scheme with a closed liquid loop. With the closed liquid loop, the two-temperature plant will be fed with gas, which is more efficient since at the optimal flow-rate ratio the gas flow molar volume is nearly twice as large as that of the liquid flow. Because of this, for the same productivity, the plant size and separation costs with the hydrogen sulphide source flow will be less than half those with the liquid source flow. The Astrakhan gas-processing facility, for one, is capable of supplying raw material to a plant with a productivity of up to 200tons of heavy water per year, and the plant performance characteristics will be far superior to those of the world's existing heavy-water plants.

The implementation of the improvements to the separation process described above should serve not only to further increase the efficiency of the two-temperature method of heavy-water production but also to extend its field of application, which, above all, refers to the isotopic purification from tritium at nuclear power plants.

In the cold column, the optimum pressure and temperature for the tritium concentrating processes will be identical to those for the H–D mixture separation ($P_1 = 2.1\text{MPa}$; $T_1 = 303\text{K}$), since these parameters are determined by the conditions of crystalline hydrate formation.

From the results of calculations performed with regard to energy costs and capital investment, it follows that the optimum temperature in the hot column for the separation of H–D and D–T mixtures is close to that for heavy water production. This is why, for the same separation conditions, Table 3.11 presents the separation factor and maximum extraction degree related to the region of low concentration of heavy isotope for all three binary mixtures.

Undeniably, the two-temperature method, as the most efficient, is of interest for the solution of the most important problems of tritium elimination such as purification of water flows, or of heavy-water moderation at nuclear power plants, from tritium as well as tritium elimination at irradiated fuel processing plants [88].

The utilization of the two-temperature hydrogen sulphide method for tritium extraction from nuclear power-plant water flow has been discussed in sufficient detail [3, 11]. Because of this, we shall restrict our consideration to only two examples: a solution of the largest-scale problem (tritium elimination in irradiated fuel processing), and a solution of

**Table 3.11**

Separation factor at $T_1 = 303K$ and $T_2 = 403K$ and maximum extraction degree of the two-temperature hydrogen sulphide plant (with withdrawal of second kind)

| Separated mixture | $\alpha_1$ | $\alpha_2$ | Maximum extraction degree |
|---|---|---|---|
| H−T | 3.34 | 2.36 | 0.29 |
| H−D | 2.34 | 1.82 | 0.22 |
| D−T | 1.42 | 1.29 | 0.094 |

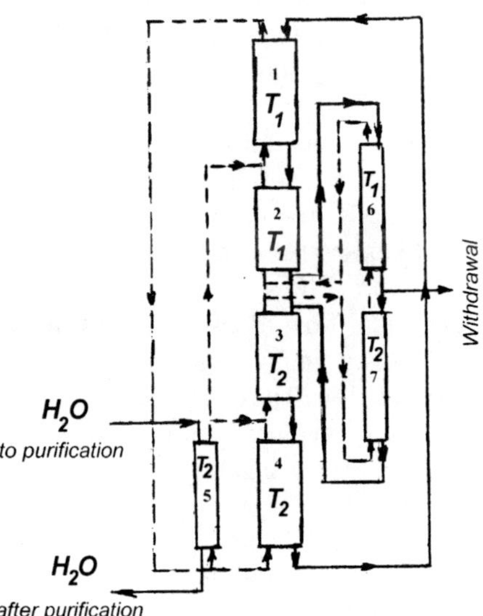


**Figure 3.24**   Scheme of two-temperature facility for purification of water flows from tritium: 1, 2, depleting and enrichment cold columns of the first cascade stage; 3, 4, enrichment and depleting hot columns of the first cascade stage; 5, column for the tritium extraction (CTE) from the purified flow; 6, 7, cold and hot columns of the second cascade stage.

the most complicated problem (purification of heavy-water moderator from tritium and protium).

In the former case, we consider a hundredfold purification of 1000kg of water per hour from tritium with the production of tritium concentrate enriched one hundredfold. The scheme shown in Figure 3.24 is based on a two-temperature plant with depletion (see section 3.1.1), operating at $P = 2.1$ MPa, $T_1 = 303$ and $T_2 = 393K$. The hot column of the plant's first stage is completed with an individual isotope exchange column for the tritium extraction from the purified flow (CTE) which reduces the admixture's purification efficiency requirements.

The calculations were done with the experimental data obtained from an experimental plant completed with a metal ring packing with a unit size of 25mm (see section 3.2.2), and with the results of determination of optimum operation conditions for a plant with depletion [3, 10, 11]. Specifically, the relative withdrawal $\theta$ was taken as 0.65 in the depletion

**Table 3.12**

Dimensions of the separation columns of the two-temperature hydrogen sulphide cascade for a hundredfold purification of 1000kg of water per hour from tritium with the production of tritium concentrate enriched a hundredfold

| Cascade stage | Column | NTU | Packing bed height, m | Column diameter, mm |
|---|---|---|---|---|
| First, depletion | cold | 68 | 27.2 | 1.0 |
| | hot | 68 | 27.2 | 1.0 |
| | CTE | 68 | 27.2 | 0.4 |
| First, enrichment | cold | 20 | 8.0 | 0.9 |
| ($K = 5$) | hot | 20 | 8.0 | 1.0 |
| Second | cold | 37 | 15.0 | 0.4 |
| ($K = 20$) | hot | 37 | 15.0 | 0.4 |

and enrichment sections of the cascade first stage, whereas at the second stage the value $\theta = 0.85$. The calculation results are presented in Table 3.12.

The energy consumption is chiefly associated with heating and humidification of flows fed to the hot columns and accounts for some 150kW·h per one ton of purified water.

Since the purifying unit dimensions are generally determined by the depletion section, they can be considerably reduced by the use of just a two-temperature concentrating unit for the tritium withdrawal from the closed water loop [3, 11].

As discussed in chapter 2, the drawback of the heavy-water moderator purification by the low-temperature rectification of deuterium is the necessity of transferring deuterium and tritium from water to the gas phase, which requires establishing a CTE (or electrolysis) unit, and leads to the increase of heavy water holdup in the plant and to an increase in the separation energy consumption.

Heavy-water direct purification from tritium and protium without transferring the extracted isotopes to the gas phase can be realized with such systems as $D_2O$–$D_2S$ and $D_2O$–$D_2$, where heavy water serves as a working substance.

In the former case, heavy-water isotopic purification requires two two-temperature plants: the first for tritium extraction and the second for protium extraction. The necessity of utilizing two plants is dictated by a sharp separation degree dependence on the flow-rate ratio, since at a $\lambda$ value optimal for tritium extraction, the protium concentration varies only slightly, and the $\lambda$ value optimal for concentrating protium does not lead to a significant change in the tritium concentration.

The separation plant is shown in Figure 3.25. The heavy water flow is fed to column 1 and enriched with tritium in its passage through the column. After the withdrawal of the tritium-enriched water, the flow is supplied to hot column 2. Issuing from column 2, the liquid flow with reduced tritium content determined by the extraction degree is delivered to the second two-temperature plant for protium extraction.

Since this plant is designed for enrichment with light isotope, upper column 3 must be hot, and lower column 4 cold. The highest protium concentration will be observed in he liquid and gas flows between the hot and cold columns. A small quantity of liquid is withdrawn as waste, and the major portion is supplied to cold column 4, after which the heavy water purified from tritium and protium is delivered to stripping column 5 and returned to the heavy-water circuit of the reactor. Considering that the protium-extraction plant must operate within the medium-concenration range, where the equilibrium lines of cold and hot columns are not straight, it is necessary to change the flow-rate ratio in the bottom of the cold column and in the head of the hot column [3, 11]. For this purpose, it will suffice to withdraw a portion of the liquid flow from some point of the hot column and return it to a point of the same isotopic composition of the cold column. Depending on the desired degree of protium enrichment, from one to three bypass lines are required for the liquid flow. The $D_2S$ gas flow in each two-temperature plant is closed, and its circulation is performed by gas blowers 6 and 7.

At the plant gas pressure of 2–2.2MPa, the temperature in the cold columns equals 303K. Since the optimum temperature for the separation of H–D and D–T mixtures in the

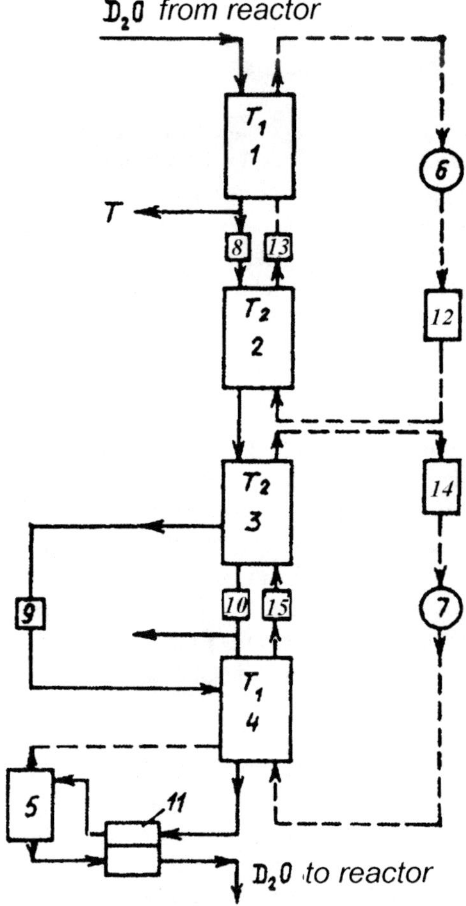


**Figure 3.25**  Scheme of plant for heavy-water isotopic purification by two-temperature hydrogen sulphide method.

hot column differs only slightly, both hot columns can operate at a uniform temperature (403–415K).

Heat exchangers 8 (for the heating of liquid), and 9–10 (for the cooling of liquid), as well as heat-recovery heat exchanger 11 of the stripping column, are shown in Figure 3.25. The gas is heated in heat exchangers 12 and 15, and cooled in coolers 13 and 14. If necessary, either of the heat recovery schemes dicsussed in section 3.2.3 can be employed. For the initial fill-up of the plant with gas, and for the compensation of gas losses, a $D_2S$ production unit is required (i.e. by catalytic deuteration of sulfur).

The heavy water flow passing through the separation plant is determined by a more complex task of tritium separation. In this case the two-temperature protium-concentration plant will operate with a rather low relative withdrawal $\theta$, which allows use of comparatively short columns  In addition, owing to the difference in optimum flow-rate ratio values in the "tritium" plant ($\lambda_0 \approx 1.35$) and "protium" plant ($\lambda_0 \approx 0.5$ at the cold column inlet), the cross-section of the protium-extraction columns, which is generally determined by the gas flow, will be smaller by a factor of 3.5–3.0 than that of the two-temperature tritium-extraction plant.

The utilization of metal mesh regular structured packing (see section 3.2.2) allows reduction in the size of the columns. Moreover, the tritium holdup in the two-temperature plant is considerably reduced due to a far thinner film of flowing down liquid for this packing, which lowers the radiation hazard in an emergency. And owing to a low hydraulic resistance of the packing, the electric power consumption for the gas circilation becomes non-essential. According to calculations [3, 11], the electric power consumption here is an order of magnitude lower than for heavy-water retarder purification by deuterium cryogenic rectification.

Now we dwell on those drawbacks of the two-temperature hydrogen sulphide method that are characteristic for the process of heavy-water moderator isotopic purification. As opposed to cryogenic rectification, its utilization is confined only to the initial concentrating of tritium owing to operating system radiolysis. Because of this, subsequent concentrating must be done by other techniques (i.e. in gas–solid systems, or by cryogenic rectification) to minimize losses of heavy water.

Operating system radiolysis leads to possible tritium self-radiation-induced formation of deuterium peroxide, deuterium, and radical products. The interaction between deuterium peroxide and sulfur deuteride can result in the columns clogging with sulfur formed by

$$D_2O_2 + D_2S \rightarrow 2D_2O + S. \tag{3.92}$$

In addition, proper allowance should be made for the fact that the heavy water supplied from the reactor for isotopic purification contains deuterium peroxide formed under exposure to the reactor's neutron, beta, and gamma radiation. At a deuterium peroxide concentration of $10^{-4}\text{mol/D}_2O$, its reaction with $D_2S$ will lead to the formation of sulfur in amounts that are one thousandfold as great as those accumulated due to the operating system self-radiolysis. If, before the arrival of the heavy water at the separation plant, it is saturated with sulfur deuteride, and the process in the column is performed over soluble additives, inhibiting the precipitation of the formed sulphur, the tritium concentration at the enriched end of the two-temperature hydrogen sulphide plant can amount to 10–18TBq/l.

Certain difficulties can be associated with the initiation of bypass flows in the "protium" plant because of the necessity to delicately adjust the liquid flow.

Finally, a thorough purification of the heavy water from $D_2S$ is required since in the reactor, the interaction with thermal neutrons by the reaction $^{34}S(n,\gamma)^{35}S$ leads to the production of the isotope $^{35}S$, which is liable to contribute, in a way, to the heavy-water circuit radioactivity.

## 3.3  HYDROGEN–AMMONIA AND HYDROGEN–AMINE SYSTEMS

### 3.3.1  Preliminary remarks

Heavy water production with the use of the $NH_3$–$H_2$ and $CH_3NH_2$–$H_2B$ systems has its origins in the following reactions:

$$NH_{3(lq)} + HD_{(gas)} \leftrightarrow NH_2D_{(lq)} + H_{2(gas)};  \tag{3.93}$$

$$CH_3NH_{2(lq)} + HD_{(gas)} \leftrightarrow CH_3NHD_{(lq)} + H_{2(gas)}.  \tag{3.93}$$

Their major advantages are high values of the separation factor $\alpha$ and a pronounced temperature dependence of $\alpha$ which is described for the reaction (3.93) by eq. (3.95), and for the reaction (3.94) by eq. (3.96):

$$\ln\alpha = (545.7/T) - 0.559;  \tag{3.95}$$

$$\ln\alpha = (653.9/T) - 0.824.  \tag{3.96}$$

Both of these reactions necessitate the application of catalysts. In 1950 it was found that potassium amide $(KNH_2)$ [89] dissolved in liquefied ammonia catalyzes the reaction (3.93). A more recent investigation indicated that potassium amide catalyzes the reaction (3.94) as well. Extended kinetics studies demonstrated that the catalyst of the reaction (3.94) is potassium methyl amide $(CH_3NHK)$ produced in the reaction of $KN_2$ with amine [90]. The creation of catalysts made it possible to develop efficient mass-transfer facilities. This cycle of studies was completed with the design and construction of pilot and semi-industrial plants, which allowed estimation of the main performance characteristics of the investigated methods of heavy water production. Simultaneously with pilot studies, development work was proceeding on process flowsheets (single-temperature scheme, two-temperature scheme, scheme with feed column, etc.). During the studies, it was established that the serious disadvantages of the use of the two-temperature schemes are a low exchange rate in the cold column and a high content of the ammonia vapor in the hot column leading to an increase in the column volumes, to a drop in the effective separation factor, and to an increase in the heat-exchange equipment cost. For the single-temperature scheme, a substantial energy input is required for the flow conversion (ammonia cracking). Because of the complexity of flow conversion, the hydrogen–amine system can be employed only in the two-temperature processing schemes. Compared with the ammonia–hydrogen system, this system offers several advantages: better kinetic performance, higher $\alpha$ values, and lower pressure of methylamine vapors.

A common disadvantage of both systems is a relatively low output (65–70tons of $D_2O$ per year) resulting from limited raw material resources. Since the chief hydrogen users are the ammonia synthesis plants, the amount of $D_2O$ produced is determined by the quantity of the nitrogen–hydrogen mixture (NHM) used in the ammonia production. This initiated a series of investigations aimed at developing process schemes which obviate the limitation on source material due to the ammonia–water isotope exchange:

$$NH_{3(gas)} + HDO_{(lq)} \leftrightarrow NH_2D_{(gas)} + H_2O_{(lq)}. \tag{3.97}$$

The exchange proceeds very quickly, which allows water to be utilized as the source flow.

Pilot research and design studies conducted in Canada, Germany, U.S.S.R., France, and Italy demonstrated that both systems involved can successfully compete with the water – hydrogen sulphide system, because of several indisputable benefits [9, 55, 64, 92–94]. Apart from the above-mentioned high $\alpha$ values and a pronounced temperature dependence of the separation factor, there are also the absence of corrosion and non-toxicity of working substances. Both of these systems have been industrially implemented in India [55].

### 3.3.2   Heavy water production by isotope exchange in hydrogen–ammonia systems

The first industrial facility for heavy water production by hydrogen–ammonia isotope exchange was put into operation in Mazingarbe, France, in 1957 [93]. The construction and commissioning of the facility was made possible by extensive studies conducted under the aegis of the French Atomic Energy Commission with the participation of Sulzer Corporation (Switzerland). For the process realization, a single-temperature flow conversion alternative, shown in Figure 3.26, was chosen. The NHM from the ammonia synthesis plant used as the feed flow, having passed through the purification system 1 (Figure 3.26) arrives at the depletion section (2a) of the isotope exchange column (IEC). Here, the feed (source) flow merges with the gas mixture flow, issuing from the enrichment column (2). Moving countercurrently to the potassium amide dissolved in liquid ammonia, which refluxes IEC, the NHM hydrogen gives up its deuterium to ammonia according to eq. (3.93) performed at $T = 263K$ and a pressure of 35–40MPa.

Repeatedly performed in special mass-exchange units, the reaction results in an increase of the deuterium concentration in the ammonia issuing from IEC (one hundred times compared with the deuterium concentration in the feed flow), and the deuterium concentration in NHM leaving IEC becomes lower than its initial concentration. Deuterium-depleted NHM flow arrives at ammonia synthesis unit 3 used as the upper flow-conversion unit (UFCU). A portion of ammonia produced in this unit (in an amount equivalent to the NHM source flow) is delivered to the user, and the balance proceeds to washover string 4.

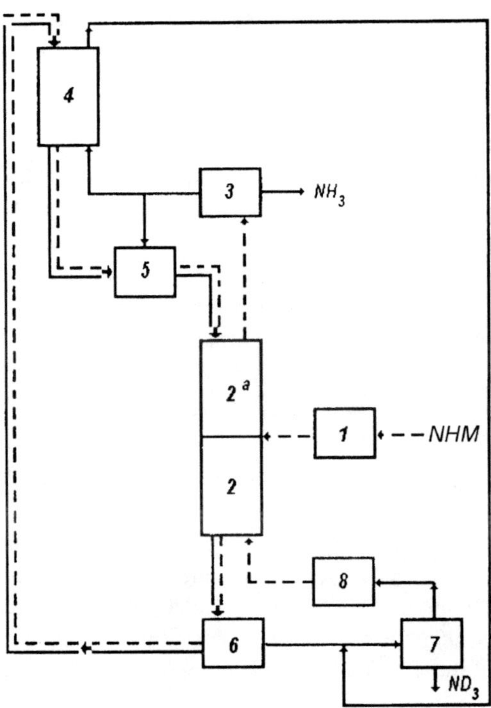

**Figure 3.26**  Scheme of Mazingarbe facility (France): 1, unit for the NHM purification from admixtures; 2, 2ª, enrichment and depletion sections of the isotope exchange column (IEC); 3, ammonia synthesis unit; 4, washover string, or the phase isotope exchange column; 5, condenser–mixer; 6, evaporator; 7, ammonia rectification unit; (8), ammonia cracking unit;----NHM flow; _____ NH₃ flow; ---- flow of potassium amide dissolved in liquid ammonia.

Issuing from the CHEX column, the deuterium-enriched flow of potassium amide dissolved in liquid ammonia arrives at evaporator 6 where it is stripped to a higher amide concentration. Gaseous ammonia formed in the evaporator is delivered for rectification, and the product potassium amide enters washover string 4 where the isotope exchange reaction between the product and a portion of gaseous  $NH_3$  obtained in the UFCU is performed. As already noted, the deuterium concentration in the liquid ammonia is are hundredfold higher than the hydrogen heavy isotope concentration in the source mixture. That is why in the process of isotope exchange performed by

$$NH_2D_{(lq)} + NH_{3(gas)} \leftrightarrow NH_{3(lq)} + NH_2D_{(gas)};  \qquad (3.97a)$$

$$KNHD + NH_3 \leftrightarrow NH_2D + KNH_2  \qquad (3.98)$$

the deuterium is extracted from the catalyst and liquid ammonia into the gaseous ammonia. Since the isotope equilibrium constants of eqs. (3.97a) and (3.98) are close to unity, the deuterium content in gaseous ammonia leaving the washover string approximates the deuterium content in ammonia obtained in the evaporator. Consequently, both flows are combined into a single flow delivered for further concentration by rectification. The solution issuing from the washover string has a deuterium content practically identical to that in $NH_3$ obtained in the UFCU. So both flows are combined into a single ammonia flow supplied to the IEC as a reflux. Washover string 4, therefore, plays an important part by performing catalyst circulation in the column without isotopic dilution.

Returning from the rectification unit, gaseous ammonia is supplied to the UFCU where ammonia cracking 8 is done with the production of NHM, which is then delivered to the CHEX column 2 bottom.

The rectification unit produces near-pure deuterammonia ($ND_3$), which is decomposed over a catalyst into nitrogen and $D_2$ to obtain heavy water by deuterium oxidation. The annual capacity of the Mazingarbe facility is 20 tons of $D_2O$.

For mass-transfer, injection plates developed by Sulzer Corporation were used in the isotope exchange column [94, 95], allowing for a greater interphase surface and a high mass transfer efficiency (the plate performance factor exceeds 30% with an output rate of $5,000m^3$ stp of NHM per hour).

Ammonia decomposition in the UFCU was done at a temperature of 823–825K and a pressure of 5.5–6.0MPa in the presence of a catalyst developed specifically for this purpose, which offered a high catalytic reactivity over a long period of operation. The $KNH_2$ preparation was conducted *in situ* by dissolving potassium metal (99.8% metal purity) in ammonia, with the ammonia–potassium reaction catalyzed by iron ions

$$K + NH_3 \rightarrow KNH_2 + \tfrac{1}{2}H_2. \tag{3.99}$$

The potassium amide is very responsive to oxygen-containing admixtures ($H_2O$, CO, $CO_2$, $O_2$) destructive to catalysts. The interaction products of $KNH_2$ with the admixtures are ammonia-insoluble and hinder the normal operation of mass-transfer units and other equipment. The occurence of the admixtures results in the consumption of potassium (some 0.6kg per 100tons of NHM passed through the unit [95]).

The commissioning and two-year trouble-free operation of the mazingarbe facility allowed valuable data on the system's technological features to be obtained and for experience to be gained, which were later used in the construction of larger facilities.

The plant capacity depends on the hydrogen resources (i.e. on the plant's capacity for ammonia synthesis) and on the hydrogen's deuterium content. And the deuterium content is determined by the production technology (electrolysis, methane conversion) and can vary from 0.010 to 0.0135at.% [2, 3]. The hydrogen production technology influences the impurity composition of NHM as well which, before the arrival at the CHEX colomn, must be subjected to thorough purification by oxygen-containing admixtures.

To make the interconnection between heavy-water plant and ammonia synthesis plant less stringent, and to enhance extraction, the C.J.B. company advanced a scheme with a feed column [8, 9] shown in Figure 3.27. Following the purification system, the source flow (NHM) arrives at the source column, where, moving countercurrently to the potassium-amide-containing liquid ammonia (the absorber's refluxer), it gives up the deuterium and goes on to the ammonia synthesis column. In the plant's two-temperature section, the deuterium content in the ammonia increases by a factor of 23 compared with that in the source flow. The subsequent concentration is done by ammonia rectification. The availability of the source flow makes it possible to substantially decrease the volume of the initial concentrating unit, enhance the degree of the product extraction from the source flow (up to 70%), and significantly reduce the adverse effect of the operation instability of the ammonia synthesis plant on the heavy-water plant capacity. References [9, 99] give a comprehensive analysis of the schemes with the source flow.

References [97–99] put forward a two-temperature scheme with depletion (Figure 3.27b) and present the process performance characteristics determined on the basis of the pilot-plant operation. It is worth noting that the diameter of the plant's isotope exchange columns was 400mm, with a maximum gas consumption rate of 12,000nm$^3$/h. The plant was coupled with an ammonia production facility with a daily output of 1,500tons of $NO_3$, and the design diameters of the plant columns ranged from 2,000 to 2,200mm. Because of this, the authors [98, 99] believed that the results of the pilot-plant operation can be taken to be a correct performance evaluation of a heavy-water production plant with an annual capacity of 75–85tons of $D_2O$.

The plant's important feature is the utilization of sieve trays with overflow in the hot and cold isotope exchange columns in lieu of costly and complicated injectors, which, because of a high hydraulic resistance, require special pumps to transfer liquid ammonia from one tray to another.

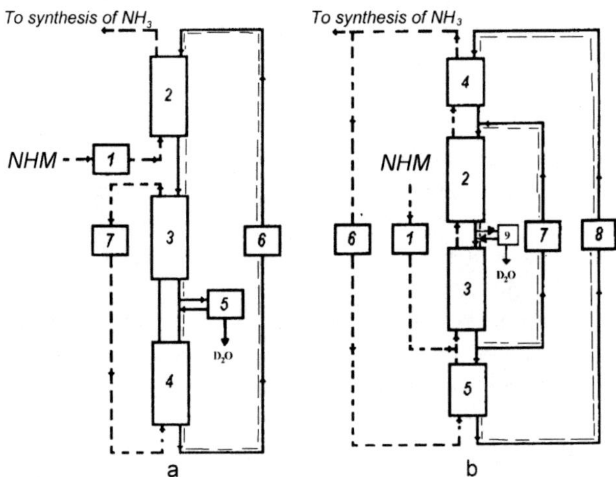


**Figure 3.27**   Types of two-temperature schemes using ammonia–hydrogen system: (a) scheme with the source column: 1, NHM purification unit; 2, source column; 3 and 4, cold and hot isotope exchange columns; 5, final concentration unit; 6, pump to transfer the potassium amide dissolved in liquid ammonia; 7, circulation gas compressor (gas blower); (b) a scheme with depletion: 1, NHM purification unit; 2 and 3, cold and hot columns of the plant enrichment section; 4 and 5, cold and hot columns of the plant depletion section; 6, gas blower; 7 and 8, pumps to transfer the potassium amide dissolved in liquid ammonia; 9, final concentration unit (flows are designated in the same manner as in Figure. 3.26).

According to data and estimates presented in the literature, the efficiency of sieve trays accounts for 3–10% and depends on the tray hole size, free cross-section content, temperature, pressure, as well as on the catalyst concentration and the presence of accelerating additives, for which amines were utilized [96, 100–102].

Based on the results of the two-temperature plant operation, the temperatures in the hot and cold columns were chosen. A temperature range of 253–263K and a pressure range of 30.0–35.0MPa were selected as the most favourable conditions for the operation of the cold column, since the mass transfer efficiency depends on the concentration of the hydrogen dissolved in ammonia, and with due regard to the fact that the hydrogen solubility in ammonia decreases with a drop in temperature, and that it is not economically expedient to increase the pressure in the column. Nor is it expedient to increase the hot column temperature beyond 333K because of a drastic rise in the partial pressure of ammonia vapor, which leads to an enlargement in the diameter of the exchange columns and an increase in energy expenditure for the gas flow circulation.

A project developed by UHDE, a Krupp company [98] and implemented in India, incorporated a three-stage two-temperature section of the plant (the scheme is shown in Figure 3.28) with the deuterium concentration at the third stage equal to 18at.%.

According to this scheme, the source flow (NHM), after purification unit 1, passes through source exchange column 2 where deuterium transfers by eq. (3.93) to the flow of $KNH_2$ dissolved in liquid ammonia, which refluxes the cold (enrichment) isotope exchange column 5 of the first stage of the plant's two-temperature section.

The solution issuing from the first-stage hot column 6 is supplied to reflux hot column 4 of the depletion section, and, upon leaving the column, delivered to reflux the source column 2. The NHM flow depleted of deuterium in source column 2 is sent for ammonia synthesis. The solution issuing from cold column 5 is supplied to the second-stage cold column 7 as a reflux, and the solution issuing from the third-stage enrichment column 9 enters, as indicated above, special-purpose unit 11 where a portion of the solution is stripped forming two flows: a gaseous ammonia flow and a flow of the solution with a heightened potassium amide content (mother solution). The gaseous ammonia is delivered to column 12 for the phase isotope exchange with the water refluxing the column. The ammonia, leaving column 12, is returned to unit 11 where it is condensed, immixed with the mother solution and delivered to the third-stage hot column 10 as a reflux. The water from column 12 enters rectification column 13 from the still of which heavy water with a concentration of 99.8at.% is withdrawn.

In the scheme under study, the first-stage columns were 2000mm in diameter and 40m in height. The availability of the source column and depletion section made it possible to increase extraction degree to 80% and to achieve an annual capacity of 60tons of $D_2O$. As indicated above, an important feature of the plant was that the sieve trays were utilized as a means of mass-transfer in the isotope columns. For the convenience of CHEX column mounting and disassembly (in case of an emergency) the trays were integrated into special packages (cartridges). Each cartridge had a height of 4–5m and comprised from 20 to 25 sieve trays. The cartridges were inserted into the CHEX colomns with a diameter ranging from 0.8 to 2.2m. The column was completed with 4 or 5 cartridges, and the sieve tray hole size varied from 0.8 to 1.2mm.

As for India, a single-temperature plant has been built [100]. The processing scheme was based on the flow chart of the Mazingarbe plant described above. The Indian plant output

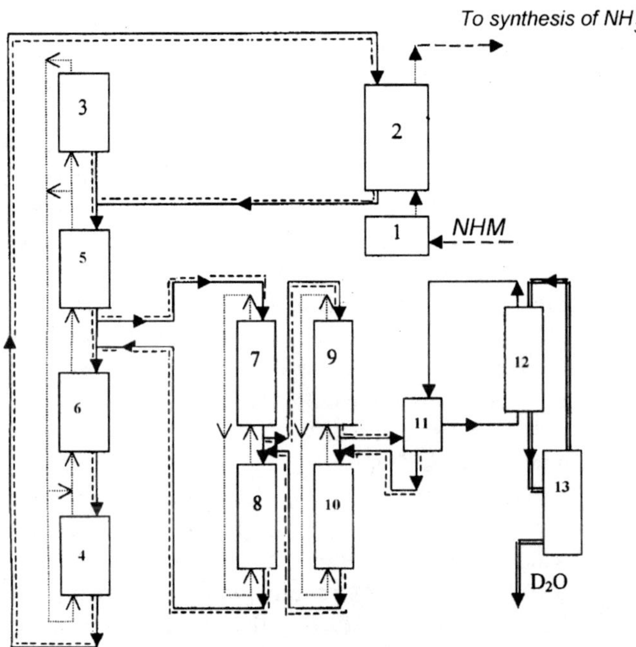

**Figure 3.28**    Scheme of UHDE two-temperature plant: 1, purification unit; 2, source column; 3, 4 cold and hot columns of the first-stage depletion section; 5, 7, 9 and 6, 8, 10, cold and hot columns of the first, second, and third sections of the plant enrichment section, respectively; 11, ammonia evaporator-condenser; 12, ammonia–water isotope exchange column; column for final concentration by the water rectification; water flow (other flows are designated in the same manner as in Figure 3.26.)

(50tons of $D_2O$ per year) exceeded that of the Mazingarbe plant. In addition, some changes were introduced into the CHEX columns and UFCU operation modes. The depletion section temperature was 248K, and that in the enrichment section 278K. These temperatures allowed the plant size to be optimized and the extraction degree to be enhanced. Contrary to the Mazingarbe plant, the Indian plant had two isotope exchange stages, which allowed the deuterium content in $NH_3$ used in rectification to be increased 2%, thereby reducing the ammonia rectification unit volume. Instead of injectors, sieve trays with holes of 0.2mm diameter were employed at the second stage of exchange. The UFCU operation conditions were also modified: $T = 873K$ and $P = 14.0MPa$ (instead of 773K and 5.0MPa in Mazingarbe), with the use of a new Danish-developed catalyst. These improvements made it possible to reduce losses due to incompleteness of the phase conversion in the cracking unit [103].

As indicated earlier, the deuterium content in the NHM hydrogen is lower than in water. It is associated with the fact that simultaneously with the hydrogen production by the methane catalytic conversion, the hydrogen–water isotope exchange proceeds by

$$HD + H_2O \leftrightarrow H_2 + HDO. \tag{3.100}$$

As the reaction equilibrium is right-displaced, and the $\alpha$ value significantly exceeds unity, the hydrogen produced is more depleted of deuterium (by a factor of about 1.7–1.9) than the natural hydrogen. This leads to a considerable drop in the heavy-water plant capacity. To compensate for the drop, the methane conversion scheme was modified [100, 103]. In the modified scheme, the deuterium-enriched water is returned to the production cycle, where it is mixed with fresh water (with a natural isotope composition of 0.0145at.%) in the ratio of about 1:9. As a result of such mixing, at the conversion temperature the isotope composition of the water vapor comes close to the equilibrium composition. That is why during the conversion the hydrogen composition varies only slightly (up to about 0.0135at.%) [100].

It is also pertinent to note that in India, pilot plants were built with columns of 250mm diameter, where all engineering problems arising from the source flow (NHM) substitution for water were solved. In the opinion of the authors [101–103], this relieves the hydrogen–ammonia system of the main trouble: the dependence of the heavy-water plant output on the capacity of the ammonia synthesis plant. The pilot plant's operation revealed several material drawbacks, the most important of which are:

- moderate isotope exchange rate at low temperatures, which determines a poor tray efficiency. In this case, for the realization of a single theoretical plate at $T = 240–250K$ and $P_{H_2} = 20MPa$, a column with a height of 5–8m is required;
- sharp temperature dependence of the ammonia partial pressure leading to a decrease in the effective separation factor, which in turn results in an increased size of exchange columns and enhanced thermal load on the heat-exchange equipment;
- a limitation on the temperature difference between the cold and the hot columns, which leads to a decrease in the two-temperature scheme extraction degree, an augmentation of the plant specific volume, and an increase in consumption coefficients.

In addition, several engineering difficulties were revealed: a high aggressivity of potassium amide solutions in ammonia toward conventional materials utilized for sealing and encapsulation; cyanides formed during the reaction between $KNH_2$ and carbon oxide, which lead to a necessity for wastewater treatment; transportation of heavy (about 270tons) sizeable columns, etc. These difficulties taken together force us to view more critically the prospects for the $NH_3$–$H_2$ system's widespread use. The authors [64] stated their belief that this system offers no decisive advantages over the $H_2O$–$H_2S$ system in terms of its application in Europe and the U.S.A. At the same time, the production performance characteristics obtained from the operation of plants established in India demonstrate a fair competitive ability of the ammonia–hydrogen system. According to reference [103], for example, electric power consumption of heavy-water production accounts for 1600kWh per one kilogram of $D_2O$, steam consumption for about two tons per kilogram of $D_2O$, and potassium consumption for about 100kg per one ton of heavy water.

### 3.3.3 Hydrogen–amine system utilization for deuterium enrichment

In 1962 it was found that hydrogen exchange with $CH_3NH_2$ (monomethylamine, MMA) proceeds (under comparable conditions) 35 times faster than that with ammonia [90]. The same paper ascertained that the activation energy of the hydrogen–MMA exchange reaction is half as high as that of the hydrogen–NH$_3$ exchange reaction. Potassium methy-lamide obtained by the metal dissolving in MMA was employed as catalyst. This finding lent impetus to intensive studies on the kinetics of hydrogen exchange with various amines. It was found that during methylamide catalysis, it is only nitrogen-linked hydrogen atoms that exchange. This is why trimethylamine, of which nitrogen-linked hydrogen atoms are replaced by methyl radicals, is of no interest. Thus, in deciding on a particular amine, there is a need to take into account the ratio between the amine molecular weight and the number of hydrogen atoms able to be involved in the isotope exchange reaction. In addition, consideration must be given to the amine vapor pressure, and to the amine solubility of the hydrogen. The latter is of prime importance for the mass-transfer process since, according to research [104–106], the value of the liquid phase mass-transfer volume factor ($\beta_X a$), having dimensionality mol/m$^3$·s, in the case considered can be determined by

$$\beta_X a = K_H \cdot P_{H_2} \cdot \sqrt{Dk} \cdot a, \tag{3.101}$$

where $K_H$ is the Henry's law constant, mol/m$^3$ atm; $P_{H_2}$ is the hydrogen pressure, atm; $D$ is the value of the methylamine-dissolved hydrogen diffusivity; $a$ is the specific interphase surface, m$^2$/m$^3$; and $k$ is the isotope exchange rate constant depending on the catalyst concentration and nature.

The MMA and diaminoethane ($H_2N-C_2H_4-NH_2$) have practically equal ratios between their molecular weights and numbers of exchanging hydrogen atoms (15.6 and 15.0, respectively). The hydrogen solubility in diaminoethane, however, is seven times smaller than that in the MMA, which excludes diaminoethane industrial application. Notwithstanding the fact that the $\alpha$ value in the hydrogen exchange with other amines is higher than that in the hydrogen exchange with MMA, the latter, from the above considerations, was chosen for pilot testing.

The investigations conducted in Israel, France, Germany, Russia, and Canada revealed that at a methylamide concentration of over 0.1 mol/l, the process rate is complicated with diffusion resistance concentrated in the liquid phase. It was also verified that the liquid ammonia solubility of the potassium methylamide is an order of magnitude lower than that of the KNH$_2$. A special cycle of research was devoted to selection of the optimal catalyst. Table 3.13 presents the results of research, according to which the exchange rate ($R$), having dimensionality kmol/(m$^3$·h), can be described in the process kinetic region by

$$R = k_0 \cdot K_H \cdot P_{H_2} [\text{cat}]^m \cdot \exp\left(-E_a \Big/ RT\right). \tag{3.102}$$

The value of the specific rate constant ($k_0$) depends on the catalyst nature and concentration.

The table shows the values of the activation energy ($E_a$) and of the quantity $m$, as well as the values of the catalyst deactivation constant measured for various methylamides at

**Table 3.13**

Kinetic performance of hydrogen–MMA isotope exchange catalyzed with different methylamides: the catalyst initial concentration is 0.03mol/l; the hydrogen pressure is 0.1MPa [41]

| Methylamide | $E_a$, kJ/mol | $m$ | $T$, K | $k_g \times 10^2$, h$^{-1}$ |
|---|---|---|---|---|
| Lithium | 29.7 | 0.4 | 253–200 | 3.8 |
| Natrium | 28.4 | 0.5 | 223 | 5.3 |
|  |  | 0.7 | 213 |  |
|  |  | 0.9 | 203 |  |
| Potassium | 22.5 | 1.0 | 240–190 | 6.7 |
| Rubidium | 22.2 | 1.0 | 240–190 | 14 |
| Caesium | 22.0 | 1.0 | 240–190 | 17 |

$T = 343$K and at a pressure of 0.1MPa. According to research [107], the $k_g$ values were calculated by

$$k_g = \frac{C_0 - C_\tau}{\tau},\tag{3.103}$$

where $C_0$ and $C_\tau$ are the methylamide concentration in MMA at zero time and $\tau$ hours later, respectively.

The catalyst deactivation is caused by

$$2CH_3NH_2 + CH_3NHK \rightarrow 2H_2 + NH_3 + CH_3N = CHNKCH_3;\tag{3.104}$$

$$CH_3NHK + H_2 \rightarrow CH_3NH_2 + KH.\tag{3.105}$$

The high values of the deactivation constant can be lowered by increasing the hydrogen pressure [108]. The pressure increase, though, results in a decrease in the potassium methylamide solubility which, in turn, leads to the exchange rate reduction.

Extensive studies revealed that catalyst stabilization can be achieved at a hot column temperature not exceeding 315K, a hydrogen pressure of about 50–70MPa, and with lithium methylamide added to the potassium methylamide solution. The optimum results are obtained with the mole ratio between lithium methylamide and potassium methylamide equal to unity.

At the cold column temperature the catalyst is fairly stable, so the operating temperature selection in this column is constrained only by the isotope exchange rate. Most authors [100, 102, 108, 109] consider a temperature range from 223 to 233K as optimal for the cold column.

The catalytic activity reduction is due as well to the presence of a variety of trace contaminants in the hydrogen (or NHM) source flow. Reactions between trace contaminants and potassium methylamide result in the degradation of the latter. The reactions' products are poorly soluble in methylamine, which leads to the mass-exchange equipment clogging.

Below are given the most typical reactions of potassium methylamide with ammonia and oxygen-containing admixtures:

$$NH_3 + CH_3NHK \rightarrow CH_3NH_2 + KNH_2, \tag{3.106}$$

$$KNH_2 + H_2O \rightarrow KOH + NH_3, \tag{3.107}$$

$$H_2O + CH_3NHK \rightarrow KOH + CH_3NH_2, \tag{3.108}$$

$$2O_2 + CH_3NHK \rightarrow KCN + 2KOH + CH_3NHOONHCH_3, \tag{3.109}$$

$$CO + CH_3NHK \rightarrow HCONKCH_3, \tag{3.110}$$

$$CO_2 + 2CH_3NHK \rightarrow CH_3NHCOONHCH_3. \tag{3.111}$$

$$\searrow$$

$$CH_3NHCOOK$$

It must be emphasized that the oxygen presence causes the potassium cyanide formation (eq. (3.109)). This poses the problem of the decontamination of the plant's industrial waste.

The need for NHM deep cleaning from admixtures brings about an increase in the cost price of the heavy water. What is more, operational difficulties arise associated with the necessity to withdraw the products of reactions (3.106–3.111), which results in the deterioration of the cost-efficiency of the methylamine–hydrogen system in the industrial production of heavy water.

As noted above, an advantage of the amine–hydrogen system is a higher exchange rate at low temperatures (compared with the ammonia–hydrogen system). According to [110], for example, the isotope exchange rate ($R$) in the amine–hydrogen system at a temperature of 253K and 233K equals 334 and 112kmol/(m$^3$·h), respectively. For the ammonia–hydrogen system in the same conditions (with a catalyst concentration of 0.2mol/l, and at a hydrogen pressure of 0.1MPa), the corresponding rates are considerably lower, namely 36 and 7kmol/(m$^3$·h). Using the $R$ value, one can determine the value of the chemical component of the transfer unit height $h_{IE}$, which, according to [110], is calculated by

$$h_{IE} = \frac{\sqrt{\beta} \cdot \ln \beta}{2(\sqrt{\beta} - 1)} \times \frac{G}{S \cdot R}, \tag{3.112}$$

where $\beta = \alpha_1/\alpha_2$; $\alpha_1$ and $\alpha_2$ are the separation factor values in the cold and hot columns, respectively; $G$ is the hydrogen flow, kmol/h; $S$ is the cross-section of the isotope exchange column, m$^2$; and $R$ is the exchange rate, kmol/(m$^3$·h).

At the cold column temperature $T_1 = 253K$, hot column temperature $T_2 = 333K$, hydrogen flow rate $G = 225kmol/h$ ($5000nm^3/h$), the column cross-section $S = 1m^2$, and at a hydrogen pressure of 1.0MPa (under these conditions, $R$ for the cold column is equal to $3.3 \times 10^3 kmol/(m^3 \cdot h)$), the $h_{1E}$ value will equal 8cm (!). The value of the chemical component of the transfer unit height, hence, becomes quite small. To realize this system's advantage, it becomes necessary to develop more efficient mass-transfer facilities, making it possible to sufficiently reduce the diffusion resistance in the liquid phase.

Apart from the above-mentioned injection plates (see section 3.3.2), sieve trays with small-size holes ($d < 0.5mm$), and high overflow weirs were tested. The combined mass-transfer units – sieve trays with a thick wire mesh layers atop, proved to be quite efficient. Here, liquid and gas move in parallel flows and, when the overflow weir border is reached, the liquid (as foam) spills down to the subjacent tray, and the gas arrives at the superjacent sieve tray. Under steady-state hydrodynamic conditions, at $T = 223K$, the efficiency of such units is about 20–22%, with an average contact time of 6–7s and interphase surface size of about $1300m^{-1}$.

Successful testing of the new mass-exchange units allowed for development of pilot plants and appraising performance characteristics of the hydrogen isotope separation method based on the utilization of the amine–hydrogen system. The system can be industrially applied with the use of the same initial concentration flow charts as the two-temperature schemes with depletion and source column discussed above (section 3.3.2). Table 3.14 gives some performance characteristics of the ammonia–hydrogen and amine–hydrogen systems, reported at different times.

An important distinctive property of the schemes using the amine–hydrogen system is an auxiliary unit for the methylamine vapor removal from NHM exiting the heavy-water plant. The removal can be efficiently performed in adsorber with activated carbon (or with 5A-type zeolite). The removal involves additional investment and requires extra power input to compensate for the pressure loss in the adsorber and for the sorbent recycling.

The performance characteristics obtained in different countries with regard to local conditions differ significantly in the capital cost value (from USD11 million to 20 million, in 1978 prices, for plants with an annual capacity of 65–100tons) and in electric power consumption.

**Table 3.14**

Performance characteristics of two-temperature schemes of $D_2O$ production with the use of ammonia–hydrogen and methylamine–hydrogen systems [87, 96, 110]

| Characteristics | Operating system | | | |
| --- | --- | --- | --- | --- |
| | $NH_3$–$H_2$ | | $CH_3NH_2$–$H_2$ | |
| Output, tons per year | 54 | 100 | 54 | 64 |
| Cold column temperature, K | 248 | 248 | 248 | 223 |
| Hot column temperature, K | 338 | 333 | 333 | 343 |
| Pressure, MPa | 30 | 30 | 35 | 6,5 |
| Electric power consumption, kW·h per 1kg of $D_2O$ | 500 | 560 | 326 | 750 |
| Steam consumption, tons per 1kg of $D_2O$ | 1.8 | 2.4 | 1.2 | 1.3 |
| Volume of columns at operating pressure, $m^3$ | 268 | 1000 | 125 | 600 |

From Table 3.14 it follows that capacity for similar capacity (54tons of $D_2O$ per year), the amine–hydrogen system consumes half as much steam and electric power as the ammonia–hydrogen system. Moreover, the volume of exchange columns is also found to be half as large (at a uniform pressure). A more thorough analysis [109] showed that in relatively small-scale production (40–65tons of $D_2O$ per year), the amine–hydrogen system outperforms in some parameters (steam and electric energy consumption) not only the ammonia–hydrogen system, but also the water – hydrogen sulphide system. But for a large-scale production (400–800tons of $D_2O$ per year), it is necessary to feed the scheme with water, which leads to a considerable increase in expenditure. According to Canadian Nuclear Association estimates, these expenditures reduce the advantages of the amine–hydrogen system to zero. In summary it should be mentioned that work which has been done in India in the past few decades which makes it possible to more optimistically estimate the future prospects of the $CH_3NH_2$–$H_2$ system. Due regard must also be had for the fact that there will be no need for high-capacity plants to compensate for $D_2O$ losses in the operable heavy-water reactors.

## 3.4  WATER–HYDROGEN SYSTEM

### 3.4.1  Historical review

The water–hydrogen chemical isotope exchange (CHEX) became the subject of investigations in the early 1940s when a demand arose for heavy-water production necessary for nuclear weapon development. During the search for the most suitable heavy water production technologies, water–hydrogen CHEX received the same attention as other chemical exchange systems (water–hydrogen sulphide, hydrogen–ammonia). The reason had to do with the system's attractive characteristics, and, above all, with a high separation factor and a low corrosivity of working substances. But because of the necessity to catalyze the water–hydrogen chemical isotope exchange, preference was given to the water – hydrogen sulphide system. Nevertheless, the water–hydrogen system has found practical use in water vapor – hydrogen exchange:

$$H_2O_{(vap)} + HD_{(gas)} \leftrightarrow HDO_{(vap)} + H_{2(gas)}, \qquad (3.113)$$

performed over molecular hydrogen activation catalysts well known by that time (vapour phase catalytic exchange, VPCE). This process, together with water electrolysis, was used in the production of the first detectable amounts of heavy water in Rjukan, Norway, in the late 1930s – early 1940s, and in the initial concentration of deuterium at a larger plant in Trail, Canada [111]. The VPCE process found another application in protium and tritium removal from the moderator and coolant of nuclear reactors. The VPCE process is used at large plants in Grenoble, France, and in Darlington, Canada, utilizing hydrogen cryogenic rectification for coolant isotopic purification. The experience of long-term operation of these plants was summarized, in particular, in references [112, 113]. Simultaneously, exploratory studies into the two-temperature scheme use for the separation process in this system were carried out. It was suggested utilizing the exchange in the water–ammonia–hydrogen ternary system, or to perform the hydrogen – liquid water exchange over a carbon-supported platinum catalyst at a pressure of up to 20MPa [111]. Many publications were devoted to the search for effective homogeneous catalysts for $H_2$–$H_2O$ isotope exchange. Among them, it is necessary to mention sodium and potassium hydroxides and many various complex compounds of transitive metals [110, 114–118]. Let us notice that all research has not resulted in recommendations for practical use in the decision of any large tasks of hydrogen isotope separation.

New efforts on the utilization of the water–hydrogen system began in the mid-1970s. The stage activities are aimed at developing efficient heterogeneous catalysts with hydrophobic properties. Owing to the surface being unwettable by water, the catalysts retain a high catalytic reactivity in the process of the molecular hydrogen activation in the presence of liquid aqueous phase, and the exchange reaction proceeds, in fact, in two steps:

$$HD_{(gas)} + H_2O_{(vap)} \leftrightarrow H_{2(gas)} + HDO_{(vap)}, \tag{3.114}$$

$$\underline{HDO_{(vap)} + H_2O_{(lq)} \leftrightarrow H_2O_{(vap)} + HDO_{(lq)}}, \tag{3.115}$$

$$HD_{(gas)} + H_2O_{(lq)} \leftrightarrow H_{2(gas)} + HDO_{(lq)}, \tag{3.116}$$

the first of which is catalytic (CTEX), and the second requires no catalyst (phase isotope exchange, PHEX). The advent of active hydrophobic catalysts made it possible to obtain the counter-current flow of phases in the water–hydrogen isotope exchange in multistage separation plants, which allowed liquid aqueous phase – hydrogen exchange (liquid phase catalytic exchange, LPCE) to tackle several new problems.

The principal types of catalyst now in use are discussed below. It might be well to point out, however, that a common property of the catalysts is a moderate thermal stability at temperatures over 373K. That is why it is inefficient to practically realize in this system the two-temperature scheme of separation, which requires a significant temperature difference between the cold and the hot columns [11]. Because of this, a single-temperature scheme is employed, and the separation plants have either one (lower) or two (lower and upper) flow conversion units. Electrolyzers of various types (water-alkaline, or with solid-state polymeric electrolyte, SPE) serve as lower flow conversion units with a considerable input of energy (from 4 to 5.5kWh/m$^3$ of $H_2$) [119] required for hydrogen production from water, irrespective of the electrolyte type. That is why the plants mentioned in the literature, based on the hydrogen–water isotope exchange over hydrophobic catalysts in combination with electrolyzers as flow conversion units (in various sources the process is referred to as CECE, or combined electrolysis and catalytic exchange, and ELEX, or electrolysis and exchange), are intended to perform rather small-scale functions. For the most part, these are the functions of processing various tritium-containing water and gas flows produced at the nuclear fuel cycle facilities. The first small-scale plants using the CECE process were developed in the late 1970s in the USA and Canada. The Canadian plant (Chalk River Nuclear Laboratory, Chalk River, Ontario) was designed for the separation of protium–deuterium isotope mixture [120], and the American plant (Mound Laboratory, Miamisburg, Ohio) for separation in the protium–tritium system [121].

### 3.4.2  Isotope equilibrium

As noted above, one of the advantages of the water–hydrogen system lies in the high values of the separation factor for various isotope mixtures. By way of illustration, Figure 3.29 presents a comparison between separation factors in various chemical exchange systems (protium–deuterium isotope mixture, region of low deuterium concentrations). As can be seen, the $\alpha_{HD}$ values for $H_2O$–$H_2$ at all temperatures are far higher than those for the $H_2O$–$H_2S$ system and second only to the methylamine–hydrogen system values. It should be stressed, however, that along with several other drawbacks, the $CH_3NH_2$–$H_2$ and $NH_3$–$H_2$ systems are catalytic, as is the $H_2O$–$H_2$ system. From this point of view the $CH_3NH_2$–$H_2$ and $NH_3$–$H_2$ systems, as opposed to the $H_2O$–$H_2S$ system, do not offer an advantage over the $H_2O$–$H_2$ system.

Table 3.15 gives the coefficients in the equation to calculate separation factors for various binary isotope mixtures ($\alpha_{HD}$, $\alpha_{HT}$, $\alpha_{DH}$, $\alpha_{DT}$)

$$\ln \alpha_{AB} = a + b/T + c/T^2 + d \ln T, \qquad (3.117)$$

generalizing the data reported by various authors [41] (in conformity with our notation system, $\alpha_{BA}$ is used for the deuterium–protium system instead of $\alpha_{AB}$ for the protium–deuterium system).

Employing the presented data, it can be shown that in the protium–deuterium binary mixture, the separation factor, at a temperature of, say, 300K, will vary, depending on the

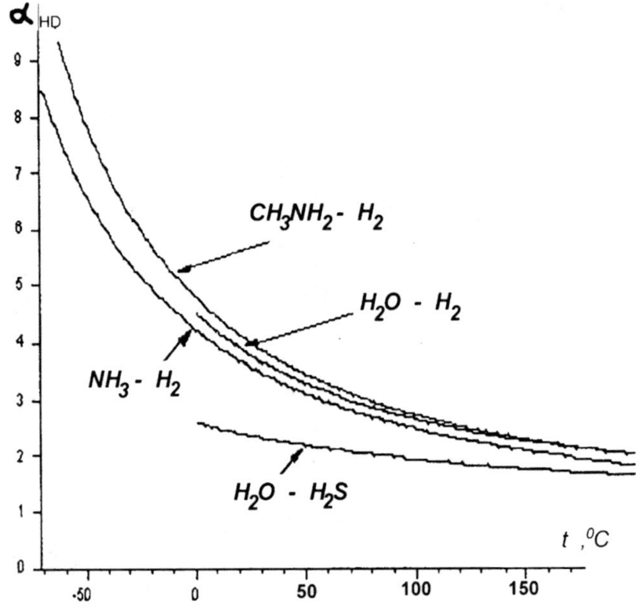


**Figure 3.29**   Comparison of values of $\alpha_{HD}$ for various chemical exchange systems.

**Table 3.15**

Values of coefficients in (3.117) for various binary isotope mixtures

| Isotope mixture | $a$ | $b$ | $c$ | $d$ |
| --- | --- | --- | --- | --- |
| Protium–deuterium [a] | $-0.2143$ | 368.9 | 27,870 | — |
| Deuterium–protium [b] | $-0.180$ | 317.2 | 27,308 | — |
| Protium–tritium | $-2.426$ | 718.2 | 24,989 | 0.292 |
| Deuterium–tritium | $-0.1974$ | 211.1 | — | — |

[a] in the book [41] a reference to a paper on the $\alpha_{HD}$ determination (Table 3.4, p. 61) is incorrect. The correct paper is: J. Rolston, J. Hartog, J. Butler, *J. Phys. Chem.*, 1976, 89, 1064.
[b] in the *J. Phys. Chem.*, 1982, 86, 2494, J.H. Rolston and K.L. Gale give other values of coefficients in (3.117): a = - 0.3600;  b = 413.8; c =15,465. The calculation by (3.117) gives practically the same separation factor values (the deviation does not exceed 2%).

deuterium concentration, from 3.76 (in the region of low deuterium concentration) to 3.25 (in the region of high deuterium concentration) according to (3.118) and (3.119), respectively:

$$H_2O + HD \leftrightarrow HDO + H_2,\qquad(3.118)$$

$$HDO + D_2 \leftrightarrow D_2O + HD.\qquad(3.119)$$

The difference is due to the departure of isotope distribution in the water and hydrogen molecules from equiprobability in the homomolecular isotope exchange reactions (HMEX). The values of HMEX equilibrium constants:

$$A_2 + B_2 \leftrightarrow 2\,AB,\qquad(3.120)$$

$$A_2O + B_2O \leftrightarrow 2ABO,\qquad(3.121)$$

where A and B are various hydrogen isotopes (protium, deuterium, tritium), at a temperature of 300K, are represented in Table 3.16. Notice that at an equiprobable isotope distribution the equilibrium constant values must be equal to 4 for all reactions.
Table 3.16

As shown in [11], for the protium–deuterium system, the equilibrium constant value at any deuterium concentration can be calculated by

$$\alpha_{HD} = \alpha_{HD}\{[1 + \tfrac{1}{2}([HDO]/[H_2O]) \\ (4/K_{HDO})]/[1 + \tfrac{1}{2}([HDO/H_2O)]\} \\ \{[1 + (\tfrac{1}{2}\alpha_{HD})([HDO]/[H_2O])]/[1 \\ + (\tfrac{1}{2}\alpha_{HD})([HDO]/[H_2O])(4/K_{HD})]\}.\qquad(3.122)$$

**Table 3.16**

HMEX equilibrium constants for water and hydrogen at a temperature of 300K

| Equilibrium constant | Exchanging isotopes | | |
|---|---|---|---|
| | A = H, B = D | A = H, B = T | A = D, B = T |
| $K_{AB}$ | 3.268 | 2.579 | 3.812 |
| $K_{ABO}$ | 3.848 | 3.699 | 3.972 |

The concentrations of various molecular forms of water (HDO and $H_2O$) required for the $\alpha$ calculation can be determined with the use of the known entity $K_{HDO}$

$$K_{HDO} = [HDO]^2/([H_2O] \cdot [D_2O]) \tag{3.123}$$

and the expression for the deuterium concentration calculation

$$x = (\tfrac{1}{2}[HDO]+[D_2O])/([H_2O]+[HDO]+[D_2O]). \tag{3.124}$$

The simultaneous solution of the above equations leads to the following quadratic equation:

$$x \cdot ([H_2O]/[HDO])^2 - (\tfrac{1}{2}-x) \cdot [H_2O]/[HDO] - (1-x)/K_{HDO} = 0. \tag{3.125}$$

The values of [HDO] and [$H_2O$] at a given $x$ concentration can be obtained by solving (3.125).

Owing to the self-radiolysis of water, the water–hydrogen system utilization for the separation of tritium-containing isotopic mixtures is possible only in the region of tritium micro-concentrations. But in such mixtures, the deuterium concentration may vary over a wide range. The deuteruim concentration variation in ternary mixtures leads to a very sharp change in the tritium distribution coefficient $\alpha_T$. In a marginal case of protium–deuterium and deuterium–tritium binary mixtures, for example, the $\alpha_T$ values change from 6.76 to 1.66 at a temperature of 300K. According to the data in reference [11], the value of the tritium distribution coefficient at any deuterium concentration can be calculated by:

$$\alpha_T = \alpha_{HT} \frac{1+\left(\dfrac{K_{DTO}}{K_{HTO}K_{HDO}}\right)^{1/2}\dfrac{[HDO]}{[H_2O]}}{1+\dfrac{1}{\alpha_{HD}}\left(\dfrac{K_{DT}}{K_{HT}K_{HD}}\right)^{1/2}\dfrac{[HDO]}{[H_2O]}} \cdot \frac{1+\dfrac{1}{\alpha_{HD}}\dfrac{[HDO]}{[H_2O]}+\dfrac{1}{K_{HD}}\left(\dfrac{1}{\alpha_{HD}}\dfrac{[HDO]}{[H_2O]}\right)^2}{1+\dfrac{[HDO]}{[H_2O]}+\dfrac{1}{K_{HDO}}\left(\dfrac{[HDO]}{[H_2O]}\right)^2} \tag{3.126}$$

In this equation, the index marks of HMEX equilibrium constants correspond to various isotopic mixtures (see Table 3.16).

### 3.4.3   Hydrophobic catalysts of the isotope exchange process

A catalyst suitable for use in the hydrogen-liquid aqueous phase isotope exchange reaction was first obtained in the early 1970s by coating a thin silicone polymer layer on to the industrial hydrophilous $Al_2O_3$-bonded Pt catalyst [123]. Later, also in Canada, other catalysts with hydrophobic properties were developed: porous Teflon-bonded platinum and composite carbon-bonded platinum in a Teflon matrix (Pt–C–Teflon) [124, 125]. A correlation between the relative specific activity of these catalysts is presented in Table 3.17.

As can be seen from the data in Table 3.17, the activity rating of Pt–C–Teflon catalyst is more than 100-fold higher than that of a non-hydrophobized catalyst.

Further progress in the development of platinum catalysts of this type is linked to the search for the best technology of catalyst preparation. The process includes many stages (supporting medium preparation, its coating with a platinum compound, platinum reduction, composite material preparation, etc.), and changes in the conditions of any stage strongly affect the catalyst's activity rating and efficiency. The optimization of the catalyst preparation technology was done in Belgium [126–128], Romania [129–131], and India [132]. Thus, domestically made catalysts of such types are owned by Canada, Belgium, India, and Romania.

A catalyst of another type was developed in Japan and in Russia. Styrene copolymers with divinylbenzene (SDVB) serve as the platinum-supporting medium of the catalyst [133–135]. As was reported in several papers, the catalyst shows a higher activity rating than the Pt–C–Teflon catalyst [130, 136]: in pilot testing in the protium–deuterium system, the Pt–SDVB catalytic activity decreased by a factor of about 5 after 100hours of contact with the liquid aqueous phase compared with the Pt–C–Teflon catalyst which kept the activity rating unchanged [130]. It should be mentioned, though, that it is the only known paper presenting such data, which is inconsistent with data obtained from Pt–SDVB catalyst utilization in Japanese and Russian laboratory-scale and pilot plants. In particular, during the seven-year operation of these plants in Russia for the separation of various isotopic mixtures, including tritium-containing ones, the catalyst's activity rating was never observed to decrease [137–140]. From our point of view, the reason for such data disagreement is that the above remarks about the Pt–C–Teflon preparation requirements are fully true for the Pt–SDVB catalyst, and the catalyst preparation technology used by Romanian scientists was improper.

**Table 3.17**

Relative specific activity of some hydrophobic catalysts[125]

| Catalyst | $k_{Pt}$ [a], $m^3H_2/(s \cdot kg$ of Pt) | Relative activity |
|---|---|---|
| 0.5% Pt on $Al_2O_3$, untreated | 0.0011 | 1 |
| 0.5% Pt on $Al_2O_3$, treated with silicone | 0.046 | 42 |
| 0.4% Pt on porous teflon | 0.064 | 58 |
| 0.4% Pt–C–Teflon | 0.146 | 133 |

[a] $k_{Pt}$, catalytic activity related to 1kg of platinum.

Mention should be made of investigations intended to obtain catalysts containing metals other than platinum (rhodium, palladium, nickel) serving as an active component [141–144]. The work, however, has not culminated in the discovery of a catalyst competitive with the platinum-containing ones. In particular, the nickel-containing catalyst was demonstrated to have a specific catalytic reactivity being two orders of magnitude lower than that of the platinum-containing catalysts [144].

It can be stated therefore that all over the world there exist two types of catalyst usable for isotope exchange in the liquid aqueous phase-hydrogen system: Pt–C–Teflon and Pt–SDVB.

A great many papers [41, 145–157] have been devoted to the influence of temperature and pressure on catalytic reactivity. Notice that, in most cases, the dependences obtained by the authors differ quantitatively, although their nature is common for both catalyst types. The distinctions are associated with the fact that the catalysts involved are porous and most of the active platinum is amassed in the interior pores. What this means is that even if the catalyst reactivity study is conducted under the conditions described by eq. 3.114 (hydrogen – water vapor isotope exchange), the quantity under measurement represents a function of the efficiency of two processes: the catalytic isotope exchange and the diffusion of reagents from catalyst's pores into the vapor-gas flow core and conversely. The change in conditions (temperature and pressure) and in the catalyst's standard reactivity, which can vary owing to a different platinum content on the carrier, diversely affects the efficiency of each catalyst. In addition, a change in temperature only, or in pressure only, leads to a change in the partial pressure ratio of the reagents themselves, which, in turn, influences the overall process efficiency. The interpretation of experimental data becomes all the more complicated when research is performed under conditions of liquid aqueous phase-hydrogen counter-current and the quantity being measured is affected by one more process – phase isotope exchange between water vapor and liquid aqueous phase – see eq. (3.115). Nevertheless, certain general regularities can be deduced from the studies done with the use of highly active catalysts.

In the temperature dependence of the isotope exchange efficiency (the mass-transfer volume factor, $K_{0YV}$, m$^3$ H$_2$/(m$^3$·s), serves most commonly as the parameter to be measured) in a temperature range of 293 to 360K two regions are retraced where the value of activation energy observed changes from over 25kJ/mol (in a temperature range of 293–333K) to 1–5kJ/mol (in a temperature range of 323–360K) [145–147, 154]. (It is necessary to notice, that in the papers on the water–hydrogen mass transfer, the mass transfer volume factor is frequently denoted by $K_{0YV}$, instead of the previously adopted notation $K_{0Y}$). As reported by Yu. Sakharovskii et al. [148], even a negative temperature dependence of $K_{0YV}$ is seen over the high-temperature range. Such dependence results from the above-mentioned reasons, and in the first place, from the contribution of chemical and diffusion components to the process efficiency, as well as from the influence of the ratio between the values of partial pressure of water vapor ($P_{H_2O}$) and hydrogen ($P_{H_2}$). As an example, in Table 3.18 are given the temperature dependences of $K_{0YV}$ at a constant value of $h = P_{H_2O}/P_{H_2}$, and conversely, the dependences of $K_{0YV}$ for $h$ at a constant temperature [41]. The measurements were taken in the conditions of the water vapour-hydrogen counter-current passing through the catalyst bed according to eq. (3.114).

**Table 3.18**

Dependence of $K_{0YV}$ ($m^3H_2/(m^3 \cdot s)$) on temperature and $h$ value: Pt–SDVB catalyst

| $T$, K [a] | 333 | 338 | 343 | 348 | 353 | 358 |
|---|---|---|---|---|---|---|
| $K_{0YV}$ | 16 | 18 | 19 | 19 | 20 | 19 |
| $h$ [b] | 0.18 | 0.31 | 0,33 | 0.44 | 0.60 | 0.88 |
| $K_{0YV}$ | 16 | 19 | 22 | 25 | 28 | 33 |

[a] $h = 0.31$.
[b] $T = 353K$.

**Table 3.19**

Influence of pressure ($P$) on $K_{0YV}$ value at a constant linear velocity of vapour-gas flow ($u$) [41]: Pt–SDVB catalyst; $u = 0.31\,m/s$; $T = 333K$, the ratio of hydrogen and water flows in the counter-current column $\lambda = 1$

| $P$, MPa | 0.16 | 0.5 | 0.68 | 0.75 | 0.90 | 1.11 |
|---|---|---|---|---|---|---|
| $K_{0YV}$, $nm^3H_2/(m^3 \cdot s)$ | 0.38 | 0.36 | 0.39 | 0.37 | 0.42 | 0.42 |

From the data presented in Table 3.18 it follows that at $h$ = const, $K_{0YV}$ is practically temperature-independent (the activation energy being observed is close to zero), and an increase in the ratio between the values of partial pressure of water vapor and hydrogen in the examined $h$ range leads to a doubling of $K_{0YV}$.

The pressure dependence of $K_{0YV}$ is reported in many sources [146, 149, 154–156]. From these papers it follows that the $K_{0YV}$ value decreases by the power law with the superscript varying over an interval of 0.3–0.6, and, when the linear velocity of vapor-gas flow remains constant with pressure, $K_{0YV}$ is scarcely affected by pressure changes (Table 3.19).

In closing the section, it should be noted that according to several publications [41, 110, 157], when highly active catalysts (with the exchange rate constant $k > 5$–$10s^{-1}$) are used, the chemical component contribution to the efficiency of the process by (3.116) is minor, and the process mass-transfer characteristics are determined by the phase isotope exchange process (3.115). This point is given more attention below in the description of various contactors utilized for the CHEX realization in this system.

### 3.4.4   Types and mass-transfer characteristics of contactors for multistage isotope exchange

As follows from the foregoing, the water–hydrogen isotope exchange by eqs. (3.114–3.116), with the use of hydrophobic catalysts, requires, at the first stage, a reaction in the vapor-gas phase over the catalyst's active surface (CTEX process, eq. (3.114)). For this purpose, the catalyst's surface must be hydrophobic. On the other hand, the PHEX reaction requires a developed surface of contact between water and water vapor. Because of this, the contactors for this purpose always represent a combination of elements to carry out CTEX and PHEX reactions efficiently. The main types of contactor are shown in Figure 3.30.

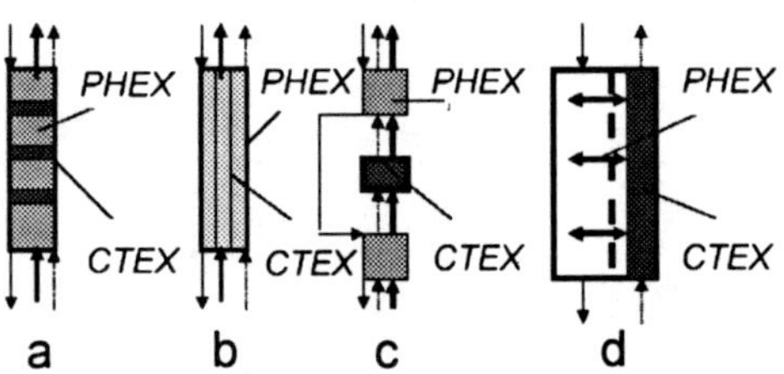


**Figure 3.30** Schemes of various Types of contactor: ⟶ liquid aqueous phase flow; ⟶ hydrogen flow; ⟶ water vapor flow.

Figure 3.30a shows the scheme of the most-used dump contactor. The contactor comprises alternating layers of hydrophobic catalyst over which the CHEX process is performed, and a hydrophilic metal packing, or their uniform blend. Figure 3.30b shows a column section filled with a regular packing material. Such a contactor arrangement is destined to enhance the column capacity. To this end, different modes of regular packing utilization were tested: from filling up of voids formed by rolled corrugated and plain strips with a granular catalyst [150,151], to the catalyst application onto a hydrophobic film which, in turn, was rolled together with a hydrophilic corrugated strip [152, 153]. Spatially divided CHEX and PHEX processes are realized in the contactor shown in Fig. 3.30c [154, 158–161]. The CHEX process in such contactors proceeds between water vapor and hydrogen, with liquid aqueous phase coming from one PHEX zone to another and bypassing the CHEX zone. Finally, Fig. 3.30d represents yet another type of contactor where the complete spatial division of the liquid aqueous phase and hydrogen flows is achieved by the use of a membrane permeable to water vapor. As reported in references [161–165], such polymeric materials as polyacrylamide or Nafion were utilized for the membrane fabrication. Below, some mass-transfer properties of all these contactor types are discussed.

Experimental data on the first type of contactor are the most widely covered in the literature. These contactors are above all characterized by the fact that the volume ratio between catalyst and packing considerably affects the overall process efficiency: capacity and mass-transfer characteristics. The influence of ratio on the capacity is due to the fact that the hydrophobic nature of the catalyst leads to a disturbance of hydrodynamic regime of the gas–liquid contact on the catalyst granules which, under counter-current conditions, results in a decrease of the column capacity. As an illustration, we can use the data, which studies the dependence of the hydraulic resistance ($\Delta P$) of a column filled with a uniform blend of Pt–C–Teflon and a metal hydrophilic packing material at a volume ratio $V_{cat}:V_{pac} = 1:2$ and 1:3 and at a fixed water flow rate, on the hydrogen flow rate $G_{H_2}$ [145] (see Table 3.20).

It can be seen from Table 3.20 that at $V_{cat}:V_{pac} = 1:2$ the column is flooded even at a hydrogen flow rate of 2.8m³/h, whereas at $V_{cat}:V_{pac} = 1:3$ the hydraulic resistance remains rather low even at a hydrogen flow rate of 5m³/h. Based on the obtained data for $\Delta P$ calculation under various experimental conditions within the region of the linear dependence of $\Delta P$ on $G_{H_2}$, the authors suggested the following equation

$$\Delta P/H = 2.3\, G_{H_2,SP} L_{SP}^{1.38}, \tag{3.127}$$

true for specific values up to $L_{SP} \leq 21.1 \,\mathrm{mol/(s\cdot m^2)}$. Unfortunately, the paper [145] does not specify the temperature at which the tests were performed. It can be presumed, though, that the temperature was equal to 313K. At the same time it should be noted that the $\Delta P/H$ value depends on the temperature, since variations in temperature lead inevitably to a change in the amount of water vapor passing through the column together with hydrogen and, consequently, to a change in the total vapor-gas flow rate at a fixed hydrogen flow rate.

In view of a substantial drop in capacity at $V_{cat}:V_{pac} = 1:2$, Belgian scientists, when passing from laboratory-scale plants with columns of 2–3 cm diameter to a pilot plant with a column diameter of 10cm, changed the volume ratio between catalyst and hydrophilic packing from about 1:2 to 1:3 [145–147]. On the other hand, at Russian plants utilizing the Pt–SDVB catalyst, the volume ratio for the columns of the same diameter remained unchanged [139, 148, 149] and equalled 1:4. Notice that in these papers a layer structure of the column filling with catalyst and packing is used, and a hydrophilic packing proportion is larger than that reported in the papers discussed before.

Table 3.21 presents the results of an investigation into the influence of the packing-catalyst volume ratio on the mass-transfer efficiency [41]. The tests were done with the use of the Pt–SDVB catalyst and a spiral-prismatic metal packing material.

As is evident from Table 3.21, a fourfold decrease in the catalyst amount (from 80% to 20%) reduces the mass-transfer coefficient only by a factor of 1.5, while a further decrease of the catalyst amount down to a volume fraction of 10% results in a drastic decrease in $K_{0YV}$. Similar results are reported by A. Bruggeman *et al.* [147], where

Table 3.20

Dependence of $\Delta P/H$ on $V_{cat}:V_{pac}$: column diameter, 10cm; column height, about 7m; liquid flow rate, 7.4l/s [145]

| $V_{cat}:V_{pac}$ | $\Delta P/H$, $10^3\,\mathrm{Pa/m}$ [a], at $G_{H_2}$ (m³/h) equal to | | | | |
|---|---|---|---|---|---|
| | 1 | 2 | 3 | 4 | 5 |
| 1:2 | 5 | 16 | | Column flooding | |
| 1:3 | 1.4 | 2.0 | 2.8 | 3.8 | 4.8 |

[a] the authors of the present study are of the opinion that the multiplier $1 \times 10^3$ was erroneously omitted in the quantities $\Delta P/H$ in the original paper.

Table 3.21

Influence of packing-catalyst volume ratio on $K_{0YV}$ value: $T = 338\mathrm{K}$; $\lambda = 1$; $G_{H_2SP} = 5.4\,\mathrm{mol/(s\cdot m^2)}$

| $V_{cat}:V_{pac}$ | 4:1 | 1:1 | 1:4 | 1:10 |
|---|---|---|---|---|
| $K_{0YV}$, m³ H₂/(m³·s) | 1.2 | 0.9 | 0.8 | 0.4 |

**Table 3.22**

Influence of observed rate constant $k$ of reaction (3.114) on $h_{eq}$ with the use of Pt–SDVB catalyst

| $T$, K | 313 | | | 333 | | | | 344 | | 355 | |
|---|---|---|---|---|---|---|---|---|---|---|---|
| $k$, s$^{-1}$ | 1 | 10 | 15 | 1 | 5 | 7 | 10 | 10 | 20 | 10 | 20 |
| $h_{eq}$, cm | 70 | 25 | 22 | 46 | 23 | 20 | 18 | 16 | 15 | 15 | 14 |

the Pt–C–Teflon catalyst was used in the tests. Here, at $T = 313$K, $\lambda \approx 4$, and $G_{H_2, SP} = 10$mol/(s·m$^2$), a change of the ratio $V_{cat}:V_{pac}$ from 1:2 to 1:3 led to only a 20% decrease in $K_{0YV}$ (from 1.99 to 1.55m$^3$H$_2$/(m$^3$s)). The same paper demonstrates that a considerable enlargement in the column size (from 3 to 10cm diameter, and from 1.15 to 6.8m height) results in a decrease in $K_{0YV}$ by no more than 15%. In tests performed at $V_{cat}:V_{pac} = 1:10$ [158] in a column of 8cm diameter and 179cm height with the use of a Degussa-produced Pt–C–Teflon catalyst (type F181 G/D, 1% by weight of platinum), $K_{0YV}$ values obtained at $T = 353$K, $\lambda = 1$ and $G_{H_2, SP} = 8.5$mol/(s·m$^2$), do not exceed 0.5m$^3$ H$_2$/(m$^3$s).

Since the isotope exchange in the water–hydrogen system proceeds in two stages (3.114, 3.115), of interest are data on the influence of the catalyst reactivity and hydrophile packing type on the overall mass-transfer efficiency. Table 3.22 presents HETP values obtained from an isotope exchange column with the use of Pt–SDVB catalysts of different specific catalytic reactivity and the same spiral-prismatic packing (SPP) at $V_{cat}:V_{pac} = 1:4$ (protium–deuterium system, low deuterium content region, $\lambda = 1$, $_{SP} = 5.4$mol/(s·m$^2$) [157].

From the above data it follows that at a catalyst reactivity $k > 5$–$10$s$^{-1}$, the $h_{eq}$ value is little affected by $k$, with a considerable enhancement at $k < 5$s$^{-1}$. By this is meant that for low-activity catalysts, the chemical component (3.114) begins to contribute significantly into the overall mass-transfer efficiency. On the other hand, tests are reported with the use of the Pt–C–Teflon catalyst ($V_{cat}:V_{pac} = 1:2$) of which the results demonstrate the mass-transfer dependence on the hydrophylic packing type [147]. At a temperature of 313K, $\lambda = 1$, and $= 5.4$mol/(s·m$^2$), replacing a packing with developed surface by a glass pellet packing with a small overall surface leads to a near tenfold decrease in $K_{0YV}$: from 2.24 to 0.27m$^3$H$_2$/(m$^3$·s). The change in $K_{0YV}$ indicates that when active catalysts are used, the PHEX process is the governing factor in the overall process efficiency.

Of interest is a comparison, presented in papers [166, 167], between mass-transfer efficiency values in the presence of Pt–C–Teflon and Pt–SDVB catalysts, performed experimentally under comparable conditions. As can be seen from Table 3.23, the Pt–SDVB catalyst utilization ensures somewhat higher mass-transfer efficiency, but in the case of layer-by-layer loading leads to a decrease in the column capacity.

To summarize the analysis of the dump-type contactor efficiency, worthy of mention is reference [166], where a comparison of isotope separation efficiency of protium–deuterium and protium–tritium mixtures is made (see Table 3.24). The study was carried out for protium isotope mixtures with low deuterium and tritium content using the Pt–SDVB catalyst and metal SPP with unit size 2mm $\times$ 12mm $\times$ 0.2mm at $V_{cat}:V_{pac} = 1:4$ for different column loading techniques (uniform and layer-by-layer loading).

From the comparison between values of $h_{eq,D}$ and $h_{eq,T}$ given in Table 3.24, it is obvious that in all tests the values of $h_{eq,D}$ are lower than those of $h_{eq,T}$. It should be recognized that

**Table 3.23**

Hydraulic resistance $\Delta P/H$ and $h_{eq}$ for isotope exchange column with the use of different catalysts: protium–tritium mixture; $T = 333K$

| Catalyst | Pt–C–Teflon, $V_{cat}{:}V_{pac} = 1{:}3$ uniform loading | | | | Pt–SDVB, $V_{cat}{:}V_{pac} = 1{:}4$ layer-by-layer loading | |
|---|---|---|---|---|---|---|
| $G_{H_2,\,SP}$ mol/(s·m²) | 10 | | 20 | | 10 | |
| $\lambda$ | 2 | 4 | 2 | 4 | 2 | 4 |
| $h_{eq}$, cm | 28 | 30 | 34 | 29 | 20 | 18 |
| $\Delta P/H$, Pa/m | 2,500 | 2,040 | — | — | 2,620 | 2,160 |

**Table 3.24**

Dependence of mass-transfer characteristics of isotope exchange column on hydrogen flow rate: $\lambda = G_{H_2}/L_{H_2O} = 4$, $T = 333K$

| $G_{H_2,\,SP}$ mol/(m²·s) | $h_{eq,D}$, m | $h_{eq,T}$, m | $h_{0Y,D}$, m | $h_{0Y,T}$, m | $K_{0YV,D}$, mol/(m³·s) | $K_{0YV,T}$, mol/(m³·s) |
|---|---|---|---|---|---|---|
| 8 | 0.17 | 0.20 | 0.20 | 0.18 | 40.2 | 44.6 |
| 10 | 0.16 | 0.20 | 0.19 | 0.18 | 53.1 | 55.8 |
| 10 [a] | 0.17 | 0.21 | 0.19 | 0.19 | 53.1 | 53.1 |
| 20 [a] | 0.33 | 0.42 | 0.38 | 0.38 | 53.1 | 53.1 |

[a]  with catalyst-packing uniform blend.

with the parity of HTU values for the separation processes under study ($h_{oy,D} = h_{oy,T}$), the calculation gives the ratio $h_{eq,D} / h_{eq,T} = 0.79$. From the comparison between experimental and calculated values of $h_{eq,D} / h_{eq,T}$, it was concluded that the assumption about the HTU parity for the separation of the protium–deuterium and protium–tritium mixtures is true.

The coincidence of HTU values and difference between HETP values for two isotopic mixtures imply that in actual processes of detritization of both tritium and deuterium containing mixtures, allowance should be made for the HETP change with deuterium concentration. In the process of separation of such mixtures HETP will vary even within a single separation plant owing to the establishment of deuterium concentration profile through the column height. Figure 3.31 shows the variation in HETP calculated over the whole deuterium concentration range (in the region of tritium micro–concentrations), at a temperature of 333K, and $\lambda$ changing from 1 to 4. From the data presented it is clear that with a variation in the concentration of deuterium mixed with protium, the $h_{eq,T}$ value changes most abruptly at $\lambda = 4$ (by a factor of 1.9), whereas at $\lambda = 2$ and $\lambda = 1$ it changes by a factor of 1.76 and 1.66, respectively.

Hence, the available data on dump-type contactors reported in the literature allow us to infer that an approach to the optimum separating layer formation in the separation columns for both types of catalyst has been worked out in sufficient detail both in column capacity

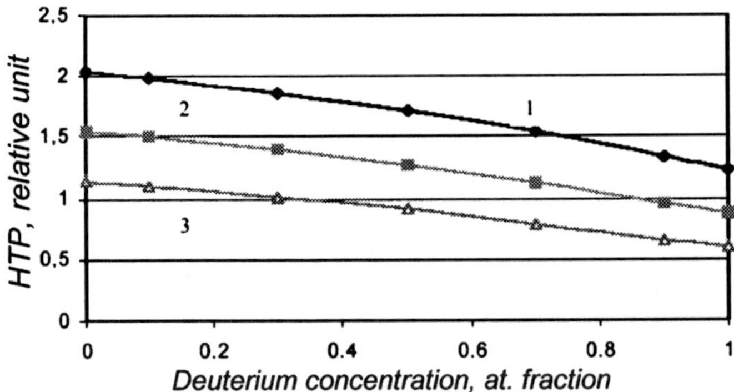

**Figure 3.31**   Dependence of the ratio $h_{eq,D}$ / $h_{eq,T}$ on deuterium concentration in separated mixture: 1, 2, 3,   at $\lambda$ = 1, 2, 4, respectively.

terms and in terms of mass transfer efficiency, with the HETP values achieved in columns of up to 10cm diameter falling in the range 15–30cm.

Experimental data on the efficiency of other contactor types used in water–hydrogen chemical isotope exchange are far less common in the literature.

A. Kitamoto *et al.* [152] present the results of experiments performed on various embodiments of a regular packing material (see Figure 3b). The catalyst represented a 0.2mm-thick Teflon film treated with an SDVB polymer and coated with dispersed platinum (1.0–1.4% by weight). The film was rolled together with hydrophilic wire mesh 0.05–0.1mm thick through a corrugated metal divider (with a crimp height of 0.6–0.75mm) made as well from wire mesh (see Figure 3.32). Some experiments were performed with the packing in which the catalyst film was complemented with one or several additional layers of hydrophilic wire mesh. The value of $K_{OYV}$ obtained in a column with a diameter of 30cm at a temperature 333K, $G_{H_2, SP}$ = 17.9mol/(m$^2$·s), and $\lambda$ = 0.15 accounts for about 85mol/(m$^3$·s), and the use of a packing with additional hydrophilic wire mesh layers increases the value to 134m$^3$H$_2$/(m$^3$·s). These values of the mass-transfer coefficient are somehow higher than those obtained with the use of dump type contactors (see Table 3.24). But attempts to find in the available literature any data on a wider use of this type of contactor in various separation plants have been unsuccessful.

The contactors of the third type (Figure 3.30c) are rather complicated. In addition, they are characterized by high values of liquid holdup and hydraulic resistance [161]. It should be recorded, though, that such a design of contactors provides a principal means of using catalysts, which are non-hydrophobic in nature, for isotope exchange in the water–hydrogen system. The above-mentioned paper [158], for example, presents the results of a comparative study on the efficiency of the protium–tritium mixture separation in a column of 8cm diameter comprising three elements filled with catalyst interspersed with three elements with hydrophilic packing material (each element is 40cm in height), with the use of Degussa-produced hydrophobic and hydrophilic catalysts. The column had upper and lower flow conversion units. Even though the conditions of the process realization were, on the whole, unsuccessful (ineffective catalysts and packing material), the separation efficiency in the presence of hydrophilic catalyst was even slightly higher than that with

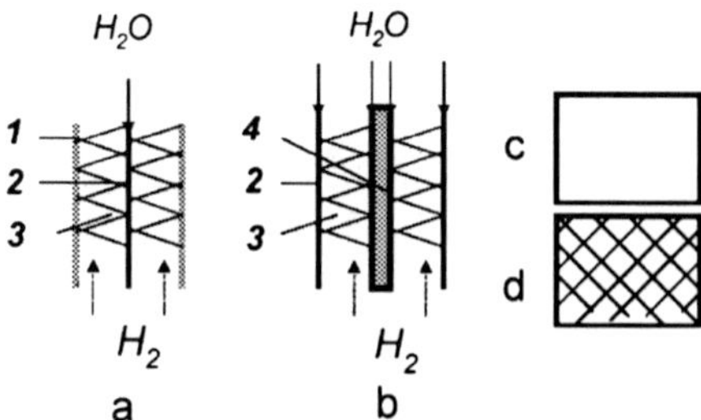



**Figure 3.32**  Schematic sketch of regular packing material: a, packing with one wire mesh layer; b, packing with two wire mesh layers; c, catalyst film; d, catalyst film with wire mesh; 1, hydrophobic film of catalyst; 2, wire mesh; 3, corrugated divider; 4, catalyst film sandwiched between two wire mesh layers.

hydrophobic catalyst under comparable conditions (at $T = 353K$, $\lambda = 1$, and $G_{H_2} = 17.9 m^3/h$, for example, the obtainable separation degree was 52 and 47, respectively).

The main idea forming the basis for the development of the contactor shown in Figure 3.30d is the separation of the catalyst from liquid aqueous phase by a water-vapor-permeable membrane, which allows use of non-hydrophobic catalysts as well. The feasibility of carrying out the separation process in such a contactor with the use of the $Pt/Al_2O_3$ catalyst has been demonstrated [162]. In addition, an advantage of the membrane-type contactors (MTC), as against the contactors of the third type, is that the CHEX and PHEX processes occur here simultaneously: the CHEX process proceeds in the MTC vapor-gas space, and the PHEX process is done through the membrane separating it from the liquid space. What this means is that there is no need for the use of additional PHEX units and for the repeated separation of water and hydrogen flows. According to the paper, the values of mass-transfer coefficients $K_{0Y}$ in the MTC with the MF-4SK membrane (perfluorochemical strong-acid cation resin membrane – the Russian clone of the Nafion membrane) per membrane unit surface at $T = 358K$, $P = 0.1MPa$, $G_{H_2,SP} = 4.3 mol/(m^2 \cdot s)$, and $\lambda = 1$, account for $0.010 m^3 H_2/(s \cdot m^2)$ of membrane surface, which, for the utilized MTC construction, corresponds to the value $K_{0YV} = 58 mol$ ($m^3$ of MTCs) (by an MTC volume is meant the overall volume of the contactor's liquid and catalytic space), and at $T = 400K$ and $P = 0.3MPa$ these values are equal to $0.014 m^3 H_2/(s \cdot m^2)$ of membrane surface and $81 mol$ ($m^3$ of MTCs), respectively (regardless of the catalyst type – either hydrophobic Pt–SDVB or hydrophilic $Pt/Al_2O_3$) [164, 165, 168].

As is evident from the above values, their order is close to those characteristic for the dump contactors. Notice that a high efficiency of the mass transfer in MTC in the presence of the thermostable $Pt/Al_2O_3$ catalyst offers possibilities for the MTC use in two-temperature plants.

One more variation of the contactor design realization is reported [169–172]. This contactor amounts to an electrochemical cell with the MF-4SK membrane serving as a solid polymeric electrolyte. The catalyst (fine dispersed platinum) is coated immediately to the membrane surface. An electric potential (lower than that of the water decomposition) is applied on both sides of the membrane, and a portion of the hydrogen flow is carried from the contactor's anode chamber to the cathode one, through which the water flow passes. The CHEX process takes place in the anode chamber, while the PHEX process proceeds through the membrane and its intensity is further enhanced due to the fact that three or four water molecules are carried to the anode chamber with each hydrogen ion. At $T = 343K$, $P = 0.1MPa$, $G_{H_2, SP} = 2.5mol/(m^2 \cdot s)$ and $\lambda = 1$, the mass-transfer coefficient value per membrane unit surface accounts for $0.019m^3 H_2/(s \cdot m^2)$: that is, about one-fifth that in the MTC.

In conclusion, mention should be made of some papers devoted to the development of mathematical models of the CHEX process with the use of contactors of the first, third, and fourth types [168, 173–178].

The aim of the papers is to search for the conditions ensuring the most effective management of the CHEX and PHEX processes in the contactors. In particular, for contactors of the first type, the objective consists of determining an optimum ratio between the heights of catalyst and hydrophilic packing beds [173].

### 3.4.5   Utilization of isotope exchange in the water–hydrogen system for hydrogen isotope separation

Nowadays, consideration is being given to the following potential fields of application of the water-hydrogen CHEX process:

(1)   removal of tritium from various light-water wastes for subsequent environmental discharge, and reduction of tritium-containing waste materials through the tritium concentrate  production;

(2)   processing of heavy-water reactor coolant to remove protium and tritium, and to obtain conditioned heavy water from various heavy-water wastes, tritium-containing wastes included;

(3)   utilization of the CECE technology as an alternative technique at the stage of heavy-water final concentration in heavy-water production from natural resources;

(4)   elaboration of alternative process applications to recycle various tritium-containing flows within the ITER program.

The first problem is characteristic for irradiated fuel processing plants, where a near-half tritium formed in the fuel is brought into the solvents of fuel elements [179], and eventually brought out for environmental discharge. The efforts directed toward using the CECE technology to solve this problem are under way in Canada, Japan, U.S.A., Germany, and Belgium [147, 150, 155, 158, 180–183]. The process flow diagrams are shown in Figures 3.33a and 3.33b.

The distinction between schemes a and b presented in Figure 3.33 is that the former represents a separation plant with upper and lower flow conversion units (UFCU and LFCU, respectively), whereas the latter shows a plant without UFCU, and the column head is fed with water of natural isotope composition. Water purified from tritium which can be either discharged to environment or utilized for the recycling of fuel elements (in this case its tritium purification efficiency can be lower), is withdrawn upon leaving the UFCU. From the plant shown in Fig. 3.33b, the product purified from tritium (hydrogen) can be delivered to the user. The flow to be recycled arrives at the middle section of the isotope exchange column. The tritium concentration increases below the column feeding point (the column

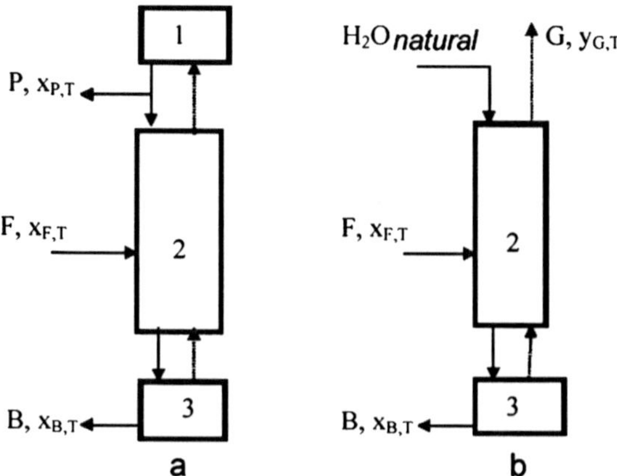

**Figure. 3.33**   Schemes of plants for detritization of light water wastes: 1, upper flow conversion unit (flame or catalytic burner); 2, isotope exchange column; 3, lower flow conversion unit (electrolyzer); F, B, P, and G, flows of source water, tritium concentrate, water purified from tritium, and hydrogen purified from tritium, respectively; $x_{F,T}$, $x_{B,T}$, $x_{P,T}$, and $x_{G,T}$ tritium concentration in corresponding flows.

enrichment section), and the tritium concentrate is withdrawn either from the electrolyzer or from the water flow at the column outlet. Above the feeding point, the tritium concentration decreases (the column depletion section), and the tritium-purified product is withdrawn from the column head. In this process, the separation degree in each column section depends on the column height. Notice that the increase of tritium concentration in water is limited by the possibility of the water self-radiolysis process [184]. As for the catalyst radiation resistance, it is reported that even a yearly stay of the Pt–C–Teflon catalyst in the water with a tritium concentration of 5Ci/kg did not reveal a change in the catalyst reactivity [130]. Data on the Pt–SDVB catalyst's radiation resistance are presented as well in another paper [41]: a gamma irradiation dose of up to 5MGr did not lead to any change in the catalyst reactivity.

In developing the technology for the solution of the problem under study, a major part is played by questions of preliminary chemical treatment of wastes before their supply to the separation plant, because the wastes can contain catalytically poisonous components, as well as exert influence on other plant units (specifically, electrolyzer as UFCU), or on the mass transfer efficiency of the separation plant. The most catalytically poisonous effect is exhibited by carbon monoxide, iodine, and its compounds, while the requirements on the content of nitric acid and tributylphosphate, in regard to their impact on the catalyst, are far milder [145, 181, 185, 186]. The technology of primary water chemistry has been detailed [145]. For water purification, processes of adsorption are suggested in preference to activated carbon (removal of organic components), ion exchange (purification from $NO_3^-$ and other ions), and distillation (lowering the content of heavy products of uranium fission).

As for contactors of the first type, therefore, it can be stated that, on the whole, all fundamental technological problems have been solved. But a practical implementation of the method as applied to the industrial waste of spent fuel treatment plants has not yet been developed beyond the scope of laboratory-scale and pilot plants. In the opinion of authors reviewing the state of the technology in Japan and Western European countries [155], the technology has not been industrially implemented for political and economic reasons. The chief cause is, undoubtedly, a high energy intensity of the process together with the need to recycle a great amount of waste (using electrolyzers as LFCU!) which will result in a rise in the cost of the whole nuclear fuel cycle.

The ways of tackling the second problem – the treatment of various tritium-containing water flows – were studied in detail by Canadian scientists. The problem stems from the necessity to remove protium and tritium, produced during the operation of CANDU-type reactors, from the heavy-water coolant. Various methods of using isotope exchange in this system to solve the problem were patented in Canada [187–189], and in countries where the energy industries are oriented to the use of CANDU-type reactors [155, 190–192]. The process alternatives are shown in Figure 3.34 [180, 193–203].

Figure 3.34a represents the scheme of a plant where the water–hydrogen isotope exchange is used only for the transfer of hydrogen isotopes from the liquid to the gas phase with the utilization of the LPCE process, and the isotope separation process is done by the

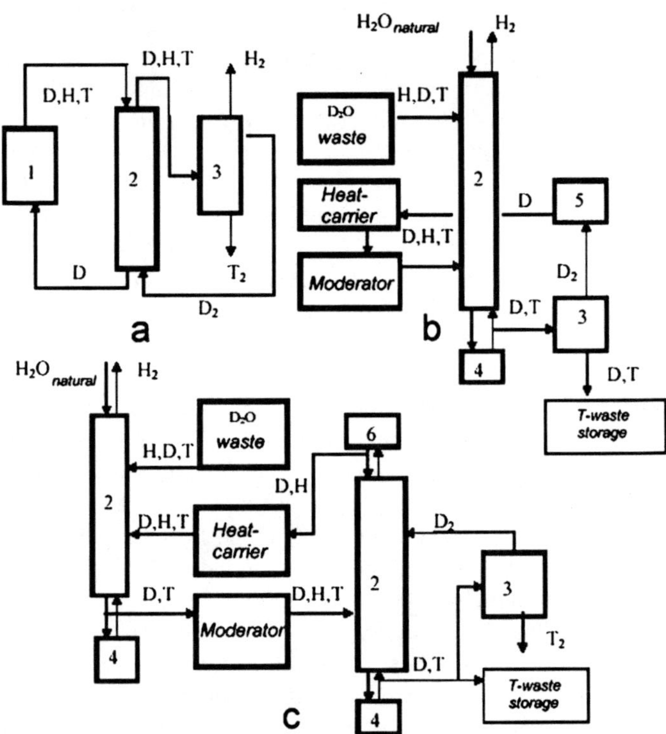



**Figure. 3.34**  Schemes of the plants for tritium removal from water of heavy-water reactors: 1, nuclear reactor; 2, isotope exchange column; 3, hydrogen cryogenic rectification plants; 4, LFCU (electrolyzer); 5, unit of hydrogen conversion into water; 6, UFCU.

hydrogen cryogenic rectification, in just the same way as is done for the heavy-water coolant purification at the Grenoble and Darlington plants [112, 113].

The schemes shown in Fig. 3.34b and 3.34c represent the plants incorporating isotope exchange units which function by CECE technology, that is, with one (Figure 3.34b) or two (Figure 3.34c) flow conversion units [180]. These columns provide not only for the isotope transfer from one phase to another, but also for the isotope mixture separation. The difference between the schemes lies in the fact that the last-named plant comprises two sequential columns in the water–hydrogen system, and the isotope mixture separation is performed even to the point of producing near-pure tritium. The scheme shown in Fig. 3.34c forms the basis of the isotope purification system of the Advanced Neutron Source (ANS) in the Oak Ridge National Laboratory (U.S.A.) [194]. In Russia, there exists the possibility of using a similar scheme for the PIK research heavy-water reactor of the V. Konstantinov Nuclear Physics Institute in Gatchina [195]. Since 1995, the isotope exchange plant forming a part of this scheme has been in pilot operation and utilized for the conditioning of various heavy-water wastes. The plant diagrammed in Fig. 3.34c is presumed to operate in two modes. The first mode is done with the use of the cryogenic rectification column, and tritium enrichment is performed as long as the tritium concentration exceeds 99%. In the second mode, the cryogenic rectification column is inactive, and the tritium enrichment in the second hydrogen–water isotope exchange column is performed until the tritium concentration in the gas ranges up to about 104ppm (which is equivalent to 300Ci/kg of $H_2O$), and the tritium concentrate is withdrawn for entombment. The practical implementation of the scheme is underway in Canada. In the late 1990s, the Chalk River Laboratories (Ontario) put into operation demonstration plants for the heavy-water coolant deprotization and detritiation based on the CECE technology (derived from CECEUD, combined electrolysis and catalytic exchange for upgrading and detritiation) [200–202]. The plants comprise isotope exchange columns with an overall height of about

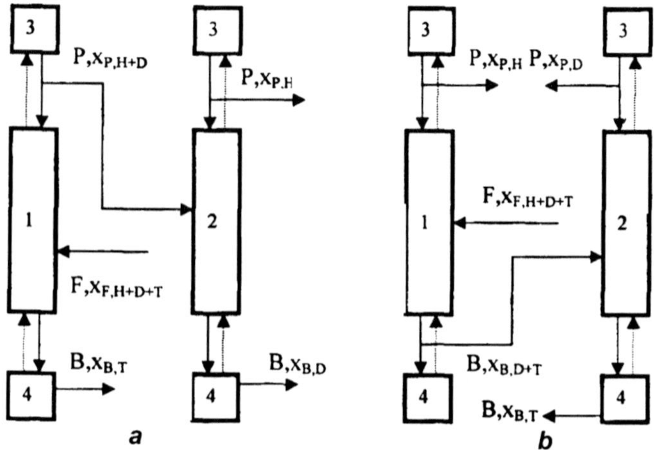


**Figure. 3.35** Variations of the process of separation of protium–deuterium–tritium mixture: a, initial detritiation of the mixture; b, initial deprotization of the mixture; 1,2, isotope exchange columns; 3, UFCU; 4, LFCU (electrolyzer).

45m, a diameter of 50mm, and an electrolyzer of about $3nm^3/h$ in capacity. The plants' operating experience allowed for the design of full-scale plants with a capacity of 300tons per year for the purification of the CANDU reactors' heavy-water coolant.

As for the problem of the purification of various heavy-water wastes from protium and tritium, we emphasize that the need for two in-series isotope exchange columns in the water-hydrogen system is typical for the sharp separation of protium–deuterium–tritium isotope mixture. The first column provides for obtaining as products either the tritium concentrate in the column bottom, and the deuterium–tritium in the column head, or the protium concentrate in the column head and the deuterium–protium mixture in the column bottom (see Figures 3.35a and 3.35b, respectively). In the second column it is practically a binary isotopic mixture that is subjected to separation. In this process, for the mixture containing less than 99 atomic per cent of deuterium, it is preferable to begin with the stripping of tritium followed by protium–deuterium mixture separation, which is dictated by the above-mentioned sharp dependence of the tritium distribution coefficient on the deuterium concentration in the mixture (see Table 3.25).

As is seen from Table 3.25, the initial stripping of tritium from the mixture with the deuterium concentration of less than 90 atomic per cent allows use of higher $\alpha_T$ values for a certain reduction of the first CHEX column volume.

In the early 1990s, a technique for processing tritium-containing heavy-water wastes was devised in the D. Mendeleev University of Chemical Technology of Russia (MUCTR). The University researchers developed and put into service pilot plants for detritiation (1994) and deprotization of heavy-water wastes (1993), allowing conditioned heavy water to be obtained from the source material containing no more than 5mCi/kg of tritium and no less than 40 atomic per cent of deuterium [135, 137, 138, 149, 196–199]. The deprotization plant is 4.5m in height, 45mm in diameter, with an electrolyzer capacity of $0.5m^3H_2/h$, and the detritiation plant is 9m in height with the same diameter and electrolyzer capacity. The above-mentioned plant for the heavy-water waste deprotization at the V. Konstantinov Nuclear Physics Institute (with an isotope exchange column of diameter 100mm and overall height 6.9m, and an alkaline electrolyzer of up to $10m^3H_2/h$ in capacity [139]), makes it possible to obtain tens of tons of heavy water per year from the source material with a deuterium concentration of 45 atomic per cent. In 1986, a similar plant was put into service in Japan. The plant recycles some 10 tons of heavy-water wastes of the Fugen nuclear power station per year [155, 191].

Noteworthy is a peculiarity of the plants for heavy-water waste detritiation by CECE technology. This peculiarity consists of utilizing the oxygen produced in the electrolyzer (LFCU) for the hydrogen oxidation in the UFCU. The oxygen issuing from the electrolyzer, in turn, is saturated with water vapor with a high concentration of tritium.

**Table 3.25**

Dependence of $\alpha_i$ values on deuterium concentration in the mixture: region of tritium micro-concentrations; $T = 328K$

| [D], at.% | 0.0147 | 10 | 50 | 90 | 99.99 |
|---|---|---|---|---|---|
| $\alpha_D$ | 3.22 | 3.07 | 2.84 | 2.78 | 2.78 |
| $\alpha_T$ | 5.40 | 5.01 | 3.40 | 1.89 | 1.57 |

If supplementary oxygen purification measures are not taken, it will transport tritium to the tritium-depleted plant end. To solve this problem, various techniques are employed in the operable plants. The scheme shown in Figure 3.36a contemplates the oxygen purification by water of natural isotopic composition, and the scheme presented in Fig. 3.36b by water detritiated in the isotope exchange column. Each technique has its advantages and shortcomings. Oxygen washing by natural water (Figure 3.36a) results in the formation of waste substances with tritium content acceptable for environmental discharge. The utilization of the scheme shown in Figure 3.36b leads to a reduction in the potential capacity of the separation plant and to a decrease in the separation process efficiency due to the mixing of isotope concentrations in unit 4.

The third potential field of CECE technology application – its utilization at the stage of the heavy-water final concentration – does not require any further consideration since the experience of the separation plant operation in Japan and in Russia for the heavy-water waste conditioning has conclusively demonstrated the system's advantages and technological effectiveness. The paper [87] compares the main characteristics of the final concentration process carried out by CECE technology and those of this process performed by the water rectification technique which is commonly used in the operable plants. The paper demonstrates that the former has significant advantages for both the separation plant volume, and power inputs. But it is highly improbable that the large-scale heavy-water plants in service today will be retrofitted with changing final concentration techniques used at present for CECE technology. The utilization of the technology is possible, however, in the future, when new heavy-water plants are constructed. This possibility, for example, is being considered in Romania [204]. In Russia this problem will also be important when a decision is made on heavy-water production from natural source materials.

Particular attention should be given to work performed in Canada on the development of the combined industrially reforming and catalytic exchange process (CIRCE) intended for heavy-water production from natural source materials [1223, 200]. The process is shown in Figure 3.37.

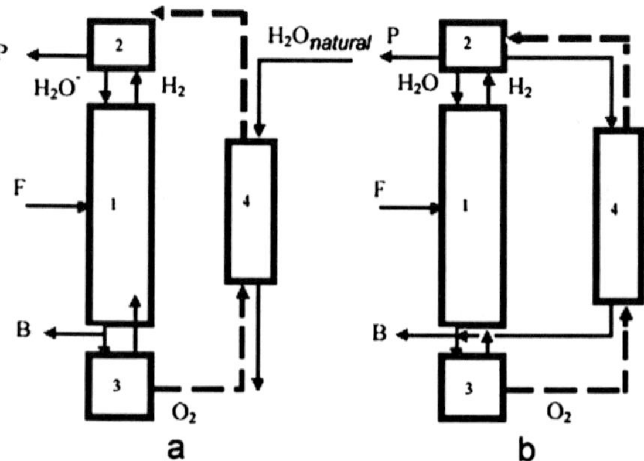


**Figure. 3.36**  Oxygen purification techniques using CECE technology for the source material detritiation: a, purification by water of natural isotopic composition; b, purification by the flow of detritiated product; 1, isotope exchange column; 2, UFCU; 3, LFCU; 4, PHEX column.

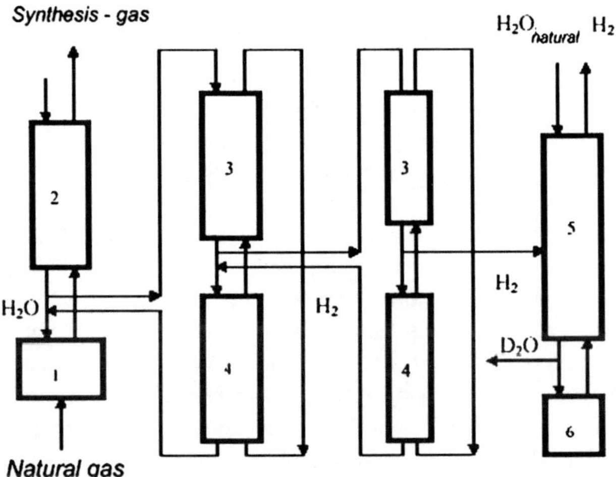


**Figure. 3.37**   Scheme of the CIRCE process: 1, steam methane reforming unit; 2 – 5, water–hydrogen isotope exchange columns; 6, electrolyzer.

Noteworthy are two features of the process. At the first stage, a reactor for the steam methane reforming serves as LFCU, which makes it possible to transfer energy costs to the cost of gas delivered from column 2 to the user. The second and the third stages of the separation cascade (columns 3 and 4) function by a two-temperature scheme (3, hot columns, and 4, cold columns) which as well obviates the need for high-energy inputs in the LFCU. The implementation of the scheme was made possible with the development in Canada of a hydrophobic catalyst maintaining thermo stability to the point of 443K [200]. It is contemplated that columns 2–4 operate at a heightened pressure (1–3.5MPa), and column 5 at a pressure close to 0.1MPa. The first three stages of the cascade are bound to produce water with a deuterium concentration of about 10 atomic per cent, and the deuterium upgrade to reactor-grade concentration ($[D] > 99.72at\%$) is done by CECE technology in column 5. The process development did not come within the province of laboratory-scale tests and, at present, the Air Liquid Canada Company (Hamilton, Canada) has begun the construction of a pilot plant with a capacity of 1ton of $D_2O$ per year.

Regarding the third class of problems, worthy of mention is another potential field of the CECE process application: production of water with a deuterium content reduced compared with that of natural water [205, 206]. This water finds use in biological and medical research. Notice that there is no need for any complementary studies to attack these problems, and, if a demand arises, the plants available in Russia (at the V. Konstantinov Nuclear Physics Institute or at the D. Mendeleev University of Chemical Technology) will allow them to be solved without any important constructional changes.

Among various problems associated with hydrogen isotope separation in the context of the ITER program, those that can be solved by the use of the hydrogen–water CHEX are as follows [207]:

- water purification from tritium for the reactor first-wall cooling (about 100kg per hour, from a concentration of 5–10Ci/kg to 0.5Ci/kg),
- purification from tritium of various auxiliary water flows (some 120kg per hour, from a concentration of 0.1Ci/kg to under 1μCi/kg),
- purification from tritium of water flows that may appear in an emergency due to the total blanket depressurization (about 10tons of water with an overall tritium concentration of 25MCi [197]).

As was initially adopted for the ITER program, the isotope purification system contemplates the use of the water and hydrogen rectification. It is suggested using a hydrogen–water vapour isotope exchange column (VPCE process) for the transfer of isotopes from water to hydrogen [208]. Several papers, though, consider the use of CECE technology instead of the VPCE process [180, 192, 197, 209–213]. The authors are of the opinion that such modification to the purification technology ensures undeniable advantages over the originally accepted technique. Moreover, it puts forward a novel ISS concept for the water flows of nuclear reactors [214]. The concept is included in a combination of the CECE process at a stage of initial concentration of tritium and purification of water up to the sanitary code requirements, and the chemical isotope exchange between hydrogen and palladium hydride (see [215], for example) at the stage of the tritium final concentration. The prospects of CECE process utilization in this field depend on the world's general trends of work on controlled nuclear fusion.

In summary it may be said that one of general technological issues peculiar to the problems of the first, second, and fourth classes and associated with the CHEX utilization for the separation of tritium-containing mixtures in water–hydrogen systems is the selection of an electrolyzer type as the flow conversion unit. The maximum tritium concentration is observed in the separation plant's LFCU; this is why of fundamental importance here is the volume of electrolyte which determines, in the first place, the separation plant accumulation time, and second, the overall tritium holdup and consequently, the plant's radiation safety. Another important criterion is the electrolyzer service life under conditions of a heightened tritium concentration. In terms of the tritium holdup, preference should be given to electrolyzers with solid polymeric electrolyte, which is characterized as well by lower specific energy inputs [119]. The performance of this type of electrolyzer was studied in sufficient detail during the operation of the plant at the Mound Laboratory (Miamisburg, Ohio, U.S.A.) [121, 158]. But reference [216] reports that at a high tritium concentration, the solid polymeric electrolyte (Nafion) is destroyed. Because of this, preference in this case should be given to other electrolyzer types, with electrolyzers based on solid oxide electrolytes being potential candidates [217, 218]. It is demonstrated that such electrolyzers are operable even in the electrolysis of near-clean $H_2O$ [218]. Moreover, H. Yamai [192] considers this electrolyzer as a basis for the development of a converter that practically combines the separation plant's upper and lower flow conversion units which may result in a very significant reduction of specific energy inputs. It should be noted, though, that alkaline electrolyzers or electrolyzers with solid-state polymeric electrolyte remain the most common type. Some features of the utilization of these electrolyzers for deprotization and detritiation of heavy-water waste have been considered [199].

## REFERENCES


1. B. M. Andreev, Ya. D. Zelvenskii, S. G. Katalnikov, *Separation of Stable Isotopes by Physical–Chemical Methods*, Energoatomizdat, **1982**, 208.
2. G. G. Philippov, K. N. Sakodynskii, Ya. D. Zelvenskii, *Khim. Prom.*, **1965**, 1, 10.
3. B. M. Andreev, Ya. D. Zelvenskii, S. G. Katalnikov, *Heavy Isotopes of Hydrogen in Nuclear Technology*, Energoatomizdat, **1987**, 456.
4. B. Brigoli, S. Villani, *Energia Nucleaire*, **1959**, 6, 784.
5. B. M. Andreev, *Khim. Prom.*, **1962**, 8, 35.
6. A. M. Rozen, *Rep. Acad. Sci. U.S.S.R.*, **1956**, 108, 122.
7. A. M. Rozen, *Materials of the Third All-Union Conference on Isotope Application, Izd. Akad. Nauk SSSR*, **1956**, 86.
8. A. M. Rozen, *Theory of Isotope Separation in Columns*, Atomizdat, Moscow, **1960**, 436.
9. K. Bier, In: *Production of Heavy Water*, (ed. G. M. Merphy), 1$^{st}$ edition, **1955**, 394, McGraw-Hill Book Co., New York.
10. B. M. Andreev, V. V. Uborskii, *TOKhT*, **1981**, 15, 5, 664.
11. B. M. Andreev, Ya. D. Zelvenskii, S. G. Katalnikov, *Heavy Isotopes of Hydrogen in Nuclear Technology*, IzdAT, **2000**, 344.
12. B. M. Andreev, G. K. Boreskov, S. G. Katalnikov, *Khim. Prom.*, **1961**, 6, 19.
13. B. M. Andreev, G. K. Boreskov, *ZhFKh*, **1964**, 38, 1, 115.
14. L. P. Portnov, G. G. Philippov, B. M. Andreev, *Production and Analysis of Pure Substances*, **1976**, 1, 83.
15. P. A. Small, *Trans. Faraday Soc.*, **1937**, 33, 820.
16. P. Geib, K. Clusius, *Fiat Rev. Germ. Sci.*, **1948**, 14, 182.
17. R. Haul, H. Benke, H. Dietrich, *Angew. Chemie*, **1959**, 71, 64.
18. D. P. Jackson, *Antifoam for Heavy Water Plant Traus*, Atomic Energy of Canada Ltd., **1969**, report AECL- 3382, 15.
19. M. P. Burgess, R. P. Germann, *AIChE J.*, **1969**, 15, 272.
20. B. M. Andreev, Ya. D. Zelvenskii, V. V. Uborskii, *Atomnaya Energia*, **1978**, 44, 240.
21. I. Bron, C. F. Chang, Z. Wolfsberg, *Z. Naturforsch*, **1973**, 28a, 129.
22. A. Van Hook, *J. Phys. Chem.*, **1968**, 72, 1234.
23. I. Kirshenbaum, *Heavy Water Physical Properties and Analysis Techniques*. Inostrannaya Literatura, Moscow, **1953**, 437.
24. F. T. Selesck, L. T. Carmichael, B. H. Sage, *Ind. Engng. Chem.*, **1952**, 44, 2219.
25. S. Sh. Byk, V. I. Fomina, *Uspekhi Khimii*, **1968**, 37, 1097.
26. B. M. Andreev, Ya. D. Zelvenskii, D. N. Maslov, *Gazovaya Prom.*, **1979**, 2, 61.
27. M. P. Burgess, *AIChE J.*, **1971**, 17, 529.
28. E. C. W. Clarke, D. N. Glew, *Can. J. Chem.*, **1970**, 48, 764.
29. G. G. Philippov, K. N. Sakodynskii, Ya. D. Zelvenskii, *Khim. Prom.*, **1965**, 1, 10.
30. B. M. Andreev, S. G. Katalnikov, *Khim. Prom.*, **1965**, 4, 28.
31. S. M. Dave, S. K. Ghosh, H. K. Sadhukhan, *Ind. J. Chem.*, **1981**, 20A, 329.
32. J. Garaund, R. Amanrich, In: *Production of Heavy Water* (ed. G. M. Murphy), **1955**, 394, McGraw-Hill Book Co., New York.

33. J. S. Spevack, U. S. Pat. 3920395 (**1966**).

34. J. S. Spevack, German Pat. 1289032 (**1970**).

35. B. M. Andreev, V. B. Petrov, V. V. Uborskii, *Rep. Acad. Sci. U.S.S.R.*, **1980**, 155, 1431.

36. E. Koldyn, *Fast Reactions in Solutions*, M. Mir, **1977**, 716.

37. *Methods of Studies on Fast Reactions* M. Mir, **1977**, 716.

38. K. F. Pavlov, P. G. Romankov, A. A. Noskov, *Examples and Problems in Processes and Apparatuses of Chemical Technology, L., Khimiya*, **1976**, 552.

39. *Brief Handook of Physicochemical Quantities*, L., Khimiya, **1974**.

40. D. Astarita, *Mass Transfer with Chemical Reaction*, L., Khimiya, **1971**, 296.

41. B. M. Andreev, E. P. Magomedbekov, M. B. Rozenkevich, Yu. A. Sakharovskii, *Heterogenous Reactions of Tritium Isotope Exchange*, Editorial URSS, **1999**, 208.

42. B. M. Andreev, N. A. Rakov, M. B. Rozenkevich, Yu. A. Sakharovskii, *Radiochemistry*, **1997**, 39, 97.

43. Ya. D. Zelvenskii, A. A. Titov, V. A. Shalygin, *Rectification of Diluted Solutions*, Khimiya, **1974**, 216.

44. B. M. Andreev, *Sep. Sci. Technol.*, **2001**, 36, 8–9, 1949.

45. A. M. Rozen. *Atomnaya Energia*, **1995**, 78, 3, 217.

46. B. M. Andreev, S. G. Katalnikov, A. V. Khoroshilov, *Proc. Mendeleev Univ. Chem. Technol. Russia*, **1998**, 171, 89.

47. W. P. Bebbington, V. R. Thayer In: *Heavy Water Production* (ed. G. Murphy *et al.*), Inostrannaya Literatura, 325.

48. J. F. Proctor, *Chem. Engng. Progr.*, **1961**, 59, 47.

49. R. J. Garvin, E. R. Norton, *Chem. Engng. Progr.*, **1968**, 64, 99.

50. H. J. Neuburg, K. T. Chuang, *Can. J. Chem. Engng.*, **1982**, 60, 504.

51. H. J. Neuburg, K. T. Chuang, *Can. J. Chem. Engng.*, **1982**, 60, 510.

52. D. A. Spagnolo, K. T. Chuang, *Ind. Eng. Chem. Process Des. Dev.*, **1984**, 23, 561.

53. D. A. Spagnolo, K. T. Chuang, *Can. J. Chem. Engng.*, **1986**, 64, 62.

54. W. Meier, *Sulzer Tech. Rev.*, **1979**, 2, 49.

55. A. K. Agarwal, In: National Sympos. on Heavy Water Technology, Bombay (India), **1989**, preprint PD-11, 5, Bhabha Atomic Res. Centr, Trombay.

56. D. A. Spagnolo, K. T. Chuang Antifoam for Heavy Water Plant Trays, Atomic Energy of Canada Limited, **1981**, report AECL-7230, 17.

57. D. A. Spagnolo, K. T. Chuang ASME Energy Sources Technology Conference and Exhibition, New Orleans, USA, 12–16 February **1984**, 84-PET-18, 16.

58. C. A. Delfino, E. A. Rojo Antifoaming materials studies in Girdler sulphide heavy water plants, Report CNEA-D-Q-FQ–100 **1988**, 20.

59. J. S. Spevak, U.S./Patent 2787526 (**1957**).

60. G. Weiss In: *Heavy Water Production* (ed. G. Murphy *et al.*), Inostrannaya Literature, 390.

61. J. F. Proctor, V. R. Thayer, *Chem. Engng. Progr.*, **1962**, 58, 53.

62. V. R. Thayer British Pat. 1363757 (**1971**).

63. W. P. Bebbington, W. C. Scotten, J. F. Proctor *et al.* Third United Nations International Conference on the Peaceful Uses of Atomic Energy, United Nations, **1964**, 12, 344.

64. J. S. Spevack In: *Technica ed Economia della Produzione di Acqua Pesante*, Roma, Comitato Naz. Energ. Nucl., **1971**, 23.

65. P. B. Lumb, L. R. Haywood, *Chem. Canada*, **1975**, 27, 19.

66. H. S. Kamath, V. Raghuraman, S. Sharma, In: National Sympos. on Heavy Water Technology, Bombay (India), **1989**, preprint PM-3, 8, Bhabha Atomic Res. Centr, Trombay.

67. R. R. Sonde, H. S. Kamath, R. K. Bhargava, S. Sharma, In: National Sympos. on Heavy Water Technology, Bombay (India), **1989**, preprint PD-2, 7, Bhabha Atomic Res. Centr, Trombay.

68. A. K. Agarwal, A. N. Verma, In: National Sympos. on Heavy Water Technology, Bombay (India), **1989**, preprint PD-12, 8, Bhabha Atomic Res. Centr, Trombay.

69. V. K. Unny, S. C. Hiremath, In: National Sympos. on Heavy Water Technology, Bombay (India), **1989**, preprint SA-3, 5, Bhabha Atomic Res. Centr, Trombay.

70. P. P. Srivastava, In: National Sympos. on Heavy Water Technology, Bombay (India), **1989**, preprint SA-5, 6, Bhabha Atomic Res. Centr, Trombay.

71. V. N. Sastry, J. S. Bhat, In: National Sympos. on Heavy Water Technology, Bombay (India), **1989**, preprint SA-1, 6, Bhabha Atomic Res. Centr, Trombay.

72. J. L. Kilby, *Chem. Eng. Progr.*, **1968**, 64, 6, 49.

73. D. R. Robert, *Chem. Eng. Progr.*, **1968**, 64, 6, 53.

74. T. A. Kletz, *What Went Wrong – Case Histories of Process Plant Disasters*, Gulf Publishing Company, **1985**.

75. A. K. Agarwal, T. Gopalakrishna, In: *Atomnaya Energia*, **1995**, 78, prep. SA-9, 10.

76. C. S. Shrivastava, N. D. Mathur, G. K. Vithal, In: *Atomnaya Energia*, **1995**, 78, prep. CA-8, 6.

77. C. Croitoru, M. Dumitrescu, T. Preda, *et al.*, 25 Years of activity in cryogenics and isotope separation in Romania. Ramnicu Valcea, Romania, **1995**. 185, 184.

78. C. Croitoru, M. Dumitrescu, J. Stefanescu *et al.*, *Chem. Canada*, **1975**, 27, 181.

79. C. Croitoru, G. Titescu, D. Smaranda *et al.*, *Can. J. Chem. Engng.*, **1986**, 64, 178.

80. M. R. Rao, K. R. Prasad, In: National Sympos. on Heavy Water Technology, Bombay (India), **1989**, preprint SA-4, 6, Bhabha Atomic Res. Centr, Trombay.

81. M. C. Agrawal, In: National Sympos. on Heavy Water Technology, Bombay (India), **1989**, preprint PD-13, 6, Bhabha Atomic Res. Centr, Trombay.

82. P. B. Kulkarmi, P. V. Joshi, S. D. Bedekar *Ind. J. Chem. Technol.*, **1997**, 4, 34.

83. H. S. Delvi, R. A. Kini, V. K. Tangri, H. K. Sadhukhan, In: National Sympos. on Heavy Water Technology, Bombay (India), **1989**, preprint MI-1, 17, Bhabha Atomic Res. Centr, Trombay.

84. P. Bruzzoni, A. L. Burkart, R. N. Garavaglia, *Carbon Steel Protection in Girdler Sulphide Plants*, Pt 3, Rep. CNEA-D-Q-FQ–86, 25 pp., April **1985**.

85. A. L. Burkart, *Carbon Steel Protection in Girdler Sulphide Plants*, Pt 4, Rep. CNEA-D-Q-FQ–92, 25 pp., April **1986**.

86. G. Venkateswaran, K. S. Venkateswarly, P. S. Joshi, In: National Sympos. on Heavy Water Technology, Bombay (India), **1989**, preprint MI-5, 6, Bhabha Atomic Res. Centr, Trombay.

87. B. M. Andreev, *Khim. Prom.*, **1999**, 4, 15.

88. B. M. Andreev, S. F. Medovshtchikov, V. V. Frunze, A. I. Shafiev: *Tritium and Environment*, M., TsNIIAtominfirm, **1984**, 65.

89. Y. Clayes, J. C. Dayton, W. K. Wilmarth, *J. Chem. Phys.*, **1950**, 18, 759.

90. K. Bar-Ely, F. S. Klein, *J.Chem. Soc.*, **1962**, 3083.

91. W. K. Wilmarth, J. C. Dayton, *J.Am. Chem. Soc.*, **1953**, 75, 4553.

92. H. K. Rae (ed.) *Separation of Hydrogen Isotopes*, ACS Sympos. series, **1978**, 184, ACS, Washington.

93. E. Roth *et al.*, Report 26 at the Second Conference on Peaceful Use of Nuclear Energy, Geneva, **1958**, 4, P/1261, 499.

94. P. Lamb, *J. Brit. Nucl. Energy Soc.*, **1976**, 15, 35.

95. S. Walter, B. Nitschke, Tecnica ed Economica della Produzione di Acqua Pesante, 173, Comitata Nazionale Energia Nucleare, Rome, **1971**, 35.

96. B. Lefrancois, L'usine d'eau lourde de Mazingarbe description et fonctionnement, ibid, 197.

97. S. Walter, U. Schindewolf, *Chem. Ing. Techn.*, **1965**, 37, 1185.

98. E. Nitschke, H. Igner, S. Walter, In: H. K. Rae (ed.) *Separation of Hydrogen Isotopes*, ACS Sympos. series, **1978**, 77, ACS, Washington.

99. Yu. A. Sakharovskii, In: *Proc. Mendeleev Univ. Chem. Technol. Russia*, **1983**, 89.

100. S. P. Mukherjecc, In: National Sympos. on Heavy Water Technology, Bombay (India), **1989**, preprint PD-3, 8, Bhabha Atomic Res. Centr, Trombay.

101. J. K. Gupta, W. S. A. Kanthiah, S. Sundaresan, preprint MI-4, 5, *ibid.*

102. T. K. Halder, H. Kumar, preprint PD-14, 11, *ibid.*

103. S. K. Gupta, W. S. A. Kantthia, R. C. Rao, Achievments of Heavy Water Plant Tuticorin in Obtaining Sustained Production and Substantial Energy Saving., *ibid.*

104. J. Ravoire, P. Grandcollat, G. Dirian, *J. Chim. Phys.*, **1963**, 60, 130.

105. P. J. Bourke, J. C. Lee, *Trans. Inst. Chem. Eng.*, **1961**, 39, 280.

106. P. V. Danquerts: *Gas–Liquid Reactions*, M., Khimiya, **1973**.

107. Yu. A. Sakharovskiy, M. B. Rozenkevich, A. S. Lobach, *Kinet. Catal. Lett.* **1978**, 8, 249.

108. W. J. Holtslander, W. E. Lockerby, In: H. K. Rae (ed.) *Separation of Hydrogen Isotopes*, ACS Sympos. series, **1978**, 40, ACS, Washington.

109. S. M. Dave, S. K. Ghosh, H. K. Sodhukhan, In: National Sympos. on Heavy Water Technology, Bombay (India), **1989**, preprint PD-8, 10, Bhabha Atomic Res. Centr, Trombay.

110. Yu. A. Sakharovskii, M. G. Slin'ko, *Khim. Prom.*, **1999**, 4, 224.

111. Ya. D. Zelvensky, (ed.), *Heavy Water Production*, Inostrannaya Literatura, **1961**, 518.

112. G. P. Pourot, *Fusion Technol.*, **1988**, 14, 480.

113. R. B. Davidson, P. Von Halten, M. Schaub, D. Ulrich, *Fusion Technol.* **1988**, 14, 472.

114. U. J. Schindewolf, *J. Chem. Phys.*, **1963**, 60, 124.

115. U. J. Schindewolf, *Ber. Bunsenges. Phys. Chem.*, **1963**, 67, 219.

116. O. Piringer, A. Farkas, *Nature (London)*, **1965**, 206, 1040.

117. Yu. A. Sakharovskii, M. B. Rozenkevich, Ya. D. Zelvenskii: *Kinetics and Catalysis*, 15, 1436 (in Russian).

118. T. E. Gogoleva, Yu. A. Sakharovsky, *React. Kinet. Catal. Lett.*, **1985**, 29, 115.

119. D. Yu. Gamburg, N. F. Dubovkina, *Handbook: Hydrogen. Properties, Production, Storage, Transportation, Application.* Khimiya, Moscow, **1989**, 672.

120. M. Hammerly, W. H. Stevens, J. P. Butler *A.C.S. Symp. Ser.*, **1978**, 68, 110.

121. M. L. Rogers, P. H. Lamberger, R. E. Ellis, T. K. Mills *In: Proc. Symp. Separation* of Hydrogen Isotopes, Montreal, **1977**, 171.

122. *Heavy Water Reactors: Status and Projected Development.* IAEA, **2001**.

123. W. H. Stevens, Canadian Pat. 907292, (**1972**).

124. J. H. Rolston, J. den Hartog, J. P. Butler, U.S. Pat. 4025560, (**1977**).

125. J. P. Butler, J. H. Rolston, W. H. Stevens *A.C.S. Symp. Ser.*, **1978**, 68, 93.

126. A. Bruggeman *et al.* Belgian Pat. 884563, (**1980**).

127. A. Bruggeman *et al.* Belgian Pat. 893715, (**1982**).

128. A. Bruggeman *et al.* U.S. Pat. 4376066, (**1983**).

129. Gh. Ionita, M. Peculea, Romanian Pat. 147684, (**1990**).

130. Gh. Ionita, I. Stefanescu, *Fusion Technol.*, **1995**, 28, 641.

131. Gh. Ionita, I. Stefanescu, In: Proc. Conf. 25 Years of Activity of Cryogenic and Isotope Separation in Romania", Romania, **1995**, 16.

132. S. K. Malhotra, M. S. Krishnan, H. K. Sadhukhan, In: Proc. 2$^{nd}$ Nat. Symp. on Heavy Water Technol., Bhabha Atomic Research Centre, **1989**, CJ2.

133. T. Sato, S. Ohkoshi, T. Takahashi, M. Shimizu, In: Proc. 6th Int. Symp. Fresh Water from the Sea, **1978**, 1, 47.

134. H. Izawa, S. Isomura, R. J. Nakane, *Nucl. Sci. Technol.*, **1979**, 16, 741.

135. Y. A. Sakharovsky, M. B. Rozenkevich, I. A. Alekseev *et al.*, RF Pat. 2050801, (**1996**).

136. Y. J. Asakura *Nucl. Sci. Technol.*, **1983**, 20, 422.

137. B. M. Andreev, Y. A. Sakharovsky, M. B. Rozenkevich *et al.*, In: Proceedings of the Third All-Russian Conference on Physical–Chemical Processes in Selection of Atoms and Molecules, Moscow, **1997**, p. 134.

138. B. M. Andreev, E. P. Magomedbekov, Yu. S. Pak *et al.*: *Atomnaya Energiya*, **1999**, 86, 198.

139. V. D. Trenin, I. A. Alekseev, S. D. Bondarenko *et al.*, *Fusion Technol.*, **1998**, 34, 963.

140. I. A. Alekseev, V. D. Trenin *et al.*, In: Proc. 20th Symp. Fusion Techn., Marseille, **1998**, 140.

141. P. V. Kurman, I. P. Mardilovitch, A. I. Trokhimets, *ZhFKh*, **1990**, 64, 711.

142. W. H. Stevens U. S. Pat. 3888974, **(1975)**.

143. I. A. Alekseev, T. V. Vasyanina, V. D. Trenin: Preprint of V. Konstantinov Nuclear Physics Institute **1955**, Gatchina, **1994**.

144. S. K. Malhotra, *et al.* In: Proc. 2$^{nd}$ Nat. Symp. on Heavy Water Techn., Bhabha Atomic Research Centre, **1989**, 3.

145. A. Bruggeman, L. Meynendonckx, C. Parmentier *et al.*, *Radioact. Waste Mngmt Nucl. Fuel Cycle*, **1985**, 6(3–4), 237.

146. L. Geens, A. Bruggeman, L. Meynendonckx *et al.*, Nucl. Sci. Techn. Separation of Tritium from Aqueous Effluents, Final Report, **1988**, EUR 11551 $_{EN}$.

147. Bruggeman A., Leysen R., Meynendonckx L. et al. Nucl. Sci. Techn. Separation of Tritium from Aqueous Effluents, Final Report, **1984**, EUR 9107 $_{EN}$.

148. Yu. A. Sakharovskii, M. V. Karpov, In: Proceedings of the Mendeleev University of Chemical Technology of Russia, **1989**, 156, 45.

149. B. M. Andreev, Y. A. Sakharovsky, M. B. Rozenkevich *et al. Fusion Technol.*, **1995**, 28, 515.

150. C. J. Sienkiewicz, J. E. Lentz *Fusion Technol.*, **1988**, 14, 444.

151. J. P. Butker, J. Hartog U. S. Pat. 4126667, **(1980)**.

152. A. Kitamoto, Y. Takashima, M. Shimizu *Fusion Technol.*, **1985**, 8, 2048.

153. A. Kitamoto, K. Hasegava, T. Masui *Fusion Technol.*, **1988**, 14, 507.

154. S. Isomura, H. Kaetsu, R. J. Nakane, *Nucl. Sci. Technol.*, **1980**, 17, 308.

155. M. Shimizu, S. Kiyota, R. Ninomiya *Bull. Res. Lab. Nucl. React. (Japan)*, **1992**, spec. iss. 1, 56.

156. J. P. Butler, *Sep. Sci. Technol.*, **1980**, 3, 371.

157. Y. A. Sakharovsky, M. B. Rozenkevich, B. M. Andreev et al., Atomnaya Energiya, **1988**, v. 85, p. 35.

158. U. Berndt, E. Kirste, V. Prech, *et al.*, Final Report on Tritium Enrichment Facility, German, KfK 4780, **1991**.

159. Y. Asakura, *J. Nucl. Sci. Technol.*, **1983**, 20, 422.

160. Y. Asakura, S. J. Uchida *Nucl. Sci. Technol.*, **1984**, 21, 381.

161. Yu. A. Sakharovskii, M. B. Rozenkevich, A. R. Korigodskii *et al. Khim. Prom.*, **1999**, 4 (251), 47.

162. A. Bekriaev, A. Markov, O. M. Ivanchuk, M. B. Rozenkevich, In: Proc. Annual Meeting of Nucl. Techn., **1998**, Munih, 337.

163. I. L. Rastunova, M. B. Rozenkevich In: Proceedings of the 5th All-Russian Conference on Physical-Chemical Processes in Selection of Atoms and Molecules, **2000**, 138.

164. I. L. Rastunova, M. B. Rozenkevich: *Khim. Prom.*, **2001**, 44, 4, 36.

165. I. L. Rastunova, M. B. Rozenkevich In: Abstracts of 6th Int. Symp. Fusion Nuclear Techn., **2002**, SWM Pos. 38.

166. A. Perevezentsev, B. M. Andreev, E. P. Magomedbekov *et al. Fusion Sci. Technol.*, **2002**, 41, 1107.

167. A. Perevezentsev, A. Bell, B. M. Andreev *et al. Fusion Sci. and Technol.*, **2002**, 41, 1102.

168. I. L. Rastunova, M. B. Rozenkevich: In: Proceedings of the 6th All-Russian Conference on Physical–Chemical Processes in Selection of Atoms and Molecules, **2001**, 167.

169. A. V. Morozov, M. B. Rozenkevich: *Zhurnal Fizitcheskoi Khimii*, **1990**, 64, 2153.

170. A. V. Morozov, V. I. Porembskii, M. B. Rozenkevich, V. N. Fateev: *Zhurnal Fizitcheskoi Khimii*, **1990**, 64, 3075.

171. A. V. Morozov, M. B. Rozenkevich: *Zhurnal Fizitcheskoi Khimii*, **1990**, 64, 2761.

172. A. V. Morozov, M. B. Rozenkevich, Y. A. Sakharovsky, In: Proc. of the 11th World Hydr. Energy Conf., Stuttgart, **1996**, 3, 2657.

173. Yu. A. Sakharovsky, D. M. Nikitin: In: Proceedings of the 6th All-Russian Conference on Physical–Chemical Processes in Selection of Atoms and Molecules, **2001**, 150.

174. O. A. Fedorchenko, I. A. Alekseev, V. D. Trenin, V. V. Uborski, *Fusion Technol.*, **1995**, 28, 1485.

175. M. J. Shimizu *Nucl. Sci. Technol.*, **1982**, 19, 307.

176. Y. J. Asakura *Nucl. Sci. Technol.*, **1983**, 20, 64.

177. T. Yamanishi, K. Okuno, *Fusion Technol.*, **1995**, 28, 1597.

178. A. Kitamoto, M. Shimizu, T. Masui, In: Proc. Int. Symp. Isot. Separation and Chem. Exchange Uranium Enrichment, Tokyo, **1992**, 497.

179. M. Benedict *et al.*, *Nuclear Chemical Engineering*, 2^nd ed., McGraw-Hill, N.Y., **1981**, 565.

180. D. Spagnolo, A. Miller, *Fusion Technol.*, **1995**, 28, 748.

181. K. Takeshita, Y. Wei, M. Shimizu *et al.*, *Fusion Technol.*, **1995**, 28, 1572.

182. R. E. Ellis, J. E. Lentz, M. L. Rogers, C. J. Sienkiewicz, Final Report, Development of combined electrolysis catalytic exchange, MLM-2952, **1982**.

183. H. J. Fick, J. Romaker, U. Schindewolf, *Chem.-Ing.-Techn.*, **1980**, 52, 892.

184. L. F. Belovodskii, V. K. Gaevoy, V. I. Grishmanovskii: *Tritium,* M., Energoatomizdat, **1985**, 248.

185. Y. Z. Wei, In: Proc. 6th Int. Symp. on Catalytic Deactivation, Ostend, **1994**, 609.

186. Y. Z. Wei, K. Takeshita, M. Shimizu *et al.*, *Fusion Technol.*, **1995**, 28, 1585.

187. R. L. LeRoy, M. Hammerly, J. P. Butler U. S. Pat. 4225402, (**1980**).

188. M. Hammerly, J. P. Butler U. S. Pat. 4191626, (**1980**).

189. J. P. Butler, M. Hammerly U. S. Pat. 4190515, (**1980**).

190. M. J. Song, S. H. Son, C. H. Jang *Waste Mgmt.*, **1995**, 15(8), 593.

191. T. Kitabata, K. Kitamura, In: Proc. 19th KAIF–JAIF Seminar of Nuclear Industry, Seoul, **1997**, 181.

192. H. Yamai, S. Konishi, M. Hara *et al.*, *Fusion Technol.*, **1995**, 28, 1591.

193. M. Hammerly *Int. J. Hydrogen Energy*, **1983**, 8, 269.

194. A. I. Miller, D. A. Spagnolo, J. R. DeVore, *Nucl. Technol.*, **1995**, 112, 204.

195. V. D. Trenin, I. A. Alekseev, S. P. Karpov *et al.*, *Fusion Technol.*, **1995**, 28, 767.

196. B. M. Andreev, E. P. Magomedbekov, Yu. S. Pak *et al.*, *Radiokhimiya*, **1999**, 14, 131.

197. B. M. Andreev, M. V. Karpov, A. N. Perevezentsev *et al.*, *Hydrogen Energetics Technol.*, **1992**, 1, 57.

198. B. M. Andreev, N. A. Rakov, M. B. Rozenkevich, Yu. A. Sakharovskii: *Radiokhimiya*, **1997**, 39, 97.

199. B. M. Andreev, E. P.Magomedbekov, Yu. S. Pak *et al.*, *Atomnaya Energiya*, **1998**, 85, 40.

200. C. J. Allan, A. R. Bennett, C. A. Fahey *et al.*, In: Preprints of 12th Pacific Basin Nuclear Conf., Seoul, Korea, **2000**, 12.

201. J. M. Miller, S. L. Celovsky, A. E. Everatt *et al.*, In: Preprints of 6th Conf. Tritium Science and Technol., Tsecuba, Ibaraki, Japan, **2002**, 5.

202. W. R. C. Graham, A. E. Everatt, J. R. R. Tremblay *et al.*, In: Preprints of 6th Conf. on Tritium Science and Technol., Tsecuba, Ibaraki, Japan, **2002**, 5.

203. Yu. A. Sakharovskii, B. M. Andreev, E. P. Magomedbekov *et al.*, theses of the Report at the 2nd Russian Conference on Radiochemistry, Dimitrivgrad, **1997**, 174.

204. Proc. Conf. 25 Years of Activity in Cryogenic and Isotope Separation in Romania", **1995**, 217.

205. L. J. Nuttall, J. H. Russell, *Int. J. Hydrogen Energy*, **1980**, 5, 75.

206. T. K. Mills, R. E. Ellis, M. L. Rogers, In: Proc. Conf. Tritium Techn. in Fission, Fusion and Isotopic Applications, Dayton, **1980**, 422.

207. R. Haange, H. Yoshida, O. K. Kveton *et al.*, *Fusion Technol.*, **1995**, 28, 491.

208. O. K. Kveton, H. Yoshida, J. E. Koonce *et al.*, *Fusion Technol.*, **1995**, 28, 636.

209. T. Yamanishi, K. Okuno, *Fusion Technol.*, **1995**, 28, 1597.

210. B. M. Andreev, A. N. Perevezentsev, V. L. Zverev *et al.*, INTER – 1L- FG – 0-9-10, **1989**.

211. B. M. Andreev, A. N. Perevezentsev, I. L. Selivanenko *et al.*, INTER – 1L- FG – 9-0-7, **1990**.

212. V. K. Kapyshev, M. V. Karpov, L. A. Rivkis *et al.*, ITER–FG–9.1-0-36, **1990**.

213. B. M. Andreev, Z. V. Ershova, M. B. Rozenkevich, *Problems of Nuclear Science and Technology*, **1990**, 2, 55.

214. B. M. Andreev, Y. A. Sakharovsky, M. B. Rozenkevich *et al.*, *Fusion Technol.*, **1995**, 28, 511.

215. B. M. Andreev, A. N. Pereventsev, I. L. Selivanenko *et al.*, *Fusion Technol.*, **1995**, 28, 505.

216. Gh. Titescu, S. Predescu In: Proc. Nat. Physics Conf., Bucharest, **1995**, 82.

217. W. Doenitz, R. Schmidberger, E. Steinheil, *Int. J. Hydrogen Energy*, **1979**, 5, 55.

218. S. Konishi, H. Yoshida, H. Ohno *et al.*, *Fusion Technol.*, **1985**, 8, 2042.

# – 4 –

# Isotope Separation in Systems with Gas and Solid Phases

---

## 4.1   ISOTOPE EQUILIBRIUM

### 4.1.1   Chemical isotope exchange reactions

Noticeable thermodynamic isotope effects observed in systems with a solid phase can be determined both by the reaction of chemical isotope exchange between two different substances and by physical sorbtion leading to the thermodynamic inequality of isotopes in the gas molecules and in the same molecules fixed by the solid phase [1–4]. The former case, to which the present chapter is devoted, is characteristic for hydrogen isotope exchange in systems involving molecular hydrogen and hydride phases of transition metals or their intermetallic compounds (IMCs). Systems with hydrides of alkaline and alkaliearth elements where significant thermodynamic isotope effects are also observed, are characterized by a very low isotope exchange rate and, thus, are of no practical interest for hydrogen isotope separation.

Irrespective of the isotope effect nature, both the ways of the counter-current isotope separation in the systems with solid phase, and main factors determining the separation plant efficiency remain common.

As follows from eqs. (1.21) and (1.22), the isotope effect in the systems under study that sorb reversibly gas is associated with the difference in isotherms of sorption of gas molecules, varying in isotope composition, by the solid phase.

The process of metal or IMC hydride phase formation is generally accompanied by the dissociation of hydrogen molecules on the surface and implantation of atoms into the crystal lattice interstices. As this takes place, on the surface both of metals and of intermetallic compounds, homomolecular isotope exchange reaction occurs with a rate several times as high as that of the interphase isotope exchange between hydrogen of the gas and solid phases [5]. In this case, a value of one separation factor would suffice to describe the isotope equilibrium in a binary mixture of hydrogen isotopes A and B (e.g. H and D). Although the gas phase may comprise three molecule species $A_2$, AB, and $B_2$ (e.g. $H_2$, HD, and $D_2$), their concentrations are bound by the equilibrium constant ($K_{AB}$) of the homomolecular isotope exchange reaction: $A_2 + B_2 = 2\,HD$ (e.g. $H_2 + D_2 = 2HD$). A significant departure of the equilibrium constant $K_{AB}$ from the value corresponding to the equiprobable distribution of hydrogen isotopes between all hydrogen molecule species ($K_{AB}^{\alpha} = 4$) results in a concentration dependence of the separation factor [6]. The mathematical formulation of the dependence is simplified because the solute hydrogen is dissociated into atoms, and the homomolecular isotope exchange reaction proceeds only in the gas phase [7]:

$$\alpha_{A-B} = a_{AB} \frac{1 + 2\dfrac{[A_2]}{[AB]}}{\dfrac{4}{K_{AB}} + 2\dfrac{[A_2]}{[AB]}}, \tag{4.1}$$

where $[A_2]$ and $[AB]$ are concentrations of molecules in the gas phase; $\alpha_{AB}$ is the separation factor over the region of the heavy isotope B low concentration. The concentration ratio $[A_2]/[AB]$ is obtained from an expression similar to eq. (125).

According to the separation factor definition by eq. (1.1), the value $\alpha > 1$ corresponds to the case when the heavy isotope is concentrated in the solid phase (hydrogen atoms occupy tetrahedral interstices of the crystal lattice).

In conformity with eq. (4.1) pertaining to a system where the solid phase is enriched with heavy isotope, the separation factor decreases with an increase in the heavy isotope concentration. The separation factor limiting values are related by the equation $\alpha_{BA} = \alpha_{AB} \cdot K_{AB}/4$. Inversely, when the heavy isotope is concentrated in the gas phase, isotope effect increases with a rise in the isotope concentration. But in this case the concentration dependence can also be calculated by eq. (4.1). A similar dependence results as well from the expression reported [8]:

$$\alpha_{A-B} = \alpha_{AB} \frac{1 + (1-y)/y\alpha_{AB}}{\dfrac{4}{K_{AB}} + (1-y)/y\alpha_{AB}}. \tag{4.2}$$

The above equation is convenient in use since there is no need to solve the quadratic equation (3.125) for the determination of the $\alpha$ concentration dependence.

In accordance with the separation factor definition, the known $\alpha$ values for two isotope pairs allow the determination of the isotope effect for the third pair. For homomolecular isotope exchange reactions, the above relation is expressed [6, 7] by $\alpha_{HT} = \alpha_{HD}\alpha_{DT}(4K_{HT}/K_{HD}K_{DT})^{1/2}$.

An experimental study of $\alpha$ versus concentration dependence was done for the H–D mixture in the $H_2$–Pd [7] and $H_2$–LaNi$_5$ [9, 10] systems characterized by the negative isotope effect. In addition, Figure 4.1 shows the experimental $\alpha_{H-T}$ values obtained at equilibrium deuterium concentration in the gas phase equal to 73 and 89at.% at a temperature of 293K [11]. As can be seen from Figure 4.1, the experimental values agree well with those obtained from the calculation by eq. (4.1).

As is evident from Figure 4.1, the dependence for the H–T mixture is sharper than that for the H–D mixture. With a decrease in temperature, the $\alpha_{A-B}$ dependence on the isotope composition of phases becomes stronger. This is due to the fact that, for one thing, $K_{HT} < K_{HD}$ in

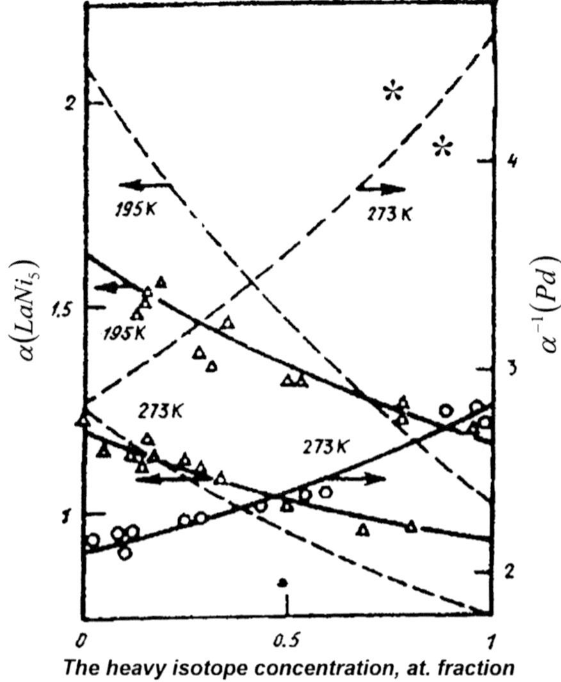


**Figure 4.1**   Dependence of separation factor for H–D (solid lines) and H–T (dotted lines) mixtures in $H_2$–$LaNi_5$ and $H_2$–Pd Systems on gas phase composition.

all cases, and for another, $K_{HT}$ and $K_{HD}$ go down with a decrease in temperature ($K_{HT}$ = 2.43 and 1.99, $K_{HD}$ = 3.19 and 2.87, at 273 and 195K, respectively).

A comparison between $\alpha_{A-B}$ versus concentration dependences for $LaNi_5$ and Pd demonstrates that in the case when the heavy isotope is concentrated in the solid phase, with a rise in the isotope concentration $\alpha$ diminishes approaching unity with possible isotope effect inversion (e.g. for $H_2$–$LaNi_5$ mixture at 273K). In the case when the heavy isotope is concentrated in the gas phase (e.g. for Pd), an increase in concentration results in a significant isotope effect enhancement, which constitutes an important advantage of such system (first of all, $H_2$–Pd system) when used at the stage of the heavy isotope final concentration. A similar concentration dependence was obtained [12] for the separation factor $\alpha_{H-D}$ at equilibrium with the hydride phase of uranium at a temperature of 500–600K. In this system, the heavy isotope is as well concentrated in the gas phase, but with a considerably lower isotope effect.

In IMC and in some metals, the hydrogen atoms can occupy both tetrahedral and octahedral interstices of the crystal lattice which complicates the $\alpha$ versus concentration dependence [13].

It has been demonstrated [7, 9] that the experimental $\alpha$ values obtained over a wide temperature range (195–323K) agree moderately with those calculated by the quantum-statistical method using a three-dimensional harmonic oscillator model. Most of the known data relate to the range of heavy isotope low concentration for which $\alpha_{AB}$ can be determined as

$$\alpha_{AB} = \frac{\varepsilon_{A_2}}{\varepsilon_{AB}} \frac{\left[1-\exp\left(-\varepsilon_A/kT\right)\right]^3 / \left[1-\exp\left(-\varepsilon_B/kT\right)\right]^3}{\left[1-\exp\left(-\varepsilon_{A_2}/kT\right)\right] / \left[1-\exp\left(-\varepsilon_{AB}/kT\right)\right]} \frac{\exp\left[1.5/kT(\varepsilon_A-\varepsilon_B)\right]}{\exp\left[0.5/kT\left(\varepsilon_{A_2}-\varepsilon_{AB}\right)\right]}, \qquad (4.3)$$

where $\varepsilon_A = \hbar\omega_A$; $\varepsilon_B = \hbar\omega_B$; $\varepsilon_{A_2}=\hbar\omega_{A_2}$; $\varepsilon_{AB} = \hbar\omega_{AB}$; $\omega_A$, $\omega_B$ is the oscillation frequency of hydrogen isotopes in the metal crystal lattice; $\omega_{A_2}$ and $\omega_{AB}$ are the frequencies of local modes of isotope species of the hydrogen molecules.

The first factor is constant, and the other two depend on the temperature. The temperature dependence of the second multiplier must be taken into account only at elevated temperatures (at lower temperatures the multiplier is equal to zero). The temperature dependence of $\alpha_{AB}$, therefore, will be determined by the third multiplier reflecting the impact of zero-point energy change $\Delta E_0$ of hydrogen atoms and molecules upon $\alpha_{AB}$ during the isotope exchange reaction. This makes it possible to simplify eq. (4.3) and write it as the following temperature dependence of $\alpha$:

$$\ln\alpha_{AB} = a + b/T. \qquad (4.4)$$

The constants $a$ and $b$ can be determined from the known relation between the interphase isotope exchange equilibrium constant ($K$) and the isobaric–isothermal potential change $\Delta G = -RT\ln K$

$$-RT\ln K = \Delta H_{AB} - T\Delta S_{AB};$$

$$\ln\alpha_{AB} = \ln 2K = \ln 2 + \frac{\Delta S_{AB}}{R} - \frac{\Delta H_{AB}}{RT},$$

where $\Delta H_{AB}$ and $\Delta S_{AB}$ are the enthalpy and entropy changes during the reaction.

Thus, if the experimental dependence is described by eq. (4.4), it is possible to determine the thermodynamic parameters of the interphase isotope exchange reaction.

An analysis of the harmonic oscillator model [14] has shown that the sharpest temperature dependence is observed in metals and IMC with local mode frequencies of hydrogen atoms over 150MeV, which generally corresponds to the hydrogen atoms localization in the tetrahedral interstices. An important temperature dependence is also observed when $\omega_H < 75$MeV. This region corresponds to the hydrogen atoms localization in the octahedral interstices. A more complicated pattern occurs in the transition range from 100 to 130MeV. According to the model, the dependence can be anomalous at a low temperature, that is, $\alpha$ increases with a rise in temperature.

Figure 4.2 presents the temperature dependences of $\alpha_{HD}$ and $\alpha_{HT}$ in the region of the low heavy isotope content calculated by eq. (4.3). The same figure shows the experimental values of $\alpha$ for some metals and IMC.

To account for the deviations from the harmonic oscillator model, a number of models was suggested: a two-parameters model [8, 15] (utilized for the isotope effect determination in the hydrogen–metal system), a three-parameters model [16] , a variational model

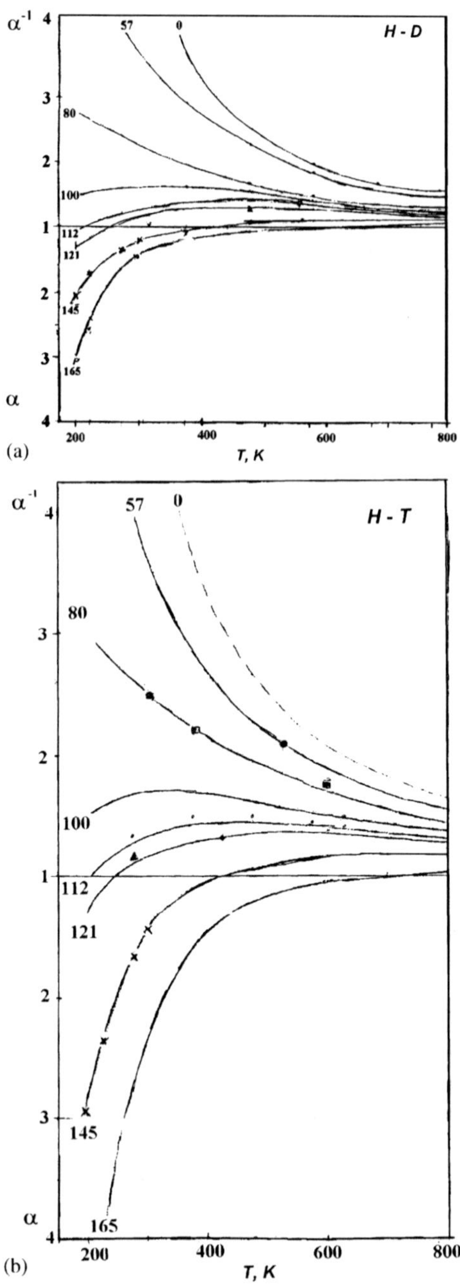



**Figure 4.2**   Temperature dependences of separation factor $\alpha_{HD}$ at different mode energies of hydrogen atom in hydride phase: ●, $Mg_2NiH_4$; ■, $TiH_2$; ▼, $UH_3$; ▲, $ZrCoH_2$; ×, $TiMn_{1.5}H_{2.5}$.

**Table 4.1**

Values of constants $a$ and $b$ of the equation for the temperature dependence of the separation factor in the systems comprising hydrogen and $\beta$-phases of metals or IMC

| Metal (IMC) | Temperature, K | H–T $a$ | H–T $b$ | H–D $a$ | H–D $b$ | Reference |
|---|---|---|---|---|---|---|
| Pd[a] | 175–330 | −0.30 | −284 | −0.023 | −203 | [7] |
| Ti | 296–663 | −0.31 | −178 | — | — | [22] |
| $TiMn_{1.5}$ | 195–296 | −1.02 | 20 | −0.75 | 295 | [23] |
| TiCrMn | 195–293 | −0.96 | 414 | −0.75 | 293 | [24] |
| $Ti_{0.8}Zr_{0.2}Cr_{1.8}$ | 195–293 | −0.99 | 416 | −0.75 | 295 | [24] |
| $LaNi_5$ | 223–323 | −1.08 | 357 | −0.53 | 193 | [9] |
| $ZrV_2$ | 250–373 | −0.88 | 394 | — | — | [20] |
| $ZrCr_2$ | 273–373 | −1.90 | 740 | −1.5 | 580 | [25] |
| $ZrMn_2$ | 239–323 | −1.02 | 424 | −0.87 | 332 | [20] |
| ZrVMn | 250–323 | −1.17 | 446 | — | — | [20] |
| $ZrMn_{2.8}$ | 239–323 | −0.81 | 337 | −0.74 | 273 | [20] |
| $ZrMn_2Cr_{0.8}$ | 250–323 | −1.06 | 440 | −0.88 | 349 | [20] |
| $ZrMn_{3.8}$ | 240–323 | −0.72 | 309 | −0.79 | 283 | [20] |

[a] Heavy isotope is concentrated in the gas phase, and the equation gives $\alpha^{-1}$.

[17] (for the calculation of the temperature dependence of $\alpha$ in the $H_2$–Pd system), as well as an averaged effective frequency model (for IMC where hydrogen atoms are capable of occupying various interstices in the crystal lattice [18]).

A considerable amount of data available in the literature on $\alpha$ for metals and IMC of $AB$, $AB_2$ and $AB_5$ types, relating generally to the region of trace tritium concentrations and comparatively low deuterium content (up to 50at.%), is generalized in monographs [19, 20] and in study [21]. The comparison presented by the authors demonstrates that the maximum $\alpha$ values can be found for compounds of the Laves phases $AB_2$. For compounds of $AB_2$ and $AB_5$ types, the $\alpha$ values will be lower due to a higher probability of the occupation of octahedral interstices. The values of constants $a$ and $b$ involved in eq. (4.4) for the metals and IMC studied over a relatively wide temperature range are given in Table 4.1.

On the basis of the data obtained for metal hydrides and IMCs of the third period elements, the authors [26] have attempted to relate the $\alpha_{HD}$ value to the number of valence electrons per metal atom ($e/Me$ value). Since the comparison was done with $\alpha_{HD}$ values obtained at various temperatures, the data have been recalculated [27], using the harmonic oscillator model, for the temperatures 173 and 273K. Figure 4.3 represents the dependences $\alpha_{HT} = f(e/Me)$, and the dependence of the hydride formation heat $\Delta H_H = f(e/Me)$. As is evident, the dependences follow an opposite pattern and have an extreme kind at the same range of $e/Me$ values.

Of practical interest are the systems with hydride phases, containing considerable amounts of hydrogen. Because of this, the above data pertain to the $\beta$-phases of metals and IMC, which allows the pressure dependence of $\alpha$ to be neglected.

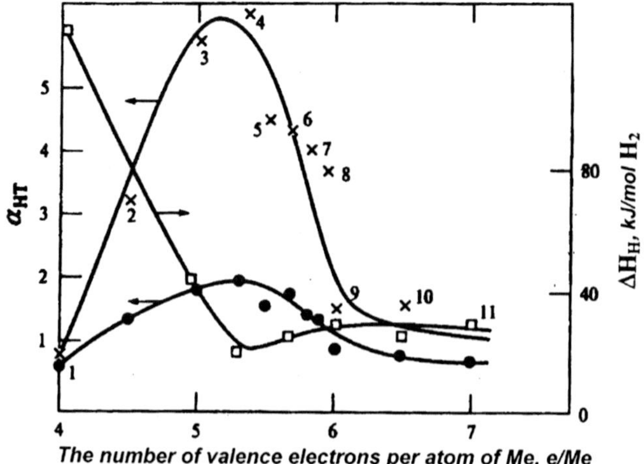



**Figure 4.3**   Dependence of separation factor $\alpha_{HT}$ and heat of hydride formation $\Delta H_H$ on number of valence electrons *e/Me* per atom of metal or IMC: $\square$, $\Delta H_H$; $\times$, $\alpha_{HT}$ at 173K; $\bullet$, $\alpha_{HT}$ at 273K; 1, Ti; 2, TiV; 3, TiCr; 4, TiCr$_2$; 5, TiMn; 6, TiCrMn; 7, TiMn$_{1.5}$; 8, TiMn$_{1.4}$Ni$_{0.1}$; 9, TiFe; 10, TiCo; 11, TiNi.

## 4.1.2   Phase isotope exchange

The isotope separation processes based on the physical gas sorption by zeolites, silica gels, activated carbons, and other sorbents, are of interest only at low temperatures when the solid phase contains sufficiently large amounts of sorbed gas.

In the physical sorption, irrespective of the sorbent and sorbed gas (sorbate) nature, the process is not accompanied by the dissociation of sorbate molecules, the homomolecular isotope exchange (HMEX) reaction does not occur, and the $\alpha$ versus concentration dependence discussed above is not observed. Another distinction of the sorption systems is a milder isotope effect compared with chemical isotope exchange reactions.

The highest values of the separation factor are seen with zeolites serving as the solid phase [28, 29]. Because of this, the zeolites will receive primary consideration.

*Separation factors of hydrogen isotopes*

For the binary isotopic mixture (A and B) in the general case, consideration should be given to the sorption of three hydrogen molecule species (A$_2$, AB, and B$_2$) and, therefore, to three separation factors $\alpha_{A_2-AB}$, $\alpha_{AB-B_2}$ and $\alpha_{A_2-B_2}$ which are related by

$$\alpha_{A_2-B_2} = \alpha_{A_2-AB}\alpha_{AB-B_2} . \tag{4.5}$$

The relation between two last-named separation factors can, in turn, be expressed [27] as:

$$\alpha_{AB-B_2} = \alpha_{A_2-AB}\frac{K_{AB}}{K_{AB}^S}, \tag{4.6}$$

where $K_{AB}^S$ is the HMEX constant of hydrogen sorbed by the solid phase.

The HMEX constant can depend on the substance phase state. The HMEX reaction constant for water molecules ($H_2O + D_2O \Leftrightarrow 2HDO$) in the condensed state, for example, is smaller than that in the vaporous state. This results in that $\alpha$ somewhat increases with a rise in the heavy isotope concentration during the water rectification. It seems likely that the same pattern is also observed for zeolites which, according to the experimental data given below, sorbs symmetric molecules (e.g. $D_2$ and $T_2$) better than asymmetric ones, that is $\alpha_{AB-B_2} > \alpha_{A_2-AB}$ (hence $K_{AB}^{TB} < K_{AB}$). Since in the sorption systems under consideration, the HMEX reaction does not occur in any phase (gas or solid), in practice only limiting $\alpha$ values can be realized, and the determination of any two separation factors involved in eq. (4.5) will suffice to fully describe the isotope equilibrium in the A–B binary isotopic mixture. In addition, it is well to bear in mind that for the enrichment over a wide isotope concentration range (e.g. for the extraction of $D_2$ from the natural isotopic mixture containing $H_2$ and HD molecules) it is necessary to perform the HMEX reaction.

The pattern discussed above is complicated by the fact that, unlike metals and IMC, the sorption systems are characterized by the selectivity of solid phase toward spin isomers of hydrogen (e.g. over zeolite NaX at 77.5K, $\alpha_{n-H_2-o-H_2} = 1.28 \pm 0.09$, and over zeolite NaA at 50K $\alpha_{n-H_2-o-H_2} = 1.68$ and $\alpha_{n-D_2-o-D_2} = 1.51$) [27]. This is why the separation factor of isotopic mixtures containing symmetric molecules is affected by the ortho–para composition of the mixtures. Seventy five percent of normal hydrogen (protium) and tritium, and 66.7% of deuterium fall on ortho-modifications. With a decrease in temperature, the para-modifications become more stable for protium and tritium, and the ortho-modifications for deuterium. It should be noted, that for tritium and deuterium the difference between normal and equilibrium ortho–para compositions become significant only at a temperature of under 50K. As for tritium, its equilibrium content of o-$H_2$ drops to 50% even at 75K.

In the isotope exchange of $H_2$ molecules with any other hydrogen molecules which are always heavier and better sorbed, the use of sorbents with agents catalyzing the hydrogen o–p conversion will lead to an increase in $\alpha$, especially at temperatures under 75K. Conversely, in the exchange between $T_2$ molecules and asymmetric molecules, the o–p conversion will depress $\alpha$, especially at $T < 50K$. And finally, in the exchange between $D_2$ molecules and asymmetric molecules, the o–p conversion will affect only slightly the isotope effect since a decrease in temperature even from 273 to 20K will lead to a rise in the number of o-$D_2$ molecules, which are sorbed worse, only from 66.7 to 80%. The largest isotope effects are observed in systems with such microporous sorbents as synthetic zeolites. Apart from the zeolite properties, the separation factor is influenced by the amount of adsorbed hydrogen (pressure) and by temperature. Most experimental data on the isotope effects have been obtained with the use of the A and X zeolites at T $\approx$ 77K and $P >$ 0.1MPa, that is with a large amount of hydrogen per sorbent volume, when $\alpha$ does not depend on the pressure (in the separation of $H_2$–HT, $D_2$–DT and $H_2$–HD–$D_2$ mixtures over NaA and NaX zeolites, within the pressure range 0.1 – 2.7MPa, the separation factor values remaining constant [30]).

The A zeolites are characterized by the highest $\alpha$ values, while X zeolites have a large hydrogen capacity (see Table 4.2).

Important isotope effects were found at temperatures under 77K for the $H_2$–HD, HD–$D_2$, and $H_2$–$D_2$ mixtures [38, 40], as well as for the $H_2$–HT and $D_2$–DT mixtures [35, 36]. The highest $\alpha$ values obtained over NaA and NaX zeolites are presented in Table 4.3, and Table 4.4 gives the coefficients $a$ and $b$ involved in the equation of the temperature dependence of $\alpha$

$$\ln \alpha = a + b/T. \tag{4.7}$$

Owing to the ferric oxide admixture, the reaction of hydrogen OP conversion can proceed on Russian synthetic zeolites, which, as noted above, should increase $\alpha$ in the exchange between $H_2$ and asymmetric molecules. Indeed, from the data presented in Table 4.3 it follows that at a low temperature the sharpest increase in $\alpha$ is seen in the $H_2$–HD and $H_2$–HT exchange reactions, and a relatively slight increase is observed in the HD–$D_2$ and $D_2$–DT exchange. If at 77K $\alpha_{HD-D_2} > \alpha_{H_2-HD}$ at all times, then at 48K, as can be seen from the Table 4.3, these coefficients are near-equal, and at 35K (when the $o-H_2$ concentration exceeds 90%) $\alpha_{HD-D_2} > \alpha_{H_2-HD}$.

It has been demonstrated [35, 36] that for the $H_2$–HT mixture over the NaX zeolite, the temperature dependence of $\alpha$ is not described by a single straight line in $\ln \alpha - 1/T$ coordinates. If over a range from 77 to 90K, where the OP hydrogen composition changes only by 7%, the experimental data fit into the straight line corresponding to the isotope

**Table 4.2**

Separation factors for hydrogen isotope species at 77.6K over NaA and NaX zeolites

| Isotope mixture | | $H_2$–HD | HD–$D_2$ | $H_2$–$D_2$ | $H_2$–HT | $D_2$–DT |
|---|---|---|---|---|---|---|
| $\alpha$ over zeolite | NaA[a] | 1.43 [31, 32] 1.47 [33] | 1.70 [31, 33] | 2.46 [31, 32] 2.50 [33] 2.55 [34] | 1.80 [30] 1.84 [35, 36] | 1.29 [35, 36] |
| | NaX[b] | 1.32 [31, 37] | 1.51 [31, 37] | 2.10 [38] 2.12 [30] | 1.95 [38] 1.70 [30] | 1.14 [38] |

[a] At $T = 77$K and $P = 0.1$MPa, the hydrogen capacity is about 90ncm$^3$/g [37].
[b] At $T = 77$K and $P = 0.1$MPa, the hydrogen capacity is 120ncm$^3$/g [39].

**Table 4.3**

The most important isotope effects over NaA and NaX zeolites

| Zeolite | Temperature, K | Mixture separation factor | | | |
|---|---|---|---|---|---|
| | | $H_2$–HD | HD–$D_2$ | $H_2$–HT | $D_2$–DT |
| NaA | 22.5 | | | 69±7 | 4.3±0.1[a] |
| | 48 | 2.5±0.2 | 2.6±0.2 | 4.7±0.2 | 1.85±0.06 |
| NaX | 35 | 3.7±0.2 | 2.9±0.2 | | |

[a] At $T = 25.2$K.

**Table 4.4**

Coefficients $a$ and $b$ in (4.7) for NaA and NaX zeolites

| Zeolite | Isotope mixture | Temperature range, K | Coefficient | |
|---|---|---|---|---|
| | | | $a$ | $b$ |
| NaA | $H_2$–HD | 48–60 | $-0.83$ | 85.3 |
| | HD–$D_2$ | 48–60 | 0.32 | 29.5 |
| | $H_2$–HT | 22–77 | $-0.95$ | 120.1 |
| | $D_2$–DT | 25–137 | $-0.27$ | 43 |
| NaX | $H_2$–HD | 35–60 | $-0.44$ | 51.7 |
| | HD–$D_2$ | 35–60 | $-0.25$ | 30.3 |

exchange reaction heat equal to 430J/mol, with a drop in temperature from 77 to 22K when the n-$H_2$ concentration enhances from 50 to 99%, the straight line slope angle grows almost twofold (the coefficients of (4.7) are given in Table 4.4). A similar pattern is seen in the $H_2$–HD isotope exchange over NaA and NaX zeolites: a weak temperature dependence of $\alpha$ over the 77–99K range, and a much stronger dependence, also described by eq. (4.7), at a lower temperature. The coefficients of eq. (4.7) for the NaA zeolite in the temperature range from 48 to 60K, and for NaX zeolite in the range from 35 to 60K are as well presented in Table 4.4 [33].

Nowadays the hydrogen cryogenic rectification method is used the most extensively for the separation of hydrogen isotope mixtures with a high tritium content. Let us compare, then, the separation factors obtained in the reactions on zeolites with those for the hydrogen rectification that are equal to 1.85 for the $H_2$–HT mixture at 20–23K, and to 1.20 for the $D_2$–DT mixture. The same mixtures over the NaX zeolite at 22.5K provide the separation factor values equal to 69 and 4.3, respectively (see Table 4.3). Even at a much higher temperature of 77.6K, the separation factor values on the NaX zeolite compare well with the above $\alpha$ values for the hydrogen rectification.

The same pattern is observed in the separation of the $H_2$–HD isotopic mixture: the rectification at 20–23K is characterized by a separation factor of 1.51, while the isotope effect on the NaX zeolite is much larger even at 35K (see Table 4.3).

The hydrogen separation factor on zeolites, therefore, is considerably stronger than that in the hydrogen cryogenic rectification that requires very complex equipment and high energy inputs.

*Isotope effects for carbon, nitrogen and oxygen*

The systems discussed below are free of the concentration dependence of the separation factor irrespective of the number of atoms of element under consideration in the sorbed molecule. This is because for the molecules containing more than two exchanging atoms of any element other than hydrogen, the HMEX constants are practically equal to limiting values conforming to the equiprobable distribution of isotopes. In the sorption of nitrogen molecules, for example, $\alpha_{^{14}N_2-^{14}N^{15}N} = \alpha_{^{14}N^{15}N-^{15}N_2}$ since the equilibrium constant $K$ of the

HMEX reaction $^{14}N_2 + {}^{15}N_2 = 2{}^{14}N^{15}N$ approximates $K^\infty = 4$. It is obvious that in this case the enrichment from the natural composition can be carried out to no more than 50at.% of the target isotope, as opposed to the HMEX systems or systems comprising molecules with a single exchanging atom (e.g. $^{12}CH_4$–$^{13}CH_4$, $^{14}NO$–$^{15}NO$) without conversion unit (to perform HMEX reactions).

In the systems under study, the isotope effects increase with a drop in temperature, but the lowest temperature threshold is limited by the danger of the gas condensation in the solid sorbent pores. The highest $\alpha$ values are seen in the systems with synthetic zeolites, just as in the hydrogen sorption.

Table 4.5 presents the separation factors of carbon, nitrogen, and oxygen isotopes on the most efficient zeolites NaX, NaA, and CaA. The isotope effects observed over several other zeolites are practically coincident with those given in Table 4.5. In the methane sorption, for one, the separation factor seen over the NaA and CaA zeolites coincides, within the limits of experimental error ($\pm 0.02$), with the value presented in Table 4.5. In the molecular nitrogen sorption over the NaX and CaA zeolites (with a capacity of about $130 ncm^3 N_2/g$), the isotope effects are as well identical, while over silica gel with its higher gas capacity (about $200 ncm^3 N_2/g$), the isotope effect is much lower (at 78K, $\alpha = 1.008$ [50]).

In none of the systems presented in Table 4.5 has the pressure influence on the separation factor been revealed, with the pressure influence (within the range from 4 to 95kPa) studied in the greatest depth in the molecular nitrogen sorption on the NaX zeolite [46] and on the CaA zeolite [45].

The $\alpha$ dependence on the temperature over the 143–273K range was studied in the methane sorption on zeolite (the separation factor varied from 1.050 to 1.011) and on the carbon oxide (attempts to reveal a selectivity in the $^{12}CO$–$^{13}CO$ mixture sorption over the investigated temperature range have not met with success).

**Table 4.5**

Experimental values of separation factor for the isotopes of light elements [41]

| Separated isotope mixture | Sorbent | $T$, K | $\alpha$ | Rectification | |
| --- | --- | --- | --- | --- | --- |
| | | | | $T$, K | $\alpha$ [49] |
| $^{12}CH_4$–$^{13}CH_4$ | NaX | 145 | 1.050 [42] | 81.6 | 1.0067[a] |
| | | 193 | 1.035 [42] | | |
| $^{14}N_2$–$^{14}N^{15}N$ | NaX | 77.3 | 1.021 [43] | 78 | 1.004 |
| | | 78 | 1.016 [44] | | |
| | CaA | 78 | 1.016 [45] | | |
| $^{14}NO$–$^{15}NO$ | NaX | 121 | 1.043 [46] | 121 | 1.027 |
| | | 146 | 1.029 [46] | | |
| $^{16}O_2$–$^{16}O^{18}O$ | NaX | 77.3 | 1.030 [43] | 83.2 | 1.0065 |
| | CaA | 77.3 | 1.020 [43] | | |
| | NaX | 77.3 | 1.018 [43, 47] | | |
| $C^{16}O_2$–$C^{16}O^{18}O$ | NaA | 195 | 1.01 [48] | | |

[a] In the carbon monoxide rectification.

In the molecular nitrogen sorption over the NaX zeolite, in addition to the temperature of 77.3K (see Table 4.5), the $\alpha$ values at 90 and 100K were determined. These values equal 1.016 and 1.010, respectively.

The temperature dependence of $\alpha_{^{14}N^{16}O-^{15}N^{16}O}$ on the NaX zeolite in the 118–195K range is characterized by

$$\ln \alpha = \frac{1035}{T^2} - \frac{2.90}{T}. \qquad (4.8)$$

Table 4.5 represents as well the separation factor values in the rectification of corresponding substances, including those for the processes that are used (or considered as prospective) in the industrial production of carbon, nitrogen, and oxygen isotopes. As can be seen from Table 4.5, the separation factors of all isotopes studied over zeolites are much higher than those in the rectification which allows to consider these systems as candidates for the separation of light elements' isotopes.

## 4.2   KINETICS OF ISOTOPE EXCHANGE AND MASS TRANSFER IN SEPARATION COLUMNS

### 4.2.1   Reactions of chemical isotope exchange

The isotope exchange in hydrogen–metal (IMC) systems can generally be described by the equation of formal kinetics presented in chapter 1. Account must be taken, however, of a large variety of special features of metals and IMC that have a profound effect on the isotope exchange kinetics. Among these features, above all, are the change of particle sizes during the process of activation (sorption–desorption) by hydrogen which leads to an increase of specific surface, and, in the case of IMC, the change of the surface composition due to segregation and partial destruction.

Intermetallic compounds produced from powdered or granulated source components require preliminary activation to convert them in a highly active state where they are capable of easily sorbing-desorbing hydrogen. The activation is accompanied by a progressive decrease of IMC particle sizes and a rise in the specific surface up to some constant magnitude. The influence of the IMC dispersity variation during the activation on the isotope exchange kinetics was studied for $LaNi_5$ and industrial IMC $Ce_{0.05}La_{0.95}Al_{0.02}Ni_{4.98}$ (CLAN) [51]. The data on the kinetics presented below relate to the activated IMCs when a further increase in the number of sorption-desorption cycles does not affect the isotope exchange rate.

Another specific feature is the solid phase polydispersity leading to a deviation of the exchange rate versus time dependence (equations of first order) from the simple exponential law [51].

An investigation into the temperature and pressure influence on the isotope exchange rate in the hydrogen isotopes – metal/IMC hydrides systems makes it possible to determine the process limiting factor and to direct the way to the process intensification. Most experimental data on the temperature and pressure influence on the isotope exchange kinetics was obtained by the static method with forced gas circulation for $\alpha \neq 1$ and $x, y \ll 1$, which corresponds to eq. (1.31). Since the hydrogen pressure determines its concentration in the solid phase, eq. (1.27) can be represented in general form as

$$R = R_o \exp(-E_a/R_g T) \cdot P^q, \tag{4.9}$$

where $E_a$ is the observed energy of the isotope exchange activation, $R_o$ is the pre-exponential factor, and $q$ is the order of the isotope exchange reaction in respect to the hydrogen pressure.

Thermodynamic and kinetic characteristics of the isotope exchange between hydrogen and some IMC hydrides are presented in Table 4.6.

It should be noted that for all IMC samples over a wide temperature and pressure range, the first-order kinetic equation eq. (1.31) remains valid up to high isotope exchange degrees ($F \leq 0.8$) which allows us to determine the $R$ value, characterizing the averaged isotope exchange rate for particles of different size, from the slope of the straight line in the coordinates $- \ln (1 - F) = f(\tau)$.

From the $AB_5$−type IMCs, of prime interest are $LaN_4Cr$ and $LaN_4Cu$, characterized not only by a high exchange rate at 195K but also by a high activation energy. The isotope exchange on such $AB_2$ type IMCs as $TiCrMn$, $Ti_{0.8}Zr_{0.2}$ and $CrMn$ proceeds as well with a high rate, but these compounds can be less efficient because of a lower $E_a$ value at elevated temperatures.

As has been shown [51, 52], in the isotope exchange between hydrogen and hydride phases of IMC activated powders the rate-determining step is the transition of hydrogen atoms from the adsorptive state into absorptive one, i.e. the diffusion of the atoms in the crystal lattice.

The familiar IMC property consisting in an increase of the crystal lattice volume (by up to 25%) in the formation of hydrides is a cause for the deformation and destruction of the IMC-made constructions. In addition, the formation of fine-grained powders during the IMC

## Table 4.6

Thermodynamic and kinetic characteristics of the isotope exchange between hydrogen and IMC hydrides [19, 20]

| IMC hydride | $\alpha_{HT}$ at 195K | $R_{HT} \times 10^4$, mol/g·min at 195K and 0.5MPa | $E_a$, kJ/mol | $\ln R_0$ | $q$ |
|---|---|---|---|---|---|
| $Ce_{0.05}La_{0.95}Al_{0.02}Ni_{4.98}H_{6.6}$ | 2.12 | 9.0 | 22 | 6.3 | 0.2 |
| $LaNi_5H_{6.6}$ | 2.12 | 6.5 | 17.4 | 3.4 | 0.3 |
| $LaNi_4CuH_{5.5}$ | 2.12 | 12.7 | 27 | 11.5 | 0.4 |
| $LaNi_4CrH_{5.0}$ | 2.12 | 25.9 | 28 | 10.1 | 0.4 |
| $TiMn_{1.5}H_{2.5}$ | 3.10 | 11.0 | 9.4 | −2.0 | 0.5 |
| $TiMn_{1.4}Ni_{0.1}H_{2.3}$ | 2.80 | 7.8 | 8.5 | −1.9 | 0.5 |
| $TiCrMnH_{0.3}$ | 3.20 | 2.1 | 8.5 | −0.9 | 0.5 |
| $Ti_{0.8}Zr_{0.2}CrMnH_{2.9}$ | 3.20 | 2.1 | 8.5 | −0.9 | 0.5 |
| $Ti_{0.8}Zr_{0.2}Cr_{1.8}H_{2.8}$ | 3.17 | 5.3 | 9.1 | −1.9 | 0.4 |
| $ZrCr_2H_3$ | 2.32[a] | 12.5[a] | 16.6 | 0.64 | 0.5 |

[a] At 273K.

activation hinders their practical application due to the removal of particles from the apparatuses and high hydraulic resistance of the powder bed. Because of this the urgent problem is to create IMC-based hydrogen sorbents of which the granules are of high mechanical strength, high resistance to destruction at multiple hydrogenation, and a high thermal and radiation resistance. The last is important for the separation of tritium-containing mixtures.

The development of granulated sorbents was investigated in many papers, of which the results are generalized in [19, 20]. The shaping of IMC powder with polymeric materials is a widely used method of the granulated sorbent production [53]. PTFE was the first binding material used for this purpose [54]. The PTFE-based sorbents, however, decompose at high temperatures (~ 450K) owing to the IMC components' interaction with fluorine atoms [55]. In addition, PTFE is the least radiation-resistant polymer [56]. Such polymeric material as polyimide provides rather high thermal and radiation resistance and required porosity.

A porous sorbent that is resistant to no less than 6000 hydrogenation cycles can be obtained by mixing IMC activated powder with a metal not forming hydrides, and by the mixture hydrogenating followed by pressing with simultaneous annealing at a temperature of 370–470K in the hydrogen environment at a gas pressure higher than the hydrogenation pressure. The hydrogen sorption–desorption rate with this sorbent is significantly higher than that with a powdered IMC [57]. Owing to the widest use of these two techniques, the mass-transfer data presented below refer to the granules of the sorbents obtained by the techniques.

A comparative analysis of the properties of sorbents prepared with the use of various polymeric materials and non-hydrogenated metals shows that the granulation technique depends on the IMC properties and on the field of application of the granulated sorbents. An important characteristic for hydrogen isotope separation, as an example, apart from the thermal conductivity, is the sorbent porosity which can vary with the thermal treatment rate at the stage of the preparation of sorbents with polymeric binders [58].

In systems with a solid phase, the mass transfer proceeds under conditions of the constant phase contact surface of which the value does not depend on the hydrodynamic mode of the reactor or separation plant operation. But, despite this fact, the mass transfer mechanisms remain complex due to the multistageness of the solid-phase substance transfer process, and the influence of the dissolved hydrogen amount on the stage rate determined by the diffusion in the solid phase.

In the hydrogen isotope exchange, the process efficiency along the whole column height can be considered constant. With regard to this simplifying property, and to the fact that the mass transfer in the gas phase [3, 52], and the hydrogen atoms diffusion in the crystal lattice (such metals and IMCs cannot be used for the isotope separation) are not the rate-determining steps, the equation similar to eq. (1.36) for the region of the heavy isotope low concentrations ($m = \alpha$) will assume the following form

$$h_{0g} \approx h_p + \frac{\lambda}{\alpha} h_{IE} = \frac{G_{sp}}{\beta_P a_{gr}} + \frac{\lambda}{\alpha} \frac{L_{sp}}{\beta_{IE} a_m}, \tag{4.10}$$

where $\lambda$ is the molar ratio of the gas flow $G$ and hydrogen flow in the sorbent $L$; $\beta_p$ and $\beta_{IE}$ are the mass-exchange coefficients due to the processes of molecular hydrogen diffusion in the sorbent pores and to the transport of hydrogen atoms from the surface to the crystal lattice interstices; $a_{gr}$ is the surface per column unit volume, $a_m$ is the mass-transfer surface determined by the geometrical surface of the IMC particles (crystallites).

The above-described approach does not take into account the HTU component due to the axial dispersion in the column. This quantity depends on the structure of flows in the column and can cause the scale-up effect consisting in the HTU rise when passing from laboratory-scale columns to industrial ones. The effective coefficient of axial diffusion $D_{eff}$ is used as a quantitative characteristic of axial dispersion. The coefficient depends on the effects of both axial dispersion determined by molecular, turbulent and convective diffusion, and of lateral irregularity of which the influence decreases with an enhancement of the lateral dispersion.

In the columns with fixed solid bed, the last effect is associated with the lateral distribution (profile) of the gas flow rates. Thus, the introduction of $D_{eff}$ accounting for all hydrodynamic effects makes it possible to describe, within one-dimensional approximation, the lateral irregularity as an increase of the axial dispersion [59]. It is apparent from the above that it is difficult to calculate the value of $D_{eff}$ which depends not only on the column geometry and operation parameters (e.g. flow rates, temperature, pressure, etc.), but also on sizes and shapes of the solid phase particles, their polydispersity type, and even on the column filling technique. That is why the experimental method of the $D_{eff}$ determination is currently the most reliable.

The influence of the hydrogen axial dispersion on HTU depending on the linear gas velocity in the column with fixed solid bed, of 1.5cm in diameter filled with solid phase particles of 2–3mm in size was studied [60, 61].

The data on the mass exchange efficiency in the columns with immobile solid bed presented below were obtained by the technique of discrete change of isotope concentration at the column input and measurement of the concentration front smearing at the column outlet [62, 63]. The results of the study of the dependence of HTU on the hydrogen specific flow for H–T and H–D mixtures at 195 and 273K with sorbents based on palladium and LaNi$_5$ are given in Figure 4.4. These results were obtained in the column of 1cm in diameter filled with particles of 1.3–1.5mm in size prepared with the use of polytetrafluorinethylene as the binding material [52, 64].

Since the mass-transfer coefficients associated with the process of the molecular hydrogen diffusion in the sorbent pores and with the transport of hydrogen atoms from the surface to the crystal lattice interstices are independent of the column hydrodynamic environment, and remain constant at a constant temperature, the corresponding $h_{0g}$ components must increase linearly with a rise in the flow rate, which is clearly demonstrated in Figure 4.4.

Owing to a relatively high activation energy (22kJ/mol for Pd and 18kJ/mol for LaNi$_5$ [52]), $\beta_{IE}$ increases with a rise in temperature and the $h_{IE}$ contribution to $h_{0g}$ decreases drastically. At 195K, the mass transfer is limited by the stage of the hydrogen implantation in the crystal lattice, which results in a $h_{0g}$ difference for the H–T and H–D mixtures (due to different values of $\alpha$). At a temperature of 273K and over, the mass transfer is generally determined by the diffusion of hydrogen molecules in the sorbent pores which leads to a weak influence of the type of exchanging isotopes on $h_{0g}$.

Irrespective of the rate-determining stage nature, the straight line slope angle, apart from the temperature, depends on the sorbent surface. At low temperatures, when the mass

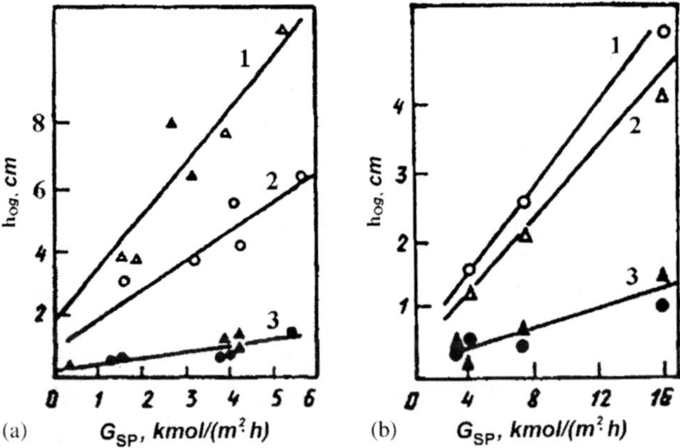

**Figure 4.4**   Dependence of HTU ($h_{0gas}$) for H–D ($\bigcirc,\bullet$) and H–T ($\triangle,\blacktriangle$) mixtures on specific hydrogen flow $G_{SP}$ in the $H_2$–Pd (a) and $H_2$–LaNi$_5$ (b) systems at $P = 0.5$MPa: 1, 2, 195K; 3, 273K.

transfer is determined by the stage with the hydrogen participation, the straight line slope angle on the HTU-gas flow dependence is determined by the ratio $1/(\alpha\beta_{IE}a_p)$. That is why with a rise in separation factor the slope angle decreases (e.g. for the H–D isotope mixture in Figure 4.4b). In the systems where the heavy isotope is concentrated in the gas phase, a rise in separation factor results in an increase of the straight line slope angle (see Figure 4.4a for the $H_2$–Pd system).

Similar results were obtained for the CLAN-based sorbents [65] and Li$_{0.8}$Zr$_{0.2}$CrMn [19, 20, 66], in a column of 1cm diameter filled with particles of 1–1.5mm in size prepared with the use of polyimide as the binding material (see Figures 4.5 and 4.6). The same sorbents were used for the study of the pressure influence on HTU.

From the dependences presented in Figure 4.7 it follows that the mass exchange efficiency increases significantly with a rise in pressure, which is especially true for Ti$_{0.8}$Zr$_{0.2}$CrMn with its higher reaction order in respect to the pressure (see Table 4.6). And finally it should be noted that the HTU value is practically unaffected by the binding agent nature which points to a high porosity of the prepared sorbent.

## 4.2.2   Phase isotope exchange

As was shown in section 4.1.2, the largest isotope effects are observed in the systems with zeolites, which receive primary consideration below. This is granulated biporous, and not powdered, sorbents that are used in the solution of practical isotope exchange problems. This is why the kinetics of phase isotope exchange in the systems with solid sorbents are generally determined by the diffusion processes in the primary and secondary porosity of the particles. The available data on the mass exchange efficiency in the columns were obtained either in the immobile solid bed by the technique of discrete change of isotope concentration at the column input and measurement of the concentration front smearing at the column outlet [62, 63], or in the moving solid phase layer of the hypersorption column (see section 4.3.1).

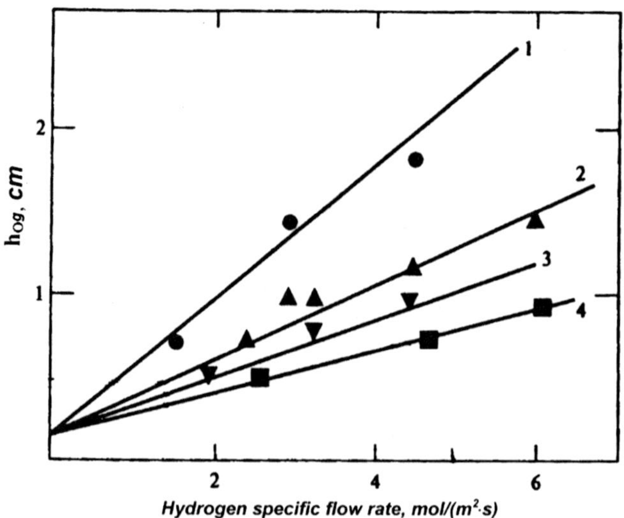

**Figure 4.5**   Dependence of HTU ($h_0$) for H–T mixture on specific hydrogen flow $G_{SP}$ in the $H_2–Ce_{0.05}La_{0.95}Al_{0.02}Ni_{4.95}$ system (with polyimide as binder) at $P = 0.5$MPa, and at various temperatures: 1, 228K; 2, 250K; 3, 273K; 4, 293K.

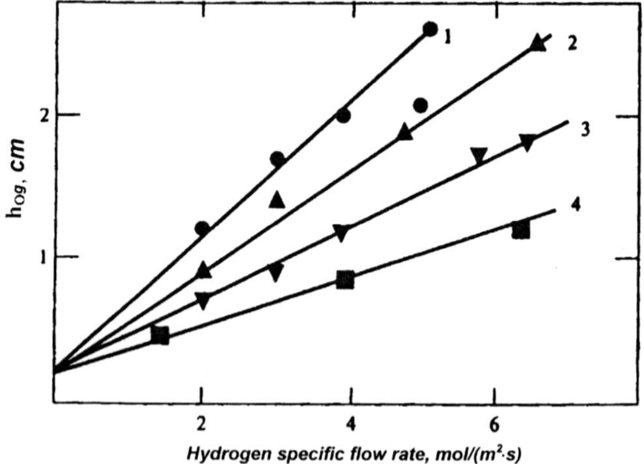

**Figure 4.6**   Dependence of HTU ($h_0$) for H–T mixture on specific hydrogen flow $G_{SP}$ in the $H_2–Ti_{0.8}Zr_{0.2}CrMn$ system (with polyimide as binder) at $P = 1.9$MPa, and at various temperatures: 1, 228K; 2, 250K; 3, 273K; 4, 293K.

The conditions of the studies are presented in Table 4.7. The results of the experiments on the separation of the $H_2–D_2$ and $H_2–HT$ mixtures on the NaX zeolite with particles 1.5 and 2.5mm in size are given in Figure 4.8, and those on the NaA zeolite with particles 1.2, 2.5, and 3.2mm in size are presented in Figure 4.9 [29, 67].

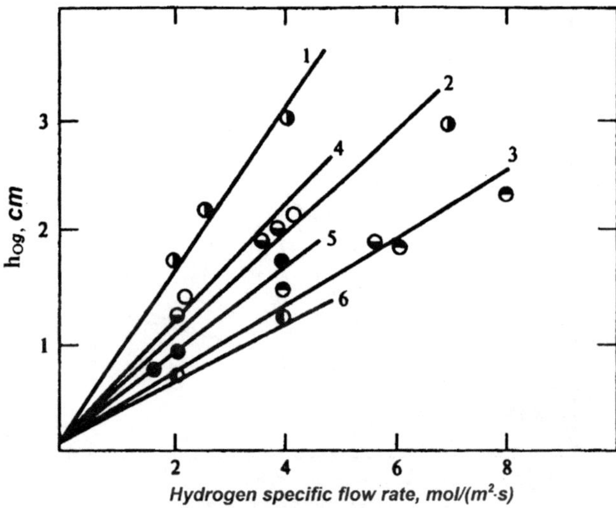



**Figure 4.7**　Dependence of HTU ($h_0$) at 228K for H–T mixture on specific hydrogen flow $G_{SP}$ in the column filled with sorbent (with polyimide as binder) on the basis of $Ti_{0.8}Zr_{0.2}CrMn$ (1, 2, 3) and $Ce_{0.05}La_{0.95}Al_{0.02}Ni_{4.98}$ (4, 5, 6) at 228K and 0.15MPa (1, 4); 0.5MPa (2, 5); and 1.9MPa (3, 6).

### Table 4.7

Conditions of experiments on HTU determination for PHEX in gas–solid systems

| HTU determination technique | Sorbent | Separated mixture | $T$, K | $P$, atm | Flow rate (maximum), $G_{SP}$, mol/(m²·s) |
|---|---|---|---|---|---|
| Concentration front | NaX zeolite | $H_2$–$D_2$ | 30.8-87.3 | 1.0 | 20 |
| smearing at the | | $H_2$–HT | 77 | 1.0 | 15 |
| immobile sorbent | NaA zeolite | $H_2$–$D_2$ | 38.5-87.3 | 1.0 | 20 |
| bed outlet | | $H_2$–HT | 77 | 1.0 | 19 |
| Separation degree | NaX zeolite | $^{16}O_2$–$^{16}O^{18}O$ | 78 | 0.04–0.17 | 5.5 |
| in the hypersorption | CaA zeolite | $^{14}N_2$–$^{14}N^{15}N$ | 78 | 0.21–0.95 | 5.0 |
| column | ACM silica gel | $^{14}N_2$–$^{14}N^{15}N$ | 78 | 0.2–0.92 | 8.5 |

Figure 4.10 shows the data on the efficiency of $H_2$–$D_2$ and $H_2$–HT isotope exchange at a temperature of 77K on the NaX and NaA zeolites for particles of various sizes. As is evident from the figure, with a decrease in particle size the HTU value drops sharply (e.g. by one half at 77K with a decrease in particle size from 3.2 to 2.5mm). In addition, the HTU values differ significantly in the $H_2$–$D_2$ and $H_2$–HT exchange. It is demonstrated [29] that such variation of HTU is due to the fact that HT, despite the same molecular mass as that of $D_2$ molecules, diffuses much faster in the gaseous hydrogen environment of the zeolite pores

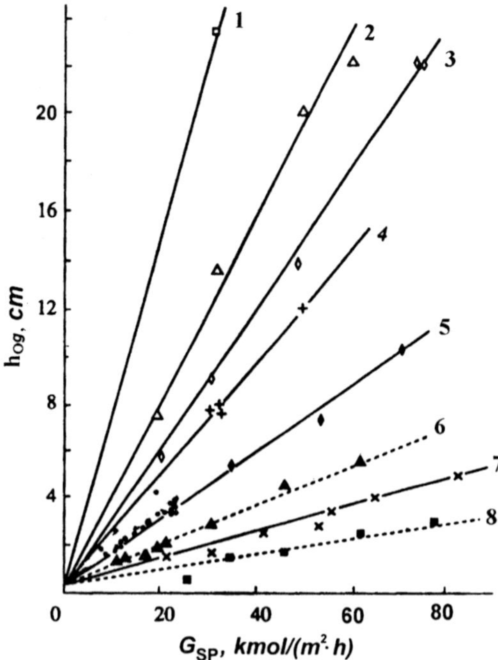

**Figure 4.8**   Dependence of HTU ($h_0$) on specific hydrogen flow $G_{SP}$ for $H_2$–$D_2$ (full line) and $H_2$–HT (dotted line) mixtures on the NaX zeolite at $P = 0.1\,MPa$ and at various temperatures: 1, 38.5K; 2, 45.5K; 3, 56.5K; 4, 66.5K; 5, 87.3K; 6–8, 77K (sorbent particle size: 1–6, 2.5mm; 7–8, 1.5mm).

than $D_2$ does. Since the HTU component determined by the external diffusion can be neglected, the mass exchange is limited by diffusion processes in zeolite pores. At 77K the rate-determining stage is the diffusion in the secondary pores of the particles, and at lower temperatures, the contribution of the diffusion in primary pores becomes significant (it accounts for about 60% for the NaX zeolite with particles 2.5mm in size at 38.5K) [29, 67].

The additivity equation of the HTU components for the region of the heavy isotope low concentrations, similar to eq. (4.10), will take the form

$$h_{0g} \approx h_{SP} + \frac{\lambda}{\alpha} h_{PP} = \frac{G_{SP}}{\beta_{SP} a_{SP}} + \frac{\lambda L_{SP}}{\alpha \beta_{PP} a_{PP}},$$

$$(4.11)$$

where $\beta_{SP}$, $\beta_{PP}$, $h_{SP}$, $h_{PP}$ are mass-transfer coefficients and HTU components, determined by the gas diffusion in the secondary and primary sorbent pores, respectively; $a_{SP}$, $a_{PP}$ are the mass-exchange surface determined by in the secondary and primary sorbent pores.

Since the HTU values on the zeolites coincide at 77K (a distinction is seen only in the H–T exchange for the particles of 1.5mm in size), the values of $\beta_{SP}$ and $a_{SP}$, according to eq. (4.11), will be uniform for the NaX and NaA zeolites, which is presumably due to the identity of the sorbents' graining techniques.

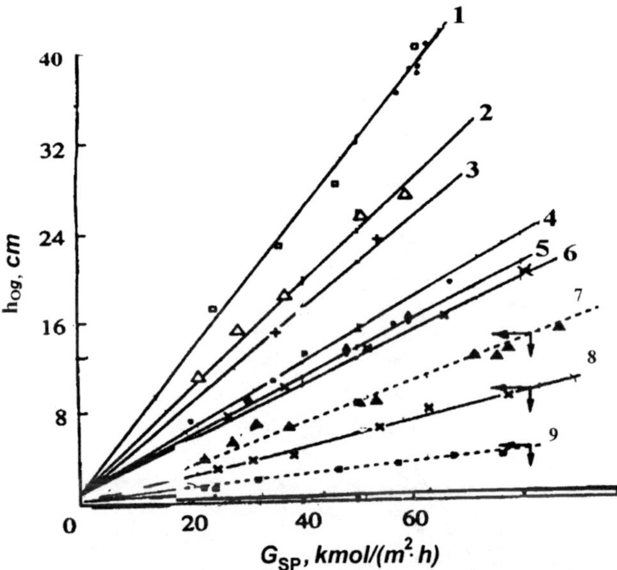

**Figure 4.9**   Dependence of HTU ($h_0$) on specific hydrogen flow $G_{SP}$ for $H_2$–$D_2$ (full line) and $H_2$–HT (dotted line) mixtures on the NaA zeolite at $P = 0.1$MPa and at various temperatures: 1, 51K; 2, 59.5K; 3, 68K; 4, 7–9, 77K; 5, 85.9K; 6, 93K (sorbent particle size: 1–6, 3.2mm; 7, 2.5mm; 8, 9, 1.2mm).

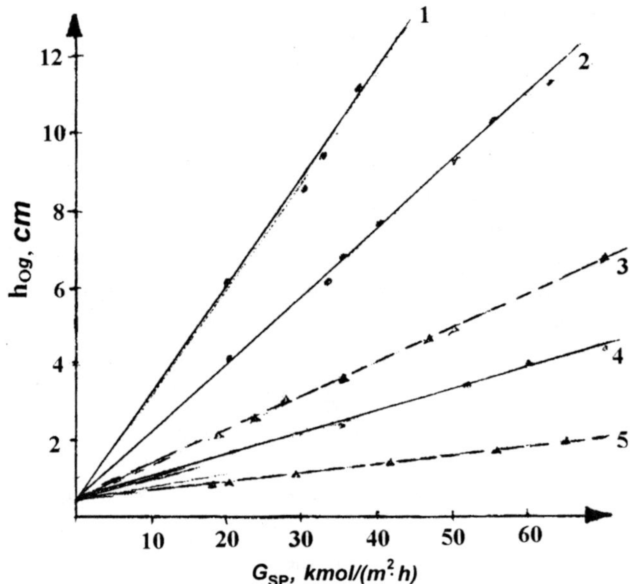

**Figure 4.10**   Influence of particle size (1, 3.2mm; 2, 3, 2.5mm; 4, 5, 1.5mm) of the NaA (●) and NaX (×) zeolites on HTU ($h_{0g}$) for the $H_2$–$D_2$ (1, 2, 4) and $H_2$–HT (3, 5) mixtures at $T = 77$K and $P = 0.1$MPa.

The pressure influence on HTU was investigated on the NaX and NaA zeolites at 77K. In the pressure measurement from 0.1 to 2MPa, the HTU value in the $H_2$–$D_2$ exchange remained constant [30, 68].

The HTU linear dependences on the gas flow obtained in the oxygen isotope separation in the molecular oxygen–NaX zeolite system ($^{16}O_2$–$^{16}O^{18}O$ separated mixture) and in the nitrogen isotope separation by the exchange of molecular nitrogen with CaA and NaX zeolites and with ACM silica gel ($^{14}N_2$–$^{14}N^{15}N$ separated mixture) are presented in Figure 4.11 [44, 46, 47, 50, 69].

The straight line 1 in Figure 4.11 represents the highest HTU values obtained on the CaA zeolite in the nitrogen separation. The results gained at a pressure of 160 and 200mm Hg are described by a uniform line. The highest mass-exchange efficiency was observed on the ACM silica gel. The dependence of HTU on the flow rate at $P = 150$mm Hg is represented by the straight line 5, and at $P = 700$mm Hg by straight line 6. The experiments on the NaX zeolite were performed at a pressure of 150 and 570mm Hg, respectively (the experimental points fell on lines 3 and 4, respectively).

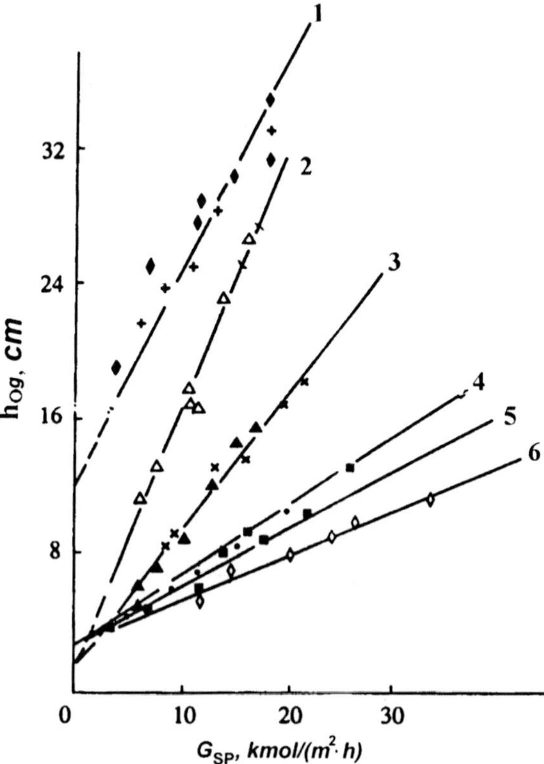


**Figure 4.11**   Dependence of HTU ($h_0$) on specific hydrogen flow $G_{SP}$ in the separation of oxygen isotopes ($\triangle$, 30mm Hg; $\times$, 150mm Hg) and nitrogen ($\blacktriangle$, 150mm Hg; $\bullet$, 570mm Hg) over zeolite NaX, and of nitrogen isotopes on CaA zeolite (+, 160mm Hg; $\blacklozenge$, 720mm Hg) and on ACM silica gel ($\blacksquare$, 150mm Hg; $\diamond$, 700mm Hg) in hypersorption column ($T = 78$K).

In the oxygen isotope separation over the NaX zeolite, the HTU values are close to those obtained in the nitrogen separation. In Figure 4.11, the experimental points at $P = 130$mm Hg fall on the same straight line 3. At $P = 30$mm Hg, the HTU values are much higher (line 2 in Figure 4.11).

Higher HTU values in the nitrogen and oxygen isotope separation in the hypersorption column, compared with those obtained for the isotope exchange in the immobile bed, are determined by the mixing effect in the column with the moving sorbent layer, as well as by the fact that the diffusion coefficients of the nitrogen and oxygen molecules are less than those of the hydrogen molecules.

## 4.3  COUNTER-CURRENT ISOTOPE SEPARATION PROCESSES

### 4.3.1  Chromatographic separation

Despite the considerable separation effects, gas–solid systems have not found an extensive practical use due to the difficulties in realizing continuous counter-current separation. This is why most studies on the isotope separation in columns are devoted to periodic separation processes and, above all, to the chromatographic separation. Chromatography can provide a higher separation degree which is of great interest in tackling analytical problems (such as the monitoring of the tritium content in the environment, or determination of the deuterium concentration at natural levels and below).

Most studies on chromatographic separation are concerned with the hydrogen–palladium system and are aimed at solving the problem of preliminary enrichment by the heavy hydrogen isotopes obtained from the water sample. When a mixture of hydrogen isotopes passes through the column filled with palladium, enrichment of the forward front with the heavy isotope occurs. Two alternative schemes of the process are suggested. According to the first scheme, the gas flow in the column is produced by the continuous hydrogen feed, and the second scheme provides for the gas flow produced by the hydrogen displacement from the palladium bed by heating a column section preliminary saturated with hydrogen.

The chromatographic separation of hydrogen isotopes in the $H_2$–Pd system was first performed by Glueckauf and Kitt [70, 71] (Table 4.8). They used a column 0.44m in height and 0.8cm in diameter, which allowed for the extraction of deuterium with 99.15% purity and a yield of over 96% from 14l of hydrogen isotope mixture containing 40at.% deuterium. Aside from the determination of $\alpha$, the chromatographic separation was used for analytical purposes, namely the preliminary enrichment of hydrogen obtained from water samples with tritium. Then Hoy [72], also with the use of the thermal desorption method, obtained, from 70l of hydrogen, a fraction of 0.5l in volume enriched with tritium, where it was possible to concentrate about 60% of the tritium contained in the initial sample. To accomplish this, a plant involving four consecutive-operating columns containing 1000, 300, 33, and 4g of palladium, respectively, was used. S. Tischtchenko and G. Dirian [73] applied the frontal chromatography technique for the enrichment of hydrogen samples with tritium. Palladium on $\alpha$-$Al_2O_3$ served as the sorbent. Initially, the column was filled with inert gas (helium) which was blown out  by the hydrogen flow during sorption. The authors succeeded in achieving a separation degree $K = 100$ in a column of 1.12m length and 10mm diameter

**Table 4.8**

Chromatographic separation of hydrogen isotopes in hydrogen-palladium system

| Num-ber | Chromatographic process | Sorbent | Palladium amount, g | Separated mixture | Separation degree | Reference |
|---|---|---|---|---|---|---|
| 1 | Displacement (thermal desorption) | Palladium black + asbestos (23% by weight) | 20 | H–D | 175 | [70, 71] |
| 2 | Displacement | Palladium black | 1400 | H–T | 90 | [72] |
| 3 | Frontal | Pd on $\alpha$-Al$_2$O$_3$ | | H–T | 100 | [73] |
| 4 | Frontal | Granulated palladium black | 136 | H–D | 37–1040 | [74] |

for an enriched fraction volume of 5cm$^3$, and in demonstrating that the separation degree dependence on the withdrawn fraction volume is exponential in nature.

In isotope separation, the Pd–Al$_2$O$_3$ sorbent is less effective than granulated palladium black (see section 4.2). Another drawback of the technique under study is due to the fact that the hydrogen sample is fed into the column preliminary filled with inert gas to avoid a pressure gradient in the column and associated with this the isotope axial dispersion. The helium injection into the column prevents the occurrence of significant pressure gradients and, as a consequence, leads to an increase in the separation degree. However, it significantly complicates the enrichment process and subsequent gas analysis since an additional analytical device is required to note the moment when the helium content at the column output reduces to zero and to take hydrogen samples (a katharometer was used by S. Tischtchenko and G. Dirian [73]). Should the need arise to take samples without the helium presence, the loss of the most enriched sample part is inevitable since the hydrogen extraction must be delayed until the completion of the helium withdrawal from the column. Owing to the pronounced exponentiality of the extracted component's concentration profile in the enriched flow, it is necessary to exactly note the moment of sampling in order to reproduce the achieved separation degree, which also somehow hampers the utilization of the technique under consideration.

The above drawbacks can be avoided if the separation process is performed at lower temperatures. In this case not only can the pressure gradient reduction be obtained in the sorption front area due to a decrease in the $\alpha$–$\beta$ transition pressure, but also the separation factor can be increased with an increase of the hydrogen capacity of palladium. The latter allows, with all other factors being equal, an increase in the sample volume or a decrease in the separation column size, thus reducing the palladium amount in the column. At lower temperatures, however, along with the above-mentioned advantages a decrease in the interphase isotope exchange rate is also observed which results in an extreme character of the temperature dependence of the hydrogen isotope separation  in the column.

**Table 4.9**

Results of experiments on chromatographic separation of protium–deuterium
mixture on palladium [74]

| Experiment number | $T$, K | $G$, mol/ $(m^2 \cdot s)$ | Enriched fraction volume, $cm^3$ | Deuterium concentration, at.% | | Separation degree, $K$ |
|---|---|---|---|---|---|---|
| | | | | In source gas | In enriched gas | |
| 1 | 299 | 0.88 | 0.62 | 2.5 | 49 | 37.5 |
| 2 | 302 | 1.10 | 0.50 | 2.5 | 52.5 | 43.1 |
| 3 | 303 | 1.10 | 0.62 | 2.5 | 54.0 | 45.7 |
| 4 | 273 | 1.22 | 1.80 | 2.5 | >98 | — |
| 5 | 273 | 1.13 | 1.74 | 0.015 | 13.0 | 997 |
| 6 | 273 | 1.22 | 1.86 | 0.015 | 13.6 | 1040 |
| 7 | 253 | 1.24 | 0.62 | 0.015 | 1.8 | 120 |

The above considerations about the temperature influence on the hydrogen isotope
separation degree in the chromatographic column were experimentally verified [74]. In
addition, the influence of the enriched fraction volume and of the gas flow rate on the
separation degree was studied in the same paper.

The experimental results are presented in Table 4.9. The analysis of these results
demonstrates the possibility of a considerable increase in the separation degree by
decreasing the temperature in the chromatographic column to 273K, as well as the possibility of reliably reproducing the enrichment by injecting the hydrogen sample into the
evacuated column (see, for example, experiments 2, 3, and 5 in Table 4.9). In conclusion
it should be noted that for the determination of the tritium content in the source hydrogen
using the results of the enriched sample analysis, it is necessary to precisely know the
separation degree achieved during the enrichment. In this connection, the preliminary
column calibration is required. The calibration, however, can be avoided by noting that
the deuterium concentration in natural objects remains practically constant (0.015at.%),
and by performing the analysis of deuterium content in the enriched hydrogen sample
together with the analysis of tritium content.


### 4.3.2  Continuous counter-current separation processes

For systems with a solid phase the possibility exists of realizing continuous counter-
current separation processes. Let us first consider variants of a simpler scheme with flow
conversion presented in Figure 4.12. The main parts of the plant are the separation column
1, desorber 2 and adsorber 3 serving for the conversion of gas and solid phase flows. If the
heavy isotope (e.g. tritium) is concentrated in the column bottom, the enriched product is
withdrawn from the gas flow leaving the desorber 2. The waste hydrogen flow is withdrawn from the gas flow before it enters the adsorber. When the heavy isotope is concentrated in the gas phase, the product is withdrawn from the column head, and the hydrogen

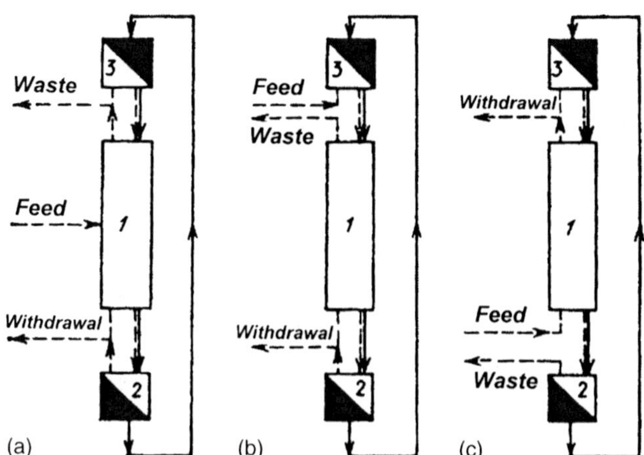


**Figure 4.12**  Schemes of separation plants using gas-solid systems: a, scheme with concentration and depletion; b, open scheme for heavy isotope concentration in solid phase; c, open scheme for heavy isotope concentration in gas phase.

released in the desorber forms the waste. Since the separation column comprises a concentration and a depletion sections (for the heavy isotope), the source flow is fed into the middle part of the column (see Figure 4.12a).

In the scheme under consideration, the flow conversion is performed at the upper end of the column (in the adsorber) through the hydrogen sorption by the solid phase, and at the lower end of the column through the gas desorption from the solid phase by way of the solid phase heating, or by other means (e.g. blowing with inert gas or gas pressure reduction over the solid phase). A coolant is supplied to the adsorber to remove the heat of hydride formation or sorption ($\Delta H_{\mathrm{H}}$). The lower flow conversion is generally done with the use of thermal desorption. This is why in the desorber provision is made for solid phase heating by a heat medium. Regenerated in the desorber, the solid phase is delivered to the adsorber. The flow conversion can be performed either immediately in the upper and lower column ends (i.e. the column comprises three sections: for separation, sorption, and desorption), or in separate units.

In the plant operation on the open scheme, only one flow-conversion unit is required: either the lower unit when using the solid phase for the target isotope concentration (Fig. 4.12b), or the upper unit when the heavy isotope is concentrated in the gas phase (Fig. 4.12c). In the former case, however, an additional adsorber is required since the column is fed by the solid phase containing isotopes to be separated, and in the latter case an additional desorber is required to produce the gas waste flow and to regenerate the solid phase which then is supplied to the upper flow-conversion unit. Thus, the separation plants using gas–solid systems are characterized by the fact that both the absorber and the desorber are required in the operation on the open scheme (though only on of them serves as the flow-conversion unit).

The main difficulties encountered in practically applying solid-phase systems for isotope separation consist in the realization of gas–solid counter-current movement. Initially, the process was performed in the column with the solid phase dense bed falling under gravity, through which hydrogen passed counter-currently. In the flow-conversion units the counter-current movement can be continued. Such counter-current separation process is often referred to as hypersorption.

The hypersorption process was originally used for $H_2$–$D_2$ mixture separation with hydrogen sorbed by activated carbon [75] and by silica gel [76]. Thereafter the process was utilized in the $H_2$–Pd system for the $H_2$–HT mixture separation [77]. In the latter case the main part of the plant is the column filled with granulated palladium sorbent (spherical grains of 1.3mm in size) [3, 77]. The velocity of the solid phase (palladium hydride) falling down under gravity is controlled by a disk feeder. The solid phase removed from the disk feeder by a special ejector enters the lower cartridge-receiver serving as a desorber. When the receiver is filled with palladium sorbent, the hydrogen is desorbed at 520–530K, and the receiver is evacuated and transferred to the column head.

The plant is capable of operating both with the upper flow-conversion unit (in the mode of the heavy hydrogen isotope concentration), and with the lower unit of the hydride phase flow conversion (in the mode of the light hydrogen isotope concentration). In the former case, upon leaving the purification system the source hydrogen isotope mixture enters the disk feeder, passes through the column from the bottom upwards, and, in the column head, is sorbed by regenerated palladium supplied from the upper receiver. The hydrogen desorbed from the palladium in the lower receiver is withdrawn from the plant. In the latter case, palladium (in the hydride form) saturated with the source mixture is delivered from the upper receiver to the column, and all desorbed hydrogen is returned to the column. The desorption degree is judged by the hydrogen waste flow issuing from the plant.

The column size, operation parameters and separation efficiency are presented in Table 4.10, where the characteristics of hypersorption plants with silica gel and activated carbon serving as the hydrogen sorbents are also shown. As can be seen from Table 4.10, the highest separation efficiency is achieved in the plant with palladium.

The hypersorption process features several drawbacks due to the solid phase movement associated with sorbent attrition, separation efficiency decrease due to the axial dispersion in the solid phase, and complexity in the sorbent batching and transporting. That is why a similar

**Table 4.10**

Characteristics of pilot hypersorption plants for hydrogen isotope separation [3, 19]

| Sorbent | Separated mixture | $T$, K | Flow rate $G_{SP}$ kmol/(m²·h) | Column size, cm | | Separation degree | HETP, cm |
|---|---|---|---|---|---|---|---|
| | | | | height | diameter | | |
| Palladium | $H_2$–HD | 294 | 1.8 | 20 | 1.5 | >122[a] | <2.5 |
| Silica gel | $H_2$–$D_2$ | 77 | 3 m/h[b] | 200 | 2.0 | 56[c] | 6.75[c] |
| Activated carbon | $H_2$–$D_2$ | 86 | 2.5–12.6 | 46 | 3.8 | 42 | 1.6 |

[a] Obtained before the attainment of the column steady state.
[b] Linear velocity of the sorbent movement in the column.
[c] Relates to the enrichment section, HETP in the depletion section equals 4.75cm.

continuous counter-current separation process in columns with the immobile solid phase bed was suggested and experimentally implemented in several variants [78]. The suggested method is possible because the counter-current movement of the phases can be achieved not only by their physical transport, but also by the displacement of the flow-conversion units (or that of the temperature zone in case of the thermal desorption) relative to the immobile solid phase. It is illustrated by Fig. 4.13 schematizing a column which comprises several identical sections. Fig. 4.13a shows the moment of the separation column operation when section 1 is the desorber and section 2 is the adsorber. The gas flow through separation sections 2–4 is produced by an elevated hydrogen pressure in the desorber and the hydrogen sorption in the adsorber. Passing through the separation sections, the gas exchanges with hydrogen contained in the solid phase, and then it is sorbed in section 5. Assuming the sections are filled with palladium sorbent, hydrogen then is enriched with the heavy isotope in the isotope exchange with the hydride phase. As soon as the sorption is completed, the flow-conversion units are displaced one section higher along the direction of the gas movement, that is section 2 becomes the desorber, section 1 serves as the absorber and sections 3–5 form the separation zone.

As a result, there occurs the counter-current movement of the solid phase $L$ relative to the gas flow $G$. Figure 4.13a schematizes the column operating in non-withdrawal mode. Figures 4.13b and 4.13c represent the open scheme alternatives similar to those discussed above and presented in Figures 4.12b and 4.12c. When the plant operates with feed and product withdrawal, the points of feed, product withdrawal, and waste discharge must be synchronously displaced together with the displacement of flow-conversion units as was discussed above for the non-withdrawal mode.

The second alternative provides for the displacement of the divisible sections in the direction opposite to that of the gas flow, that is and in this case, as soon as the sorption is completed, owing to the displacement of the sections, section 2 also becomes the desorber and section 1 becomes the adsorber.

As is evident, in addition to the schemes presented in Figure 4.13, both alternatives of the counter-current movement realization allow separation in the column with enrichment and depletion sections, operating on the principle shown in Figure 4.12a.

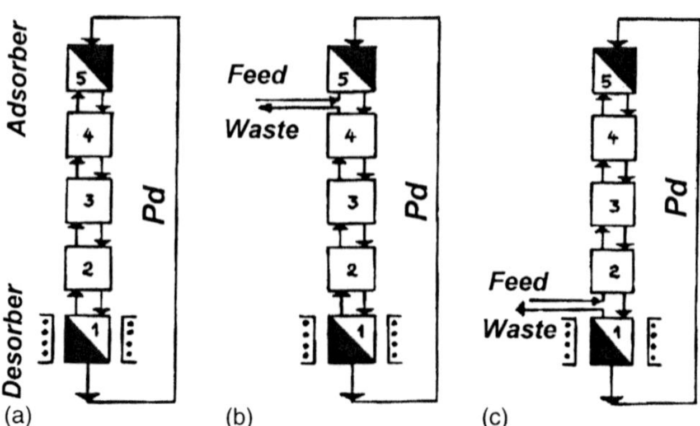


**Figure 4.13**   Realization of counter-current flow in columns with solid phase immobile relative to plant walls.

Based on the above principle, a series of separation plants with sectioned columns was developed in the D. Mendeleev University of Chemical Technology of Russia (MUCTR). Initially, it was a plant with simple physical displacement of the sections relative to the flow conversion units, operating in non-withdrawal mode [79]; later a plant was tested where the counter-current movement was realized through the displacement of sorption-desorption zones with respect to immobile sections of the column. Experiments on the H–D mixture separation using granulated palladium black with grains of 0.5–1.0mm in size as the hydrogen sorbent (Table 4.11, position 1) were performed in this plant.

The further process improvement consisted of minimizing the plant free space leading to the isotope concentrations' mixing, and in automating the plant operation. For this purpose, the sections are connected by a common flow shutoff and distribution device ensuring the plant's tightness. The rotating part of the device is attached to the column sections filled with sorbent, and the stationary part is connected to the external pipelines which are linked with the column sections. The separation plant is schematized in Figure 4.14 [80–82].

Experiments on the separation of all hydrogen binary mixtures (H–T, H–D, and D–T) on a palladium sorbent with aluminium binder were performed in this plant (Table 4.11, position 2). Table 4.12 presents data on the influence of the sorbent bed height, temperature, and hydrogen flow in the column on HETP [80–84].

The same plant was used for experiments on the separation of hydrogen, oxygen, and nitrogen isotopes with NaX zeolite serving as sorbent [82]. The data presented in Table 4.12 demonstrate that efficient isotope separation processes in the gas–solid systems can be implemented both for hydrogen and for other light elements. In the H–D mixture with palladium system, for example, with the total sorbent bed height of 30mm (three sections of 10mm each), a separation degree value equal to 517 (experiment 1) has been achieved, and for the system with zeolite this value was equal to 990 (experiment 12). For nitrogen and oxygen isotopes the separation degree value accounted for 1.142 and 1.186, respectively [82].

The separation efficiency values characterized by HETP are in close agreement for all experiments with the section's sorbent bed height of 10mm and account for 2.6–3.7mm,

**Table 4.11**

Characteristics of plants with sectioned columns for isotope separation in gas–solid systems

| Number | Number of section | Section dimensions, cm | | Separation section height, cm | Separated mixture | Separation degree | Gas flow, l/h |
|---|---|---|---|---|---|---|---|
| | | Diameter | Height | | | | |
| 1 | 8 | 0.4 | 3 | 18 | H–D | 1050 | 6 |
| | | | | 3–31.8 | H–D | >15600 | ≤20 |
| 2 | 5 | 1.1 | 1–10.6 | 3 | H–T | 1660 | 4 |
| | | | | 3 | D–T[a] | 10 | 6 |
| | | | | | | 774 | 16.4 |
| 3 | 12 | 1.5 | 6 | 54[b] | D-T | 112 | 13 |
| | | | | | | 41.7 | 5.2 |

[a] Experiments were done in the region of trace tritium concentrations.
[b] Steady state was not achieved, the separation degree value was obtained only for four sections, i.e. with the separation section height equal to 24cm.

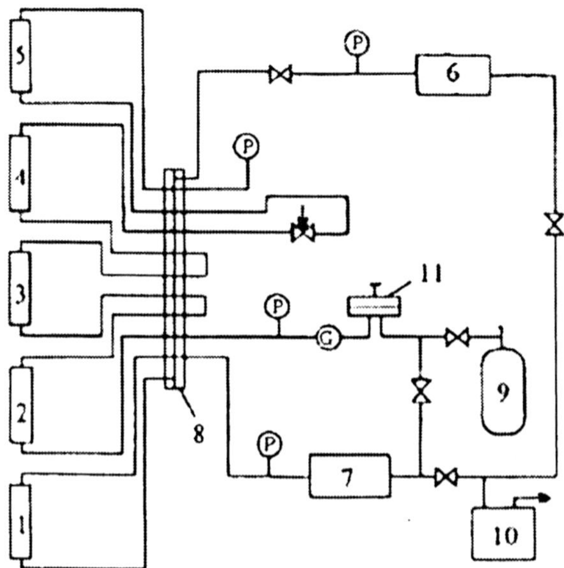

**Figure 4.14**   Scheme of plant with sectioned column for isotope separation in gas–solid systems: 1–5, sections with sorbent; 6, 7, and 9, reservoirs for product, waste, and source material; 8, shut-off and distribution device; 10, evacuation system; 11, pressure regulator.

with the separation efficiency being independent of the operation system, temperature, pressure (within the range 0.02–0.1 MPa), and flow rate which is indicative of the governing role of the gas molecules' diffusion in the solid phase particles. The mass-exchange efficiency, however, decreases considerably with an increase in the section's sorbent bed height (see Figure 4.15) which is due to the periodicity of the solid phase movement (the process departure from the true counter-current nature).

In the H–D mixture separation, a degradation in the separation efficiency was revealed at a deuterium concentration over 98 at.% [84], with the occurrence of a temperature and flow rate dependence of HETP which is characteristic for a process limited by the rate of the hydrogen chemical isotope exchange reaction. In all other cases the process is limited by the molecular hydrogen diffusion in the pores of sorbent grains.

Based on the performed developments, a pilot plant for the separation of the D–T mixture with a high tritium content and for the removal of small amounts of protium was elaborated and put into service in RFYaTs VNIIEF [85]. The sectioned column incorporates 12 sections of 15 mm diameter and 60 mm height, each filled with a palladium sorbent (the section hydrogen capacity is 0.87 l) [86, 87]. The plant operates in periodic mode on the closed scheme. The plant peculiarity consists in accumulating reservoirs at the depleted and enriched ends of the column, with the ratio of the reservoirs' volumes determined by a specific separation objective: either by the tritium concentration in the product, or by the degree of tritium extraction from the source material. The operation of such periodic plant is much simpler: after the column is filled with separated mixture it operates until the required tritium concentrations are obtained in the reservoirs. This is

**Table 4.12**

Results of experiments on separation of hydrogen, oxygen and nitrogen isotopes in five-section column

| Experiment number | Mixture | Sorbent/ Bed height, mm | $T^a$, K | $P$, atm | Gas flow, $l$/h | Heavy isotope concentration, at.%. in section 1 | in adsorber | HETP, mm |
|---|---|---|---|---|---|---|---|---|
| 1 | H–D | Pd/10 | 296 | 1 | 6.8 | 5.7 | 96.9 | 2.8 |
| 2 | H–D | Pd/20 | 296 | 1 | 7.6 | 5.7 | 97.2 | 5.0 |
| 3 | H–D | Pd/56 | 296 | 1 | 10.0 | 6.1 | 97.8 | 12 |
| 4 | H–D | Pd/10 | 276 | 1 | 3.0 | 5.7 | 96.7 | 3.7 |
| 5 | H–D | Pd/10 | 296 | 1 | 3.0 | 5.7 | 96.5 | 3.7 |
| 6 | H–D | Pd/10 | 336 | 1 | 3.0 | 5.7 | 94.0 | 3.7 |
| 7 | H–D | Pd/10 | 296 | 0.5 | 3.0 | 5.7 | 96.4 | 3.7 |
| 8 | H–D | Pd/10 | 296 | 1 | 5.1 | 5.7 | 96.7 | 3.1 |
| 9 | H–T | Pd/30 | 293 | 0.8 | 4.0 | 4.6 | 98.2 | 14 |
| 10 | H–T | Pd/30 | 293 | 0.8 | 4.0 | 2.5 | 97.7 | 14 |
| 11 | $H_2$–$D_2$ | NaX zeolite/10 | 77 | 1 | 3.4 | 96.9 | 15.8 | 3.8 |
| 12 | $H_2$–$D_2$ | NaX zeolite/10 | 77 | 1 | 8.8 | 90.0 | 0.10 | 2.6 |
| 13 | $^{16}O$–$^{16}O^{18}O$ | NaX zeolite/10 | 77 | 0.2 | 4.0 | 0.46 | 0.388 | 3.5 |
| 14 | $^{14}N$–$^{15}N$ | NaX zeolite/10 | 70 | 0.8 | 4.0 | 0.778 | 0.681 | 3.5 |

[a] Temperature in adsorber and in the plant separation section. Temperature in the desorber with the use of palladium sorbent is 493K, and with the use of NaX zeolite – 293K.

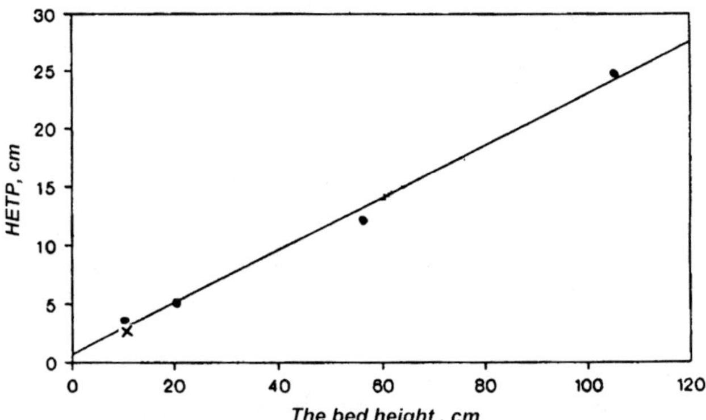

**Figure 4.15**   Dependence of HETP on sorbent bed height in Pd section at $T = 296$K, $P = 0.1$MPa for H–D mixture (●), and NaX zeolite at $P = 0.1$MPa, $T = 70$K for $^{16}O$–$^{16}O^{18}O$ (×).

why the experimental data given in Table 4.11 (position 3) refer to the unsteady state of the separation column.

As an illustration of the experiment with $K = 112$ (see Table 4.11), Figure 4.16 presents kinetic curves characterizing tritium concentration changes in the accumulating reservoirs of 100l in volume each, at a gas flow rate $L = 13$l/h (the sections' switching periodicity is

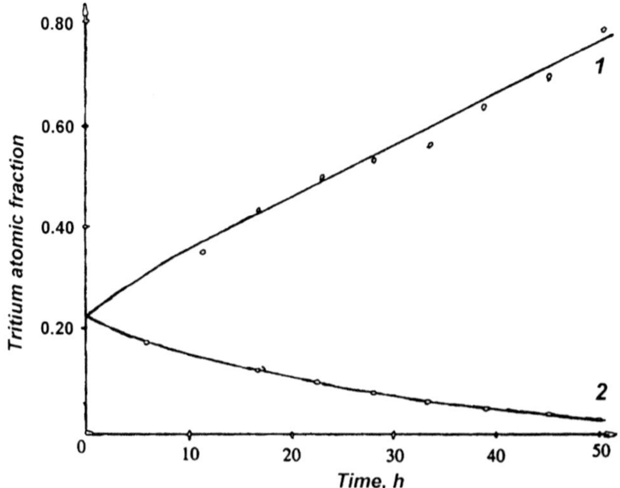


**Figure 4.16**   Tritium concentration change in accumulating reservoirs at the separation column enriched end (1) and depleted end (2).

4min, initial concentration $x_F = 23$at.%). After 50hours of the plant operation, the tritium concentration at the enriched end was 80at.% (with the gas amount of about 40l), and at the depleted end 3.45at.% (with the gas amount of about 90l). The rate of the tritium concentration growth at the plant enriched end (in the accumulating reservoir and in six separation sections) agrees well with the value of the initial isotope transfer in the column, determined by eq. (1.90)

$$j_0 = LX_F \frac{(\alpha - 1)(1 - x_F)}{\alpha - (\alpha - 1)x_F} = 0.83 \mathrm{l/h}. \tag{4.12}$$

HETP calculation was performed on the basis of data on the tritium concentration profile along the sorbent bed height in the separation column. In the separation of the initial mixture containing 23–25at.% of tritium, no more than two to three sections are sufficient to achieve the tritium concentration of about 96at.%. As an example, Figure 4.17 shows the concentration profile for an experiment where $K = 41.7$ (see Table 4.11). The plant was filled with the D–T mixture of 23.4at.% tritium concentration. The accumulating reservoirs at the depleted and enriched ends of the plant were of 100l volume each. At the sections switching periodicity of 10min the gas flow rate in the column was 5.2l/h. After 57hours, the average concentration in section 1 accounted for 18.87at.%, and in section 3 it reached 90.65at.%. The HETP value obtained from the tritium concentration profile presented in Figure 4.17 is 13mm, which agrees well with the data on the H–D mixture separation efficiency. In experiment 3 (Table 4.12), HETP was 12mm with a sorbent bed height of 5.6cm. Figure 4.17 shows as well the deuterium concentration profile for the above experiment. A steeper deuterium concentration profile is due to the fact that the H–D mixture separation factor is higher

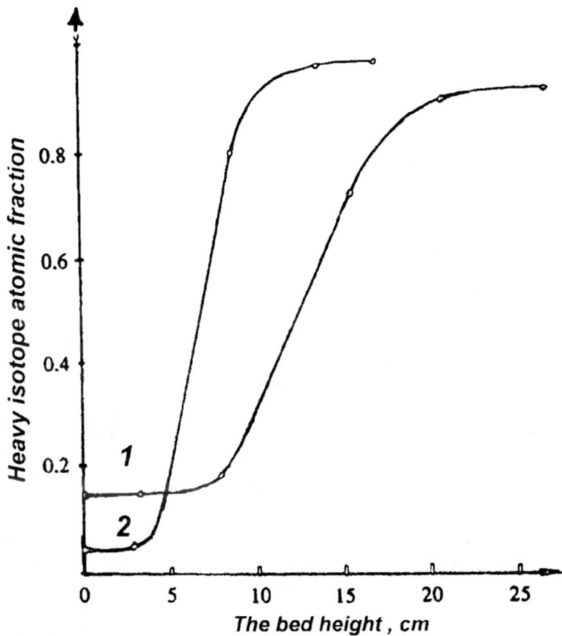

**Figure 4.17**    Dependence of tritium (1) and deuterium (2) concentration on the column separation section height.

than that of the D–T mixture (for the H–D mixture, with a deuterium concentration increase from $x \approx 0$ to 100%, $\alpha$ rises from 2.17 to 2.70, and for the D–T mixture it accounts for 1.47).

The high separation degree values at a small sorbent bed height, obtained in the plants described above, as well as their operation ease and reliability, demonstrate the possibility of the efficient application of the gas–solid systems for the separation of light elements' isotopes.

In conclusion of this section devoted to the techniques of the implementation of the continuous counter-current separation process in solid–phase systems, it should be noted that in the sectioned column an efficient nitrogen isotope separation was achieved in the systems both with liquid phase ($NH_4OH$ aqueous solution), and with KU-2 cationite [88], and that the principle of the relative counter-current movement was successfully utilized in the two-temperature separation by the displacement of temperature zones with the immobile solid phase bed [78].

## 4.4   APPLICATION OF THE SOLID-PHASE SYSTEMS FOR THE SEPARATION OF TRITIUM-CONTAINING HYDROGEN ISOTOPE SYSTEMS

The separation of tritium-containing mixtures with the use of hydrogen-solid systems excels most systems used for heavy water production. The advantage lies in the fact that the working substances are not liable to radiolysis. A relatively low capacity of the columns, however, restricts their possibilities to rather small-scale functions. Let us consider, hence, such functions as the purification of the nuclear reactor heavy-water moderator and regeneration of the D–T mixture following the  thermonuclear reactor plasma chamber as rather complicated, but not calling for the processing of high flow rates, tasks.

Estimating the IMC systems as prospective, we will consider the CHEX systems only from the point of view of Pd-containing systems. On the basis of such systems, chromatographic plants have been tested, and continuous counter-current separation processes in the immobile solid-phase bed, as well as in hypersorption columns, have been developed.

The purification of the nuclear reactor's heavy-water moderator with the use of sectioned columns filled with a palladium sorbent is considered by Andreev *et al.* [3],  and that with the use of zeolite by Alekseev *et al.* [89].

Irrespectively of the sorbent, the plant for the heavy-water moderator isotope purification must include a catalytic isotope exchange (CTEX) unit, and a column with sections intended for the extraction of tritium and protium. Figure 4.18 schematizes plants for the isotope purification with the use of palladium and zeolite. Let us consider the stages of the admixture concentration without dwelling on the CIE unit which, just as it does in the cryogenic rectification plants (see section 2.4.4), serves to transfer tritium and deuterium admixtures from the heavy-water coolant to the deuterium circulating in the closed circuit.

When palladium is used (see Figure 4.18a), deuterium from the CTEX unit comes to the recuperative heat exchanger to be cooled by the deuterium arriving at the CTEX unit, and enters the tritium concentrating column where the following isotope exchange reaction occurs:

$$D_2 + T(Pd) \leftrightarrow DT + D(Pd). \tag{4.13}$$

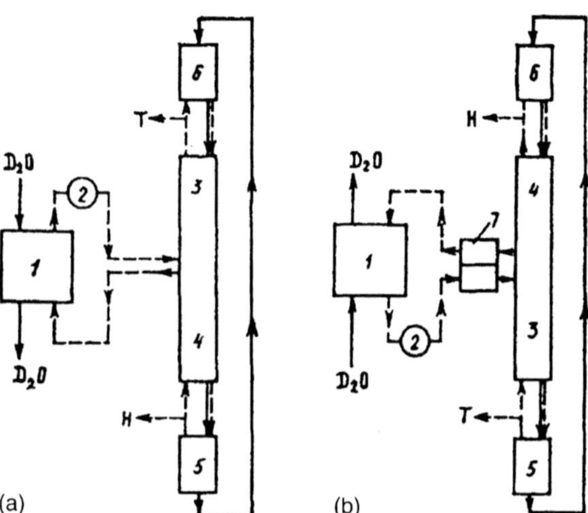


**Figure 4.18** Schemes of heavy-water retarder isotope purification plants with the use of palladium (a) and zeolite (b): 1, CTEX unit; 2, circulation compressor; 3, tritium concentrating column (section); 4, protium concentrating column (section); 5, desorber; 6, adsorber; 7, recuperative heat exchanger.

Enriched with tritium, the gas is withdrawn from the column head. In principle, owing to the $D_2 + T_2 \leftrightarrow 2\,DT$ reaction, tritium enrichment can be performed up to commercial product level. It is advisable to utilize a second column with a smaller cross-section (second order) to decrease the tritium holdup in the separation equipment serving for the final concentration (at this stage the separation factor increases to $\alpha_{DT}^{-1} = 1.53$ at 298K). The column bottom is used for the concentration of protium withdrawn from the gas flow leaving the desorber in the form of an equilibrium mixture $H_2 + HD + D_2$.

To compare the technique under study with the deuterium cryogenic rectification, Table 4.13 presents the results of the calculation for a plant of the same capacity as that of the operable plant for the heavy-water moderator purification at the Grenoble nuclear site [90]. The table gives the characteristics of the main column serving for the initial tritium concentrating (up to $5.55 \times 10^{13}$Bq/l as converted to heavy water), and for the protium concentrating up to 40at.%.

Isotope purification with the use of the $H_2$–Pd system has the following advantages:

1.  the purification is done at an indoor temperature and atmospheric pressure;
2.  the final concentration of tritium can be performed in the separation unit without a special stage of DT catalytic decomposition in a separate reactor;
3.  low energy inputs.

The energy inputs are required only to heat the palladium sorbent (in the hydride form) up to the desorber temperature (420–470K), and to supply the desorption heat. Since no special coolant is required for the cooling of the gas and palladium sorbent flows issuing from the desorber (cooling water can be utilized), the energy consumption at this stage is minor. Drawbacks of this technique such as the high cost of palladium require a search for other efficient methods with hydride phases.

In the isotope purification over zeolite, deuterium from the CTEX unit passes as well through the recuperative heat exchanger and enters the separation plant (Figure 4.18b). Depleted from tritium and protium, the gas is withdrawn from the column above the feed

**Table 4.13**

Main characteristics of the plant for heavy water isotope purification with an annual tritium capacity of $8.9 \times 10^6$GBq (at a tritium concentration in the source flow of 70GBq/$l$) and an annual protium capacity of 100$l$ of $H_2O$ (at a protium concentration in the source flow of 0.4 at.%) [3]

| Characteristic | Plant with sorbent | | Grenoble |
| --- | --- | --- | --- |
| | palladium | zeolite | cryogenic rectification plant |
| Temperature, K | 298 | 77 | 24–25 |
| $\alpha_{DH}$ | 1/2.47 | 1.70 | 1.51 |
| $\alpha_{DT}$ | 1/1.47 | 1.29 | 1.22 |
| NTP | 30 | 50 | 110 |
| Column cross-section, cm$^2$ | 1400[a] | 2100[a] | 490 |

[a] At $G_{SP} = 20$ kmol/(m$^2$·s).

point and returned to the CTEX unit. The gas enriched with protium (HD + $D_2$) is withdrawn from the column head, and the gas enriched with protium ($D_2$ − DT) is withdrawn from the desorber. In the desorber, zeolite is heated to 300K. This is why it is necessary to cool both the gas released in the desorber and supplied to the column bottom, and zeolite fed to the adsorber, to the operating temperature equal to 77K.

The separation energy inputs are generally determined by the energy spent for  cooling zeolite and deuterium flow, and for discharging the adsorption heat from the adsorber.  On the whole, the energy consumption is lower than that in the deuterium cryogenic rectification. Another advantage over the cryogenic rectification is the possibility of performing the process at higher temperatures.

Notice that the deuterium accumulation in the desorber complicates radiation safety problems during zeolite transportation to the adsorber in the hypersorption separation, as opposed to the plant with palladium. For this reason, as well as for the reason that the HMEX reactions do not occur over zeolite, it is advisable to use the hydrogen–palladium system at the second stage of the cascade.

The results of the calculation of an adsorption separation plant with the NaA zeolite are presented as well in Table 4.13. As can be seen from the table, the isotope separation in the solid-phase systems has major advantages over the cryogenic rectification. In spite of this, the hydrogen isotope separation in solid-phase systems has not yet found practical use in the heavy-water retarder purification owing to the problems associated with the implementation of the continuous counter-current separation process in the hydrogen-solid systems.

We now turn our attention to the problems of the isotope purification of the mixture issuing from the thermonuclear reactor plasma chamber. By now, a unified hydrogen isotope separation system for both water and gas flows containing tritium has been put forward. In particular, the Isotope Separation System (ISS) project provides for a system which, in addition to the water rectification system, incorporates four cryogenic rectification columns. The complexity of the scheme is because of the necessity of using the HMEX units (see section 2.4.5).

Since the isotope separation on palladium is accompanied with HMEX reactions, the ISS can be significantly simplified with a reduction of tritium amount. It can be illustrated by the stage of D–T mixture purification from protium which determines the major contribution (about 70%) to the overall tritium amount in the separation system.

The ISS ITER provides for the processing of 71.4g-mol/h of the D–T mixture (1.0at.% of H + 49.5at.% of D + 49.5at.% of T) comprising the mixture purification from protium with concurrent protium and deuterium concentrating at the other column end (at a tritium content of about 2at.%). To solve this problem with the use of the $H_2$–Pd system, one column operating on the scheme presented in Figure 4.12a is required. Based on the feed flow F and its isotopic composition, as well as on the regenerated D–T mixture composition and on the tritium concentration equal to 2at.%, the flows B (regenerated D–T mixture withdrawal) and P (protium and deuterium withdrawal), as well as the protium and deuterium concentrations in the flow P, can be  determined by the material balance equation. The B and P values are given in Table 4.14, with the following flow P isotope composition: 2.56at.% H + 95.44at.% D + 2at.% T.

The column head ensures the purification of protium from 1 to $10^{-3}$ at.% ($K_H$ = 1000). Since the tritium–deuterium ratio in the source mixture is equal to unity, in the calculation

**Table 4.14**

Main parameters of the plant for regeneration of 71.4g-mol/h of the D–T
mixture issuing from the thermonuclear reactor plasma chamber in a sectioned
column with the use of palladium sorbent [19]

| Parameter | Column head | Column bottom |
| --- | --- | --- |
| Withdrawal flow, g-mol/h | $B = 43.5$ | $P = 27.9$ |
| Gas flow $G$, g·mol/h | 197.6 | 126.2 |
| Flow-rate ratio | 1.28 | 0.82 |
| Separation factor | $\alpha_H^{-1} = 2.4$ | $\alpha_{DT}^{-1} = 1.47$ |
| NTP | 11 | 18 |
| Number of separation sections | 4 | 6 |
| Tritium amount, g | 24 | 10 |

of this column section, the value $\alpha_H^{-1} = 2.4$ at $\alpha_{HD}^{-1} = 2.01$ and $\alpha_{HT}^{-1} = 2.67$ was used (in fact, the average separation factor $\alpha_H^{-1}$ is somewhat higher owing to the ratio increase to 0.8:0.2 = 4). At a relative withdrawal value $\theta = 0.8$, the flow-rate ratio $\lambda = G/L = 1.28$, and NTP calculated by the "plate-to-plate" method is equal to 11. As can be seen from the calculation, at this flow-rate ratio and separation factor $(\alpha_{D-T}^{o}) = 1.51$, about 100 theoretical plates of separation are required for the tritium concentration from 49.5 to 80at.% ($K_T = 4.08$).

In the column bottom, the most difficult problem is the gas purification from tritium $K_T = (49.5 \times 98)/(50.5 \times 2) = 48.03$. Taking the lower range of the D–T mixture separation factor D–T $\alpha_{DT}^{-1} = 1.47$ at $\theta = 0.8$, we obtain $\lambda = 0.82$, and by the "plate-to-plate" method we obtain NTP equal to 18. At a specific flow rate $G_{SP} = 20$kmol/(m²·h) in the column head, where the gas flow is the largest, the column cross-section equals 98.8cm².

Assuming the separation section height equal to 6cm at HETP = 2cm, the column head will incorporate four, and the bottom six, separation sections. Therefore, with regard to two sections in the flow-conversion units, the column with a palladium sorbent will be of the following dimensions: 72cm in height and 11cm in diameter. At a bulk weight of the palladium sorbent equal to 1.4g/cm³ with a palladium content of 75%, a relatively small amount of palladium (about 10kg) is required to fill the column.

Table 4.14 presents as well the results of the calculation of the tritium holdup in the column. The calculation was done with regard to the tritium concentration profile along the column height, the palladium gas capacity equal to 50 cm³ stp/g of Pd, and subject to the condition that the entire sorption section (adsorber) is filled with a mixture composed of 80at.% T + 20at.% D.

As can be seen from the calculation, it is possible to solve the problem of the regeneration of the D–T mixture issuing from the TNR plasma chamber by a compact plant (total column volume is 1.7l) of a simple construction (operable at room temperature and atmospheric pressure) with the use of a relatively small amount of palladium. This makes it possible to reduce the tritium amount in the isotope separation system by at least half.

The results of the operation of the plant used at the final stage of the processing of the mixture leaving the JET reactor plasma chamber indicate a lower efficiency of the gas-chromatographic regeneration of the D–T mixture compared with the sectioned column. Such gas-chromatographic regeneration of the D–T mixture with a daily capacity of   20g-mol requires four chromatographic columns of 5l in volume each [90, 91]. Such a system of chromatographic columns, however, was tested over a wide range of tritium concentrations.

In conclusion it should be remarked that, in addition the problems considered above, the $H_2$–Pd system shows promise both for the protection of environment against tritium discharges at nuclear power plants and at irradiated fuel processing plants, and for emergencies at thermonuclear sites. Owing to the possibility of processing large flows with a relatively low tritium content, it is worthwhile not using the cryogenic rectification method, but more economic techniques not requiring the tritium transfer from water into hydrogen (GS-process, $H_2O$–$H_2$ chemical isotope exchange). It makes possible the realization of a non-waste technology on the condition that tritium isotope enrichment to product level is performed in the sectioned column with palladium sorbent. Such combined schemes for tritium recovery at nuclear-power facilities have been studied [3, 80, 93–96], as have those for radiation safety systems [96]. In addition, the final tritium concentrating in the sectioned column with palladium sorbent after the stage of TNR water coolant purification from tritium with the use of the chemical isotope exchange in the $H_2O$–$H_2$ system or of the GS–process, as well as the protium removal from tritium extracted from the blanket, was considered.

The advisability of utilizing the sectioned column with palladium sorbent for the recycling of tritium waste resulting, in particular, from the production of tracer compounds, is demonstrated in references [97, 98]. For example, an essentially laboratory-scale plant consisting of five separation sections of 10mm diameter with an overall sorbent bed height of 1.5cm (20g of palladium) allows are to obtain, in a matter of hours, 1l of pure gaseous tritium from the waste of the H–T mixture with a tritium atomic fraction of no less than 20%.

Unfortunately, isotope separation with the use of the gas physical sorption has not yet found a wide application. Small-capacity chromatographic processes find use in analytical studies and in the solution of some small-scale problems. Despite several attempts [44, 47, 50], the hypersorption isotope separation of light elements has not found industrial use due to a variety of engineering problems associated with the management of the solid phase displacement.

In closing we should dwell on the advantages and prospects of the gas–solid systems used in solving different isotope separation problems.

The advantages of the method are most obvious in the processing of the hydrogen isotope mixtures with a high level of tritium content. The efficient chemical isotope exchange techniques are unsuitable due to the radiolysis of working substances under the tritium radiation effect. The hydrogen cryogenic rectification currectly used for this purpose is complicated as regards the equipment, and has a variety of other serious drawbacks. The use of the hydrogen–palladium system makes it possible to perform the process in compact equipment under mild conditions – at room temperature and atmospheric (or lower) pressure, with tritium in the gaseous hydrogen form being less ecologically destructive than, for example, water. An important advantage is the hydrogen homomolecular isotope exchange reaction over palladium (e.g. $H_2 + D_2 \Leftrightarrow 2HD$), which allows separation in a single column over the entire isotope concentration range.

As for the separation of other hydrogen isotope mixtures, it is generally the problem of a considerably larger scale, the solution of which requires sorbents of much lower cost than that of palladium. IMCs and zeolites can prove to be such sorbents. To increase the sectioned column productivity, a novel solution must be found for the fast heating–cooling of the sections entering the flow-conversion units (since it is the rate of the gas sorption–desorption in the flow-conversion units that determines the column flow rate).

The gas–solid system utilization for the separation of other light elements (carbon, oxygen, nitrogen) also holds much promise (especially at the final concentration stage) owing to a high efficiency and environmental friendliness, while the available physico-chemical methods are either very energy-intensive, or ecologically hazardous.

## REFERENCES


1. Ya. M. Varshavsky, S. E. Vaisberg, *Rep. Acad. Sci. U.S.S.R.*, **1961**, 140, 1361.
2. J. Bigeleisen, *Proc. Natl. Acad. Sci. U.S.A.*, **1981**, 78, 5271.
3. B. M. Andreev, Ya. D. Zelvenskii, S. G. Katalnikov, *Heavy Isotopes of Hydrogen in Nuclear Technology*, M., IzdAT, **2000**, 344.
4. D. A. Knyazev, N. F. Myasoedov, A. V. Botchkarev, *Uspekhi Khimii*, **1992**, 61, 384.
5. B. M. Andreev, E. P. Magomedbekov, Yu. S Pak, A. A. Firer, *Proc. D. Mendeleev Univ. Chem. Technol. Russia*, **1987**, 147, 59.
6. B. M. Andreev, Ya. D. Zelvenskii, S. G. Katalnikov, V. V. Uborskii, *Isotopenraxis*, **1977**, 13, 12, 440.
7. B. M. Andreev, A. S. Polevoy, A. N. Perevezentsev, *Atomnaya Energia*, **1978**, 46, 1, 53.
8. V. S. Parbuzin, N. I. Malyavskii, *Zhurnal Fizitcheskoi Khimii*, **1976**, 50, 2944.
9. B. M. Andreev, E. P. Magomedbekov, V. V. Shitikov, *Atomnaya Energia*, **1983**, 55, 2, 102.
10. B. M. Andreev, V. V. Shitikov, E. P. Magomedbekov, A. I. Shafiev, *J. Less-Common Met.*, **1983**, 90, 161.
11. B. M. Andreev, A. N. Perevezentsev, I. A. Mandrykin, N. F. Myasoedov, *Radiokhimiya*, **1986**, 28, 212.
12. S. Imoto, T. Tanade, K. Utsunomiga, *Int. J.Hydr. Energ.*, **1982**, 7, 597.
13. A. B. Sazonv, E. P. Magomedbekov, *Atomnaya Energiya*, **1999**, 87, 1, 62.
14. B. M. Andreev, G. H. Sicking, *Ber. Bunsenges. Phys. Chem.*, **1987**, 91, 177.
15. N. I Malyavskii, V. S. Parbuzin, *Vestnik Mosk. Un-ta, ser. Khimiya*, **1977**, 18, 111.
16. A. V. Botchkarev, D. A. Knyazev, B. M. Andreev, E. P. Magomedbekov, *Zhurn. Fiz. Khimii*, **1997**, 71, 1428.
17. A. B. Sazonov, E. P. Magomedbekov, *Zhurn. Fiz. Khimii*, **1999**, 73, 1919.
18. A. V. Botchkarev, E. P. Magomedbekov, A. B. Sazonov, S. A. Samoylov, *Zhurn. Fiz. Khimii*, **1999**, 73, 2180.
19. B. M. Andreev, E. P. Magomedbekov, G. H. Sicking, *Interaction of Hydrogen Isotopes with Transition Metals and Intermetallic Compounds*, Heidelberg, Springer Verlag, **1996**, 168.
20. B. M. Andreev, E. P. Magomedbekov, M. B. Rozenkevich, Yu. A. Sakharovskii, *Heterogenous Reactions of Tritium Isotope Exchange*, Editorial URSS, **1999**, 208.
21. B. M. Andreev, E. P. Magomedbekov, I. L. Selivanenko, *Atomnaya Energiya*, **1998**, 84, 2, 132.
22. B. M. Andreev, E. P. Magomedbekov, A. V. Korolyov, *Zhurn. Fiz. Khimii*, **1990**, 64, 434.
23. B. M. Andreev, O. V. Dobryanin, E. P. Magomedbekov *et al., ibid..*, **1982**, 56, 463.
24. B. M. Andreev, E. P. Magomedbekov, Yu. S. Pak, M. G. Zagliev, *ibid.*, **1984**, 58, 2841.
25. I. I. Vedernikova, E. P. Magomedbekov, B. M. Andreev *et al., ibid.*, **1991**, 65, 1657.
26. J. Tanaka, R. H. Wiswall, J. Reilli, *J. Inorg. Chem.*, **1978**, 17, 498.
27. B. M. Andreev, E. P. Magomedbekov, A. S. Polevoy, *Proc. D. Mendeleev Univ. Chem. Technol. Russia*, **1984**, 130, 45.
28. V. E. Kotchurikhin, Ya. D. Zelvenskii, In: *Zeolites, their Synthesis, Properties, and Application*, M-L., Nauka, **1965**, 319.

29. B. M. Andreev, A. S. Polevoy: *Theoretical Basis of Chemical Technology*, **1995**, 29, 261.
30. B. M. Andreev, A. S. Polevoy, I. A. Alekseev, *Zhurn. Fiz. Khimii*, **1986**, 60, 698.
31. G. M. Pantchenkov, A. M. Tolmatchev, T. V. Zotova, *Zhurn. Fiz. Khimii*, **1964**, 38, 5, 1361.
32. A. M. Tolmatchev, T. V. Zotova, N. M. Eliseeva, *Zhurn. Fiz. Khimii*, **1965**, 39, i. 4, 1021.
33. V. A. Yakovlev, dissertation abstract, M. M. V. Lomonosov Moscow State Univ., **1975**, 14.
34. I. A. Alekseev, I. A. Baranov, V. A. Novozhilov, G. A. Sukhorukova, V. A. Trenin, *Atomnaya Energiya*, **1983**, 54, 409.
35. I. A. Alekseev, V. A. Trenin, *Zhurn. Priklad. Khimii*, **1993**, 66, 132.
36. I. A. Alekseev, S. P. Karpov, V. D. Trenin, *Fus. Techn.*, **1995**, 28, 28.
37. V. S. Parbuzin, G. M. Pantchenkov, *Rep. Acad. Sci. U.S.S.R.*, **1965**, 164, 4, 856.
38. A. S. Polevoy, I. P. Yudin, *Zhurn. Fiz. Khimii*, **1982**, 56, 1229.
39. A. S. Polevoy, I. P. Yudin, *ibid..*, **1982**, 56, 2015.
40. G. G. Zhun, Yu. P. Bagoy, V. P. Maletsky, *ibid.*, **1988**, 62, 11, 2990.
41. B. M. Andreev, E. P. Magomedbekov, I. L. Selivanenko, *Atomnaya Energiya*, **1998**, 84, 2, 141.
42. A. S. Polevoy, A. I. Durneva, *Zhurn. Fiz. Khimii*, **1990**, 64, 2, 440.
43. Yu. D. Zelvenskii, *Isotopenpraxis*, **1968**, 4, 388.
44. A. V. Sarukhanov, V. E. Kotchurikhin, E. P. Magomedbekov, Ya. D. Zelvenskii, *Atomnaya Energiya*, **1974**, 36, 208.
45. A. V. Sarukhanov, Ya. D. Zelvenskii, V. E. Kotchurikhin, *et al.*, *Zhurn. Fiz. Khimii*, **1975**, 49, 2108.
46. A. V. Sarukhanov, dissertation abstract, D. Mendeleev *Univ. Chem. Technol. Russia*, **1974**, 12.
47. A. V. Sarukhanov, V. E. Kotchurikhin, E. P. Magomedbekov, Ya. D. Zelvenskii, *Zhurn. Fiz. Khimii*, **1974**, 48, 1512.
48. T. I. Loseva, V. E. Kotchurikhin, Ya. D. Zelvenskii, *Isotopenpraxis*, **1970**, 13, 389.
49. B. M. Andreev, Ya. D. Zelvenskii, S. G. Katalnikov, *Separation of Stable Isotopes by physical-Chemical Methods*, M., Energoatomizdat, **1982**, 208.
50. A. V. Sarukhanov, V. E. Kotchurikhin, V. V. Uborsky, Ya. D. Zelvenskii, *Proc. D. Mendeleev Univ. Chem. Technol. Russia*, **1974**, 81, 49.
51. K. I. Blank, E. P. Magomedbekov, A. V. Krupentchenko, *Proc. Mendeleev Univ. Chem. Technol. Russia*, **1984**, 130, 80.
52. B. M. Andreev, A. N. Perevezentsev, V. V. Shitikov, *Zhurn. Fiz. Khimii*, **1981**, 55, 1993.
53. H. Buhl, S. Will, U. S. Pat. 4110425 **1978**.
54. H. Buhl, S. Will, U. S. Pat. 2550584 **1977**.
55. S. S. Badayev, L. I. Kopaneva, E. V. Lazarev, *Zhurn. Neorg. Khimii*, **1983**, 28, 1063.
56. E. Henley, E. Johnson, *Chemistry and Physics of High Energy Reactions*, Oxford University Press, Oxford, **1969**, 415.
57. M. Ron, U. K. Pat. 2126206 **1983**.
58. B. M. Andreev, A. N. Perevezentsev, Yu. N. Pisarev, S. M. Ivanov, *Izv. Akad. Nauk. Inorganic Materials*, **1987**, 23, 233.
59. P. G. Romankov, N. B. Rashkovskaya, V. F. Frolov, *Mass-Transfer Processes of Chemical Technology*, L., Khimiya, **1975**, 192.
60. B. M. Andreev, A. S. Polevoy, Izv. Vuzov, *Chemistry and Chemical Technology Series*, **1982**, 7, 889.
61. B. M. Andreev, E. P. Magomedbekov, A. S. Polevoy, *Proc. Mendeleev Univ. Chem. Technol. Russia*, **1989**, 156, 24.
62. M. S. Safonov, V. K. Schiryaev, V. I. Gorshkov, *Zhurn. Fiz. Khimii*, **1970**, 44, 975.
63. V. I. Gorshkov, M. S. Safonov, N. M. Voskresenskii, *Ion Exchange in Columns*, M., Nauka, **1981**, 224.

64. B. M. Andreev, A. N. Perevezentsev, I. L. Selivanenko, In: Proc. Int. Symp., Isotope Separation and Chemical Exchange Uranium Enrichment, Tokyo, October 29 – November 1, 1990, (ed. Y. Fujii, T. Ishida, K. Takeuchi, *Bull. Research Lab. Nucl. Reactors Tokyo Inst. Technol.*, **1992**, special issue 1, 428.

65. B. M. Andreev, E. P. Magomedbekov, G. N. Schvedova, I. I. Levin, *Zhurn. Fiz. Khimii*, **1987**, 61, 1827.

66. B. M. Andreev, E. P. Magomedbekov, A. S. Polevoy, *Proc. Mendeleev Univ. Chem. Technol. Russia*, **1989**, 156, 24.

67. A. S. Polevoy, I. A. Alekseev, *Zhurn. Fiz. Khimii*, **1992**, 66, 2698.

68. A. S. Polevoy, I. A. Alekseev, *TOKhT*, **1989**, 6, 230.

69. B. M. Andreev, A. S. Polevoy, *Zhurn. Prikl. Khimii*, **1977**, 50, 3, 570.

70. E. Glueckauf, G. P. Kitt, *Angew. Chem.* **1957**, 69, 567.

71. E. Glueckauf, G. P. Kitt, In: *Proc. Int. Symp. Amsterdam*, 1957, AOC Nier North Holland, Publ. Co. **1958**, 20.

72. J. E. How, *Science*, **1968**, 161, 464.

73. S. Tistcenko, G. Dirian, *Bull. Soc. Chim. France*, **1970**, 1, 16.

74. B. M. Andreev, A. S. Polevoy, A. N. Perevezentsev, *Radiokhimiya*, **1986**, 28, 489. (in Russian).

75. D. Basmadjian, *Can. J. Chem. Engng.*, **1963**, December, 269.

76. A. Clayer, L. Agneray, G. Vandenbusche, P. Petel, *Z. Anal. Chem.*, **1968**, 236, 240.

77. B. M. Andreev, A. S. Polevoy, *D.A.N. Gruz. SSR*, **1981**, 7, 181.

78. B. M. Andreev, G. K. Boreskov, *Zhurn. Fiz. Khimii*, **1964**, 38, 115.

79. B. M. Andreev, A. S. Polevoy, *Zhurn. Fiz. Khimii*, **1982**, 56, 349.

80. B. M. Andreev, N. A. Rakov, M. B. Rozenkevich, Yu. A. Sakharovsky, *Radiokhimiya*, **1997**, 39, 2, 97.

81. B. M. Andreev, E. P. Magomedbekov, I. L. Selivanenko, *Atomnaya Energiya*, **1998**, 84, 3, 242.

82. B. M. Andreev, A. V. Kruglov, I. L. Selivanenko, *Sep. Sci. Technol.*, **1995**, 30 (16), 3211.

83. B. M. Andreev, I. L. Selivanenko, A. I. Vedeneev, A. N. Golubkov, B. N. Tenyaev, In: Proc 4th All-Russian Conference on Physical–Chemical Processes in Selection of Atoms and Molecules, M., TsNIIatominform, **1999**, 147.

84. A. N. Perevezentsev, B. M. Andreev, I. L. Selivanenko, I. A. Yarcho, *Fus. Engng. Design*, **1991**, 18, 39.

85. B. M. Andreev, A. N. Perevezentsev, I. L. Selivanenko, B. N. Tenyaev, A. I. Vedeneev, A. N. Golubkov, *Fusion Technol.*, **1995**, 28, 505.

86. B. M. Andreev, L. Selivanenko, A. I. Vedeneev, A. N. Golubkov, B. N. Tenyaev, In: Proc. International Seminar on the Potential of Russian Nuclear Centers and MNTTs in Tritium Technologies, Sarov, 17–21 May 1999, Sarov, RFYaTS VNIIEF, **2000**, 58.

87. A. N. Golubkov, A. N. Vedeneev, S. B. Lebedev, *et al. ibid.*, 52.

88. A. V. Kruglov, B. M. Andreev, I. E. Pojidaev, *Sep. Sci. Technol.*, **1996**, 31 (4), 471.

89. N. A. Alekseev, N. A. Baranov, V. A. Novozhylov, G. A. Sukhorukova, V. D. Trenin, preprint of LIYaF-858, **1983**, 22.

90. R. Lässer, G. Jones, J. Hemmerich *et al.*, *Fusion Technol.* **1995**, 28, 681.

91. R. Lässer, A. Bell, N. Bainbridge *et al. Fusion Engng. Design.* **1999**, 47, 301.

92. R. Lässer, *et al.*, *Fusion Engin. Design.*, **1999**, 47, 333.

93. R. D. Penzhorn, J. Anderson, R. Haange, B. Hircq, A. Meikle, Y. Naruse, *Fusion Engng. Design*, **1991**, 16, 141.

94. B. M. Andreev, M. V. Karpov, A. N. Perevezentsev, M. B. Rozenkevich, Yu. A. Sakharovsky, *Hydrogen Energetics Technol.*, **1992**, 1, 57.

95. B. M. Andreev, Yu. A. Sakharovsky, M. B. Rozenkevich, E. P. Magomedbekov, Yu. S. Park *Fusion Technol.* **1995**, 28, 515.

96. B. M. Andreev, Z. V. Ershova, V. L. Zverev, A. V. Kapyshev, A. N. Perevezentsev, M. B. Rozenkevich, *Thermonuclear Fusion*, **1990**, 2, 55.
97. B. M. Andreev, E. P. Magomedbekov, Yu. S. Pak, M. B. Rozenkevich, Yu. A. Sakharovsky. In: 5th All-Russian Conference on Physical-Chemical Processes in Selection of Atoms and Molecules, M. TsNIIatominform, **2000**, 138.
98. B. M. Andreev, E. P. Magomedbekov, Yu. S. Pak, M. B. Rozenkevich, Yu. A. Sakharovsky, I. L. Selivanenko, V. V. Uborsky, *Atomnaya Energiya*, **1999**, 86, 3, 198.
99. B. M. Andreev, E. P. Magomedbekov, Yu. S. Pak, M. B. Rozenkevich, Yu. A. Sakharovsky, I. L. Selivanenko, *Radiokhimiya*, **1999**, 41, 131.

# – 5 –

# Carbon Isotope Separation

## 5.1  CARBON ISOTOPE SEPARATION BY RECTIFICATION

### 5.1.1  Isotope effect in the phase isotope exchange and the properties of main operating substances

Isotope effect based on a pressure difference of saturated vapors of isotopically substituted carbon-containing molecules has been studied by various authors for different operating substances (see Table 5.1) [1–9]. The highest values of the single stage separation factor $\alpha$ are typical for two substances, namely carbon oxide or carbon monoxide (II) CO and methane $CH_4$, for which the available experimental data are the most extensive.

Carbon oxide (II) has the critical temperature $T_{cr} = 132.92K$, normal boiling point $T_{b(n)} = 81.61K$, and melting point $T_{m(n)} = 68.09K$. The liquid density of CO at $T_{b(n)}$ $\rho_l = 790kg/m^3$. The heat of evaporation $\Delta H_{ev} = 6.04kJ/mol$.

In the CO low-temperature rectification over the region of normal boiling point $T_{b(n)}$ and melting point $T_{m(n)}$, the separation factor values fall within the range $\alpha = 1.01–1.003$, with the heavy carbon isotope $^{13}C$ concentrated in the liquid.

According to T. Johns [1], in the liquid-vapour equilibrium of the $^{12}C^{16}O–^{13}C^{16}O$ mixture over the temperature range of $68.1 – 81.2K$, the single stage enrichment factor is described by

$$\varepsilon = 78.2/T^2 - 0.394/T. \tag{5.1}$$

According to E. Ancona *et al.* [2], the enrichment factor within the temperature range of 79 to 108.0K is approximated by

$$\varepsilon = 71.251/T^2 - 0.2423/T - 8 \cdot 10^4. \tag{5.2}$$

Over a relatively wide temperature range, the data of the above-mentioned authors are somehow different. The best convergence is achieved at 79–81.6K. In conformity with eqs. (5.1) and (5.2) at the normal boiling point, $T_{b(n)}$, the separation factor is equal to 1.0069.

Methane is characterized by the critical temperature $T_{cr} = 190.77K$, normal boiling point $T_{b(n)} = 111.66K$, and melting point $T_{m(n)} = 90.68K$. At $T_{b(n)}$ the liquid $CH_4$ density $\rho_l = 426kg/m^3$. The heat of evaporation $\Delta H_{ev} = 8.2kJ/mol$.

As is seen from Table. 5.1, in the methane rectification the separation factor values obtained by the Rayleigh distillation method correlate well, but exceed in absolute magnitude those obtained by the differential method. As pointed out by E. Oziaschvili and A. Egiazarov [10], the reason has to do with an inadequate chemical purity of the substances studied in the mentioned sources [6, 7].

It is felt that the $\alpha$ values for methane within the range from the melting point $T_{m(n)}$ to the boiling point $T_{b(n)}$ vary from 1.0055 to 1.003 [11].

By results of K. Haga and H. Soh [12] the vapor pressure of $^{12}CH_4$ molecules is more than vapor pressure of $^{13}CH_4$ approximately to 0.35%, i.e. $\alpha = 1.0035$. In accordance with

$$\varepsilon = 85.0/T^2 - 0.442/T, \qquad (5.3)$$

at the methane normal boiling point $T_{b(n)}$, $\alpha \approx 1.003$. In any case, the methane rectification is characterized by a lower separation factor value as compared with that of the carbon oxide (II) rectification.

Heavier carbon-containing molecules – ethane, ethylene, etc. (see Table 5.1) – are exemplified by a significantly smaller isotope effect than that in the rectification of CO and $CH_4$. For example, the separation factor values in the benzene rectification ($\alpha = 1.00025$), or in the methanol rectification ($\alpha = 1.0002$) are negligibly small. For this reason, the carbon oxide (II) rectification is of the most practical importance for the carbon isotope separation.


## 5.1.2  Carbon oxide (II) cryogenic rectification

The properties of carbon oxide (II) allow direct use of liquid nitrogen as coolant for liquid CO production. The rectification at the liquid nitrogen temperature level necessitates careful preliminary purification of CO from the admixtures of carbon dioxide (IV) $CO_2$, oxygen, and water to the ppm grade.

The pure gaseous carbon oxide CO can be supplied by chemical plants or produced *in situ*, by the carbon reduction of $CO_2$ (e.g. in the form of activated carbon) at a temperature of 1000°C by

$$CO_2 + C \rightarrow 2\,CO. \qquad (5.4)$$

The history of carbon isotope separation by CO rectification technology accounts for about 50 years. The pilot plants were developed in various countries and at different times, and the first production cascades were put into operation in Great Britain, the U.S.S.R. and U.S.A. CO cryogenic rectification is by now a well–understood, and sufficiently reliable, process. The attraction of this carbon isotope separation technique is indicated by practically first results of laboratory studies (see Table 5.2). The data presented in Table 5.2 demonstrate that the CO rectification is characterized by relatively small HTP values equal to 2–3cm.

Along with the highest separation factor values among rectifiable working substances, the practical interest in this carbon isotope separation technique is determined by the favorable HTP values in the CO rectification.

**Table 5.1**

Isotope effect in the carbon isotope separation by rectification (experimental and calculation data)

| Operating substance | Molecules | Temperature, K | $\alpha$ | Approximation of temperature dependence of isotope effect | Reference |
|---|---|---|---|---|---|
| | | 68.1 | 1.0120[a] | | |
| | | 68.2 | 1.0109[a] | | |
| | | 75.0 | 1.0086[a] | $\varepsilon = 78.2/T^2 - 0.394/T$ | [1] |
| | | 81.1 | 1.0070[a] | | |
| | | 81.2 | 1.0104[a] | | |
| | | 79 | 1.0075[a] | $\varepsilon = 71.251/T^2 - 0.2423/T -$ | |
| Carbon | $^{12}C^{16}O$ | 84.7 | 1.0064[a] | $8 \cdot 10^{-4}$ | |
| oxide CO | $^{13}C^{16}O$ | 87.6 | 1.0057[a] | | [2] |
| | | 96.2 | 1.0043[a] | | |
| | | 108.0 | 1.0031[a] | $\varepsilon = 1.3141/T - 0.0092$ | |
| | | 68.1 | 1.0125[b] | $\varepsilon = 0.889/T - 0.0006$ | [3] |
| | | 81.0 | 1.0104[b] | | |
| | | 69.0 | 1.0131[b] | | |
| | | 81.6 | 1.0093[b] | $\alpha = 0.9954 \exp(1.447/T)$ | [4] |
| | | 101.2 | 1.0062[b] | | |
| | | 90.5 | 1.0049 | $\varepsilon = 2.3(4.1986/T + 0.0775 \times$ | [5] |
| | | 111.8 | 1.0010[a] | $\lg T - 0.1959)$ | |
| | | 90.5 | 1.0114[a] | | |
| | | 97.5 | 1.0108[a] | $\alpha = 0.0040 + 0.6689/T$ | [6] |
| Methane | $^{12}CH_4-$ | 111.8 | 1.0099[b] | | |
| CH$_4$ | $^{13}CH_4$ | 91 | 1.0054[a] | | |
| | | 96.0 | 1.0046[a] | $\lg\alpha = 36.8/T - 0.191/T$ | [1] |
| | | 104.8 | 1.0035[a] | | |
| | | 97.5 | 1.0101[b] | $\varepsilon = 1.1275/T - 0.0019$ | [7] |
| | | 111.8 | 1.0068[b] | | |
| Ethylene C$_2$H$_4$ | $^{12}C^{12}CH_4-$ $^{12}C\,^{13}CH_4$ | 120 / 169 | 1.0014[b] / 1.0016[b] | $\lg\alpha = 0.00136 - 0.0945/T$ | [8] |
| Ethane C$_2$H$_6$ | $^{12}C^{12}CH_6-$ $^{12}C\,^{13}CH_6$ | 130 / 184.5 | 1.0022[b] / 1.0012[b] | $\lg\alpha = 0.3958/T - 0.00213$ | [8] |
| Chloroform CH$_3$Cl | $^{12}CH_3Cl-$ $^{13}CH_3C$ | 307.4 | 1.0008[c, d] | | [9] |
| Benzene C$_6$H$_6$ | $^{12}C_6H_6-$ $^{12}C_5\,^{13}CH_6$ | 307.4 | 1.00025[c, d] | | [9] |
| Methanol CH$_3$OH | $^{12}CH_3OH-$ $^{13}CH_3OH$ | 307.4 | 1.0002[c, d] | | [9] |

[a] Obtained by differential method.
[b] Obtained by Rayleigh distillation method.
[c] Obtained from experiments in the column.
[d] $^{13}C$ is concentrated in vapour phase.

**Table 5.2**

Results of experiments on carbon isotope separation by CO rectification

| | Column | | | HETP, cm | Refer-ence. |
| Material | Diameter, mm | Height, m | Packing | | |
|---|---|---|---|---|---|
| Glass | 17 | 1.34 | SPP[c] 1.5 × 2 × 0.2mm | 4[a] | [3] |
| Steel | 17 | 3.7 | SPP[c] 1.5 × 2 × 0.2mm | 2.7 | |
| Steel | 25 and 17 | 12[b] | SPP[c] 1.5 × 2 × 0.2mm | ≈ 2.2 | [13] |

[a] Relatively high HTP value is due to an inadequate adiabaticity of the glass column used in experiments.

[b] Depletion section of 2m in height and 25mm diameter, concentrating section comprising two units of 5m in height each and 25mm and 17mm diameter.

[c] Spiral-prismatic packing (Levin's packing).

A plant developed by H. London [14] in Harwell, Great Britain, may be considered as the first pilot plant for CO rectification. The plant produced 0.4g of $^{13}$C per day (up to 150g of $^{13}$C per year) with a concentration of 60–70at.%, and involved a single column comprising two sections with the total NTP equal to 600.

A relatively low pressure of 4.75kPa, which corresponds to a temperature of 75.4K, was maintained in the column head. Most likely, the indicated conditions were chosen for the purpose of increasing the separation factor (in these conditions, $\alpha = 1.011$ for $^{12}$C$^{16}$O–$^{13}$C$^{16}$O molecules), and it is hardly probable that they were optimal.

Thereafter, more productive plants for carbon oxide (II) rectification were developed at Prochem, British Oxygen Company in Great Britain, at the Los Alamos National Laboratory in the U.S.A., and at the Research Institute of Stable Isotopes in Tbilisi, U.S.S.R.

The Prochem plant [15] constitutes a three-stage automated cascade of packed columns. The first stage of the cascade consists of two columns of 62.5mm diameter placed in tandem configuration. The first-stage columns feed one second-stage column of 25mm diameter. All columns are 20m in height. The plant produced 1.5kg of $^{13}$C with a concentration of about 91at.% per year. It should be noted that the CO waste flow depleted from $^{13}$C isotope was used to feed a separate rectification column employed for the production of carbon oxide (II) with a concentration of 99.9at.% of $^{12}$C isotope. Subsequently the above-described cascade was supplemented with a rectification column of 19mm diameter and 12.5m height for the $^{13}$C post-concentration from 90.9 to 97.7at.% [16].

It is pertinent to note that the CO cryogenic rectification, apart from the need for thorough purification of the carbon oxide (II) feed flow and severe requirements of the process adiabativity, is characterized by a special feature described below and associated with the CO isotopic composition.

Natural carbon oxide is a mixture comprising six isotopic species of molecules of which three have the highest content of $^{12}$C$^{16}$O, $^{13}$C$^{16}$O, and $^{12}$C$^{18}$O in the natural mixture. Owing to a small difference between the pressure of saturated vapors of $^{13}$C$^{16}$O and $^{12}$C$^{18}$O ($\alpha = 1.002$ at $T = 74.4$K [1]) on the one hand, and because of a low rate of the isotope exchange reaction between CO molecules on the other, only the lightest molecules $^{12}$C$^{16}$O are separated from the binary mixture $^{13}$C$^{16}$O + $^{12}$C$^{18}$O during carbon oxide (II) rectification. In the end, the product (heavy carbon isotope $^{13}$C) is found to be diluted with its lighter counterpart – $^{12}$C isotope – and the production of the highly concentrated carbon-13 with a concentration of over 90–95at.% proves to be very difficult without special measures.

Such measures should include the use of an additional rectification column as it was in the CO rectification process developed by Prochem Company [16], or, which is more effective, the isotope exchange reaction

$$^{13}C^{16}O + {}^{12}C^{18}O \leftrightarrow {}^{13}C^{18}O + {}^{12}C^{16}O \tag{5.5}$$

followed by the separation of lighter molecules in an individual rectification column (the plant at the Los Alamos National Laboratory in the U.S.A. [17]).

In the first case [16] the CO isotope species mixture fed into the additional rectification column had, with the total concentration of $^{13}$C equal to 90.9at.%, the following composition:

| $^{12}C^{16}O$ | $^{13}C^{16}O$ and $^{12}C^{17}O$ | $^{12}C^{18}O$ and $^{13}C^{17}O$ | $^{13}C^{18}O$ |
|---|---|---|---|
| 2.9% | 90.4% | 6.2% | 0.5%. |

After rectification, the content of the light molecules $^{12}C^{16}O$ became significantly lower:

| $^{12}C^{16}O$ | $^{13}C^{16}O$ and $^{12}C^{17}O$ | $^{12}C^{18}O$ and $^{13}C^{17}O$ | $^{13}C^{18}O$ |
|---|---|---|---|
| 0.14% | 92.4% | 2.1% | 5.36%. |

which made it possible to increase the total concentration of $^{13}$C up to 97.7at.% [16].

The second alternative to the use of additional technological operations for the production of highly concentrated $^{13}$C was first implemented in the Los Alamos laboratory in the U.S.A. (see below).

In the USA, several cryogenic rectification plants for the production of the $^{13}$C isotope with a concentration of over 90at.% were developed [17, 18]. The 1969 plant represented a three-stage cascade of packed columns with a total height of 42.7m with a depletion section 6m high (see Figure 5.1a). The lower, or third, stage of the enrichment section comprised a single column of 25.4mm diameter and 24.4m height. The middle, or second, stage include four columns of the same diameter and 6.1m high each placed in parallel, and the upper, or first, stage incorporated 12 columns operating in parallel (for each second-stage column there were three first-stage columns of the same diameter and 12m in height each, of which 6m fell on the depletion section).

The columns were filled with Helipack-type wire mesh packing material made from stainless steel. The plant's annual capacity accounted for about 3–3.5kg of $^{13}$C with a concentration of 92–93at.%. The plant characteristics are given in Table 5.3.

Transition chambers provided with electric heating elements to supply heat for vapor generation were installed between the cascade stages. The system was mounted in a vertical position which allowed for the liquid phase gravity movement, and was enclosed into a duralumin-made vacuum case of 152mm diameter with multilayer insulation made from aluminized polymer film. The plant was placed into a well of 38m depth with a steel encasement of 450mm diameter.

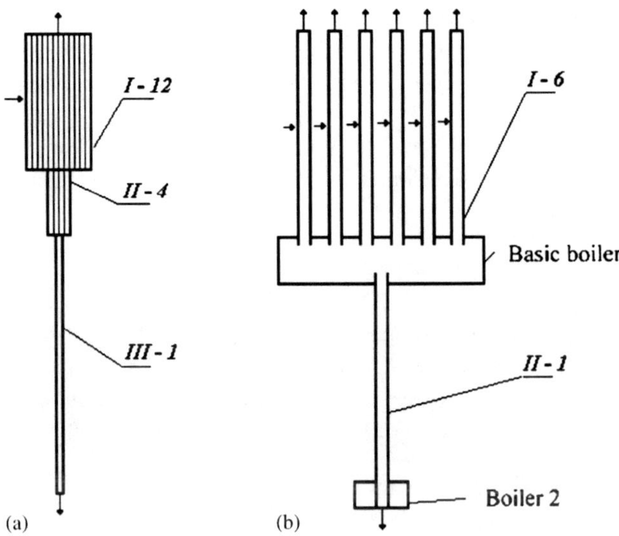


**Figure 5.1**   Schemes of the plants for carbon isotope separation by the CO cryogenic rectification at the Los Alamos National Laboratory, U.S.A.: a, 1969 plant; b, COLA plant; I, II, III, cascade stage numbers; 1, 4, 6, 12, number of parallel columns in a stage of cascade.

## Table 5.3

Characteristics of the plant for the production of $^{13}C$ by the CO rectification (Los Alamos, USA)

| Stage | | I | | II | III |
|---|---|---|---|---|---|
| | | Depletion | Enrichment | | |
| Number of columns | | 12 | | 4 | 1 |
| Column diameter $D$, mm | | 25.4 | | 25.4 | 25.4 |
| Column height $H$, m | | 6 | 6 | 6.1 | 24.4 |
| Packing type | | | Helipack, stainless steel | | |
| Stage flow rate , mol CO per day | | | 13,772 | 4,589 | 0.8125[a] |
| HETP, cm | | | 2.0 – 2.9 | | |
| Pressure $P$, kPa | head | 75[b] | — | — | — |
| | bottom | | — | — | 127 |
| Temperature $T$, K | head | 78.7 | — | — | |
| | bottom | | | | 83.6 |
| $^{13}C$ concentration, at.% | head | 0.63 | 1.1 | | |
| | bottom | 1.1 | | | 92–93 |

[a] Withdrawal flow.

[b] Condenser pressure.

To ensure the plant uninterrupted operation, the $CO_2$ admixture content in the feed flow must not exceed 0.05ppm. For this reason, the feed flow was thoroughly purified at a pressure of 0.7MPa over Ascarite, and after cooling to a temperature of 93K at $P = 0.2$MPa, was filtered.

The plant cooling at the start, and the compensation for the refrigeration loss, were ensured by the evaporation of liquid nitrogen at a daily rate of 1,000l, of which about 52% went to the carbon oxide (II) condensation. The liquid nitrogen specific consumption accounted for about 80tons per 1kg of $^{13}C$ with a concentration of 92–93at.%.

The withdrawn product approximated the following composition:

| $^{12}C^{16}O$ | $^{13}C^{16}O$ | $^{12}C^{18}O$ | $^{13}C^{18}O$ |
|---|---|---|---|
| 1.8% | 92.5% | 5.4% | 0.3% |

The $^{13}C$ concentration in the waste gas flow equal to 0.63at.% corresponded to an extraction degree $\Gamma = 43.5\%$.

The operation experience of this ingeniously designed facility demonstrated the complexity of the plant control, with difficulties in avoiding the liquid holdup fluctuations in the packing which, in turn, leads to the variation of the isotope concentration profiles along the height of the columns. Nevertheless, the HETP value was equal to 2.0–2.9cm, which is a satisfactory index for the rectification columns of this diameter.

A separate rectification column similar to the Prochem plant allowed production of more than 100kg of $^{12}C$ isotope with a concentration of 99.99at.% per year.

In 1979, as a result of the above-mentioned plant improvement, a new and more powerful plant for the production of $^{13}C$ and $^{12}C$ isotopes was developed at the Los Alamos National Laboratory ($B = 10$kg of $^{13}C$ with a concentration of about 90% per year). The plant, named COLA, is shown in figure 5.1b.

The main head unit of the system – the cascade first stage – comprised six packed rectification columns operating in parallel. The columns of 50mm diameter and 100m height were each filled with Pro-Pack packing material.

Each column had a depletion and an enrichment section, individual feed flow input and individual condenser with the depleted flow withdrawal, but the bottoms of all columns were linked by a common evaporating still. From the still, the flow was supplied by gravity to a one second-stage column as a reflux. This column, of the same size as the first-stage columns, was provided with its own evaporizer. The flow rate in each first-stage column accounted for 8400mol of CO per day, that is, eight times as high as that in the column of 25mm diameter of the 1969 plant (see Table 5.3). This rate corresponds to a twofold specific flow rate $G_{SP}$ which is due to the use of a high-capacity packing material.

The use of the Pro-pack packing material, an increase in the column diameter, and a rise in $G_{SP}$ made it possible to enhance HETP up to about 70mm, that is by a factor of 2.4–3.5. The total number of separation plates amounted to $NTP_{\Sigma} = 3000$. The withdrawal flow rate from the second-stage evaporator was equal to 2–2.5mol of $^{13}C$ per day. At a withdrawal flow rate $B = 2.2$ mol of $^{13}C$ per day, the $^{13}C$ concentration in the product accounted for 91.11at.%. The product composition was as follows: 90.82% fell on $^{13}C^{16}O$ molecules, and 0.29% on $^{13}C^{18}O$ molecules.

The degree of $^{13}C$ extraction from the source material was increased to $\Gamma = 60\%$, with liquid nitrogen serving as coolant, as before.

To obtain a product with a higher concentration of the $^{13}C$ isotope, the carbon oxide (II) produced at the COLA plant described above was passed through the catalyst layer at a temperature of 400°C to carry out the HMEX reaction (5.5). Following the reaction (5.5), the gas mixture was subjected to rectification in an additional small rectification column for purification from $CO_2$, and, at a rate of 2.5mol of $^{13}C$ per day, was delivered to the COLITA column with an overall height of 51m (24m depletion section and 27m enrichment section). The rectification in the COLITA column allowed the product to be obtained at a rate of 1.7mol of $^{13}C$ per day (or about 8kg of $^{13}C$ per year) with the $^{13}C$ isotope concentration $x_B = 99.1$at.% at the $^{13}C$ concentration in the waste flow equal to 45at.%.

Within the American National Stable Isotope program (Isotopes of Carbon, Oxygen, Nitrogen, ICON) aimed at producing stable isotopes by cryogenic rectification, the operation of COLA and COLITA plants continued successfully until the late 1980s, when the plants were mothballed because of the privatization of isotope production.

At present, the Cambridge Isotope Laboratories, Inc. (CIL) is the world's premier producer of $^{13}C$ isotope. Using CO cryogenic rectification technology, the company increased, during the past decade, the production capacity several times – from 30kg of 99% $^{13}C$ in the early 1990s up to 120kg per year by 2000. Engineering data on the company's production facilities are not available except for the total length of the columns producing about 30kg $^{13}C$ per year, equal to about 3000 m [19].

In the U.S.S.R., a CO cryogenic rectification plant for $^{13}C$ isotope production was developed at the Research Institute of Stable Isotopes (NIISI) in Tbilisi, Georgia [10, 20]. The plant annual capacity is equal to 2.5–3.0kg of $^{13}C$ with a concentration of over 90at.%.

The plant is a three-stage cascade of columns, schematized in Figure 5.2. The first stage I comprises four columns operating in parallel (see Figure 5.2) of 41mm diameter and with the packing bed height of 26.8m of which 9.7m fall on the depletion section (the $^{13}C$ concentration in the waste flow is 0.6at.%). The second II and third III stages consist of single columns. The second-stage column diameter is 36mm, with the packing bed height equal to 21m. The third-stage column is 20mm in diameter with the packing bed height equal to 27.8m.

All columns of the cascade are filled with a spiral-prismatic packing material made of copper wire 0.2mm in diameter. The packing element size is 2.6mm × 2.6mm in the first and second stages, and 2.2mm × 2.0mm in the third stage.

The plant characteristics are presented in Table 5.4 [20].

Each column is provided with its own feed input, evaporator E and condenser C. All condensers are immersed in a bath of boiling liquid nitrogen.

The liquid for the column II reflux is supplied by gravity from the first-stage withdrawal manifold. The temperature in the second-stage evaporator is 83K.

The concentrate in the vapor phase is withdrawn from the second-stage bottom and delivered to the third stage as the feed flow. The pressure in the third-stage head is 85.3kPa, and in the bottom 96.2kPa. The CO vapors from the third-stage head arrive at the intermediate condenser C' (see Figure 5.2), whence the liquid drains down for the column reflux.

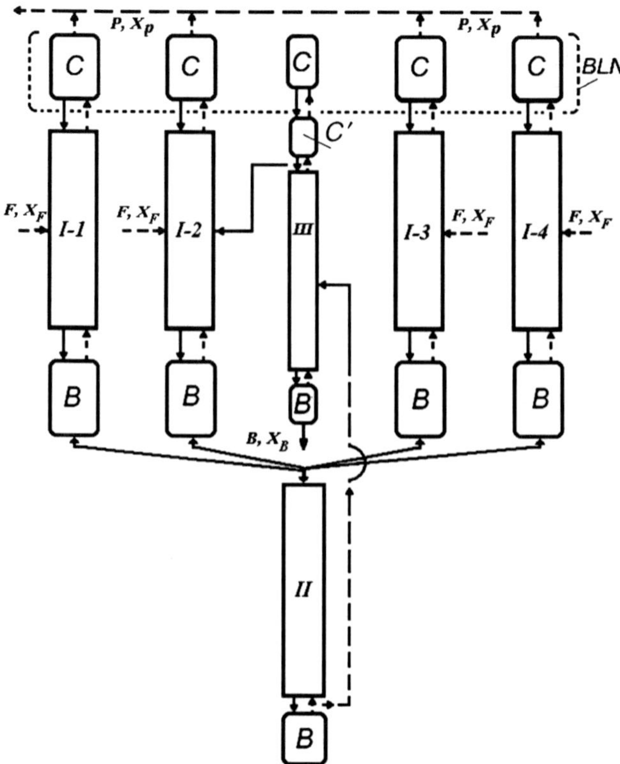



**Figure 5.2**   Scheme of the plant for $^{13}C$ production by CO rectification (NIISI, Tbilisi, Georgia) [20]: I, II, II, stage number; 1, 2, 3, 4, column number; C, condenser; C', intermediate condenser; B, evaporator; BLN, bath of liquid nitrogen.

In the intermediate condenser C', carbon oxide (II) serves as coolant at a corresponding pressure maintained with the use of a helium cushion. The steam condenser for this intermediate cooling agent (CO), in its turn, is immersed in the above-mentioned nitrogen bath.

The third stage has an enrichment and a depletion section arranged so that the upper product of the stage arrives to the feed collector of the column I-2. So, the first and the second stages are coupled by the withdrawal of second kind, and the withdrawal of first kind in the vapor phase is performed from the second stage. This linkage between the second and the third stage makes it possible to avoid a pressure increase in the latter. It allows it as well to be placed within the four-column cluster of this stage, at the same level as the first stage. All the columns are provided with a multilayer insulation and mounted in a common evacuated casing where a residual pressure of not higher that 0.12Pa is maintained.

The purification of the source carbon oxide supplied in bottles under a pressure of 15MPa is performed in a trap cooled with waste cold gaseous nitrogen, and then in a rectification purifier of 23mm in diameter and 1.5m height.

**Table 5.4**

Characteristics of the plant for $^{13}C$ production by CO rectification (NIISI [a], Tbilisi, Georgia)

| Stage | | I | | II | III |
|---|---|---|---|---|---|
| | | Depletion | Enrichment | | |
| Number of columns | | 4 | | 1 | 1 |
| Column diameter $D$, mm | | 41 | | 36 | 20 |
| Column height $H$, m | | 9.7 | 17.1 | 21 | 27.8[b] |
| Packing | | SPP[c] 2.6 × 2.6 × 0.2mm, copper | | | SPP[c] 2.2 × 2.0 × 0.2 mm. copper |
| HETP, cm | | 3.8 | | 3.5 | 2.5 |
| Pressure $P$, kPa | head | — | — | 96.5 | 85.3 |
| | bottom | — | — | 112 | 96.2 |
| $^{13}C$ concentration, at% | head | 0.6 | 1.1 | 9.1 | — |
| | bottom | 1.1 | 9.1 | 22 | 90–92 |

[a] At present, I.G. Gverdtziteli NIISI.
[b] Of which 7m is the depletion section height.
[c] Spiral-prismatic packing (Levin's packing).

The carbon oxide consumption accounts for 1300kg per year, or about 400–500kg per kilogram of $^{13}C$. The liquid nitrogen consumption for the plant cooling is about 100tons per kilogram of $^{13}C$.

In recent years, a new $^{13}C$ isotope production facility based on the CO cryogenic rectification has been developed at the Russian Research Centre Kurchatov Institute [21]. It is planned to develop a separation complex comprising four modules (cascades), with an annual capacity of about 5kg of $^{13}C$ each. It is suggested that all production modules are placed on a single constructing-and-mounting structure of 45m in height.

By this time, a three-stage cascade (the first pilot module) had been developed. The first two stages of the cascade are rated for the production of $^{13}C$ with a concentration of 25 – 30at.%, and the third 90 – 92at.%. An annual design capacity of the module is 4.5 – 5kg of $^{13}C$ [21]. The module is shown in Figure 5.3, with main characteristics given in Table 5.5.

The first stage of the cascade comprises four columns operating in parallel, with depletion and an enrichment sections connected by a common intermediate evaporator. The second and the third stages consist of single columns, with the third stage including a depletion section. The first and the second stages are coupled by the withdrawal of second kind, and the second and the third stages by the withdrawal of first kind.

All the columns are installed in a common high-vacuum jacket, where a residual pressure is maintained at a level of 0.013Pa. The columns are heat-isolated with 20 layers of a screen-vacuum heat insulator.

The liquid nitrogen flow rate accounts for 60l/h.

Special attention is given to the source CO purification from the admixtures of hydrogen, nitrogen, oxygen, argon, water, methane, and carbon oxide (IV). From the bottles, the source CO comes to a catalytic afterburner of oxygen which converts the $H_2$ and $O_2$ admixtures into

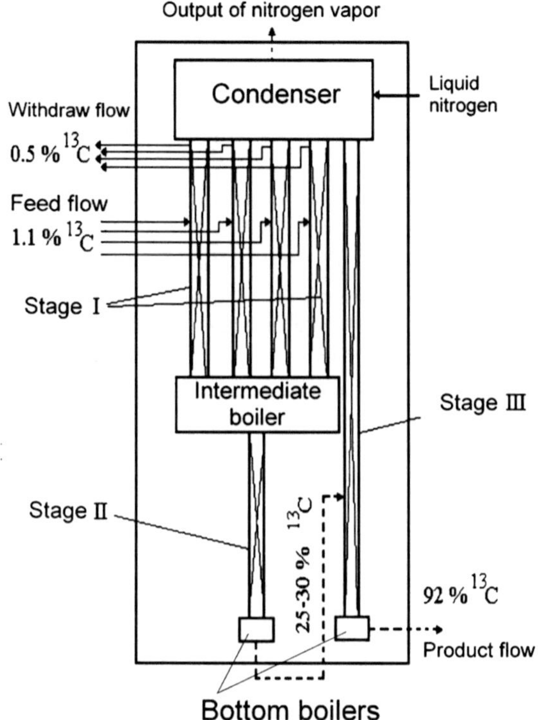


**Figure 5.3**    Scheme of pilot module for $^{13}$C production by CO rectification at Kurchatov Institute Science Centre, Russia [21].

**Table 5.5**

Characteristics of pilot module for $^{13}$C production by CO rectification (Research Centre Kurchatov Institute, Russia, 2003) [21].

| Stage | | I | II | III |
|---|---|---|---|---|
| Number of columns | | 4 | 1 | 1 |
| Column diameter $D$, mm | | 72 | 72 | |
| Column height $H$, m | | | 39[a] | 39 |
| Packing | | SPP[b] 2.6 × 2.6 × 0.2 mm, stainless steel | | SPP[b] 2.2 × 2.5 × 0.2 mm, stainless steel |
| CO flow rate, $l$/h | | 160[c] | 3[d] | 1[e] |
| Evaporator capacity, W | | 1500 | 400 | 60 |
| $^{13}$C concentration, at.% | head | 0.5[f] | — | 1–5[f] |
| | bottom | — | 25–30 | 92 |

[a] Total height of stages II and III.
[b] Spiral-prismatic packing (Levin's packing).
[c] Stage I feed flow rate (40 l of CO per hour to each column).
[d] Stage II withdrawal flow and stage III feed flow.
[e] Stage III withdrawal flow.
[f] In the waste flow.

$H_2O$ and $CO_2$. After that, the source CO is supplied to the silica gel and zeolite adsorbers, then filtered and rectified in a cryogenic rectification purifier. The purification system ensures a residual content of condensable admixtures being at a level of 1ppm.

In the immediate future it is planned to supplement the cascade with a homomolecular isotope exchange unit, and with an additional rectification column for the production of $^{13}C$ with a concentration of 99at.%.

A somewhat different technique allowing production of $^{13}C$ by cryogenic rectification has been suggested by the D. Mendeleev University of Chemical Technology of Russia (MUCTR) [22]. The technique is also based on a modular construction concept, but with different functions, capabilities, and technological characteristics of individual modules.

The production plant, for example, can comprise modules with an annual capacity of 5kg and 10kg of $^{13}C$ with 92at.% concentration (M-5 and M-10 modules, respectively), final concentration modules with an annual capacity of 4.8kg of $^{13}C$ with 99at.% concentration (MF module), and modules capable to annually produce up to 100kg of the carbon light isotope $^{12}C$ with a concentration of 99.99at.% (ML module). With a combination of the above modules, it is possible to produce the carbon isotopes $^{13}C$ and $^{12}C$ with a required capacity and quality.

The selection of technologies and execution of process calculations were performed with the use of the results of an extensive cycle of studies on the cryogenic rectification hydrodynamics and mass transfer in the columns with fine effective packing performed at the laboratory-scale and pilot plants by the Department of the Technology of Isotopes and High-Pure Substances of the MUCTR [23–26].

In conformity with available data on the optimum conditions for the carbon oxide (II) rectification, the condensate formation at a pressure of 67kPa (79K temperature) was adopted for all modules. These conditions are met at a carbon isotope separation factor $\alpha$ = 1.0075 (see eqs. (5.1) and (5.2)).

The MMR packing in the form of wire mesh rings with elements of 4–5mm in size (MUCTR clone of Dixon rings) was adopted as the contactor for the rectification columns with the highest flow rate. Compared with SPP packing, the MMR packing has a higher efficiency with the same HETP value. For the rectification columns with a lower flow rate, the SPP packing with elements of 3mm $\times$ 3mm or 2.5mm $\times$ 2.5mm in size made from wire of 0.2mm in diameter, or caprone or polypropylene thread, which is easier in manufacturing, was adopted (the use of the caprone SPP packing instead of the metallic one allows reduction of the packing weight by a factor of 6, see Table 1.1).

In the determination of the parameters of separation modules, a limitation was set on the height, namely, the packing bed height of the rectification columns must not exceed 20m.

For M-5 and M-10 modules, a three-stage scheme of continuous cascade with a depletion section (extraction degree $\Gamma$ = 80%) and staged flow reduction, comprising three packed columns (4–7 in Figure 5.4) located at the same level, was adopted.

Instead of the pressure increase by the cascade stages leading to a decrease in the separation factor, the pressure in the condensers (top refluxers) 8–11 of all columns of the module is taken to be uniform ($T$ = 79K, $P$ = 67kPa).

The movement of liquid flows in the columns is by gravity, for which purpose the flows from the column bottoms are withdrawn in the vapor phase and then condensed. The condensers are placed at a corresponding level above the columns which allows complete

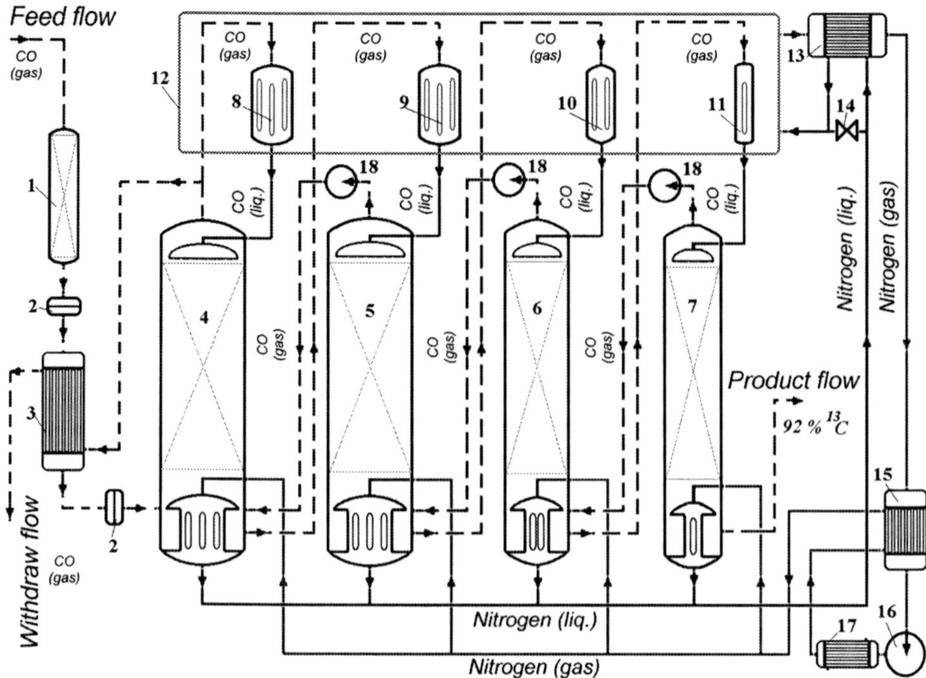


**Figure 5.4**   Scheme of four-column cascade for $^{13}$C production, M-5 or M-10 module [22]: 1, purifier with Ascarite; 2, filter; 3, heat exchanger for the cooling of CO source flow; 4, rectification column with evaporating still – cascade depletion section; 5, 6, 7, rectification columns (stages I, II, II,
respectively) – cascade enrichment section; 8, 9, 10, 11, condensers (refluxers); 12, liquid nitrogen
bath; 13, overcooler; 14, bypass valve; 15, heat exchanger; 16, compressor; 17, cooler; 18, blower.

obviation of the necessity for pumps. For the return of vapor flows from the succeeding
stages to the preceding ones, glandless blowers 18 are mounted.

For the condensate formation, the cooling of the carbon oxide condensers is done in the
nitrogen bath 12 where nitrogen boils at a near atmospheric pressure.

For the vapor flow formation in the columns, the power supply is performed on the thermal pump principle with nitrogen serving as well as an auxiliary heating medium. At a
pressure sufficient for the creation of temperature difference between condensating nitrogen and boiling carbon oxide, the gaseous nitrogen arrives at the pipes of condensers in the
bottoms of the rectification columns, and initiates the carbon oxide boiling in the interpipe
space. Liquid nitrogen formed during the process is withdrawn from below the tubular
evaporators and then throttled through the over cooler 13 to the liquid nitrogen bath to be
used as heat-transfer medium. Gaseous nitrogen from the upper part of the bath passes
through the heat exchangers 13 and 15 and arrives at the compressor 16 to form the heating nitrogen flow and, hence, the nitrogen cycle is closed. This nitrogen cycle makes it
possible to decrease the consumption of liquid nitrogen as heat-transfer medium for the
carbon oxide condensation by about half.

To reduce the heat in leak from the surrounding environment (minimize cold consumption) all columns of the modules are shielded by the EVTI type (screen-vacuum heat insulation) multilayer insulator and housed in a common (for a module) pressurized casing evacuated by the vacuum-maintenance system to a residual pressure of $10^{-2} - 10^{-3}$Pa.

The results of appraisal calculations for a module with an annual capacity of 10kg of $^{13}$C are presented in Table 5.6. The aggregate liquid nitrogen consumption rate for the cooling of condensers accounts for about 120l/h. The difference between the liquid nitrogen aggregate consumption and its supply from an external source is compensated for by the liquid nitrogen formed in the above-mentioned cycle serving for the heating of the rectification columns' evaporators.

The plant consumption rates per kilogram of $^{13}$C are given in Notice that the liquid nitrogen consumption rate at the Los Alamos plant accounts for about 80tons·(1kg of $^{13}$C)$^{-1}$, whereas the rate at the NIISI plant is 100tons·(1kg of $^{13}$C)$^{-1}$.

The isotope exchange and final concentration module (MF) schematized in Fig. 5.5 is designed for the production of highly enriched $^{13}$C (99%).

**Table 5.6**

Characteristics of M-10 module consisting of four CO cryogenic rectification column

| Num-ber | Characteristic | Column | | | |
|---|---|---|---|---|---|
| | | Depletion | Stage I | Stage II | Stage III |
| 1 | Column diameter (int), mm | 100 | 100 | 70 | 40 |
| 2 | Packing bed height, m | 20 | 20 | 20 | 20 |
| 3 | Packing type | MNR[a] | MNR[a] | SPP[b] | SPP[b] |
| 4 | Packing element size, mm | 4–5 | 4–5 | 3 × 3 | 3 × 3 |
| 5 | Packing volume, l | 160 | 160 | 77 | 26 |
| 6 | CO flow rate in the column kg/h | 47 | 47 | 10 | 1,6 |
| 7 | HETP, cm | 5 | 5 | 4 | 3 |
| 8 | NTP | 400 | 400 | 500 | 667 |
| 9 | $^{13}$C concentration, at.%: | | | | |
| | head | 0.22 | 1.1 | 5.3 | 33 |
| | bottom | 1.1 | 5.3 | 33 | 92 |
| 10 | CO consumption rate, kg/h | 0.36 | | | |
| 11 | Heating nitrogen flow rate ($P = 0.21$ MPa), m³/h | 56.5 | | | |
| 12 | Power consumption for the liquid nitrogen compression, kW | 2.2 | | | |
| 13 | Liquid nitrogen consumption[c], kg/h | 66 | | | |
| 14 | Power consumption for the compression of interstage vapour flows, kW | 0.2 | | | |
| 15 | Product flow in the form of CO, g of $^{13}$C per hour | 3 | | | |

[a] Mesh rings.
[b] Spiral-prismatic packing (Levin's packing).
[c] From external source.

**Table 5.7**

Consumption rates per kilogram of $^{13}$C produced in the form of CO with a $^{13}$C concentration of 92at.%

| Pos. | Consumption item | Rate |
|---|---|---|
| 1 | Carbon oxide, kg | 290 |
| 2 | Gaseous heating nitrogen flow rate ($P = 0.21\,MPa$), m$^3$ | $45.2 \times 10^3$ |
| 3 | Liquid nitrogen, ton | 53 |
| 4 | Power consumption (liquid nitrogen compression, vapour return), kWh | 2000 |

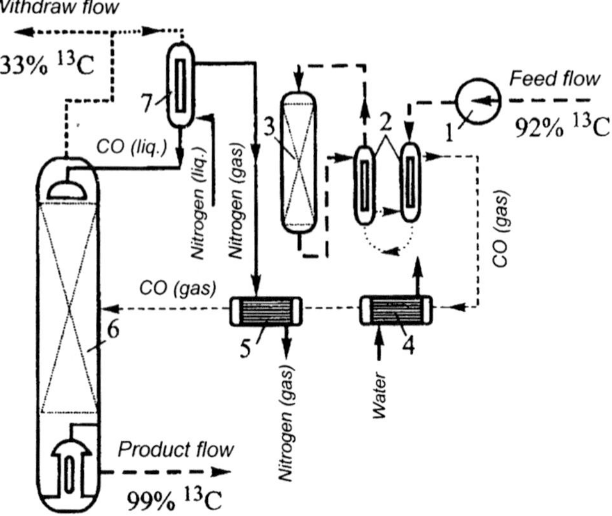


**Figure 5.5**  Scheme of MF module [22]: 1, blower; 2, heat exchanger; 3, isotope exchange reactor; 4, cooler; 5, heat exchanger; 6, rectification column with evaporating still; 7, condenser (refluxer).

The module is fed by the carbon oxide with a 92at.% concentration of $^{13}$C produced in the M-10 or M-5 modules. The blower 1 delivers the CO feed flow through the heat exchanger 2 to the catalytic isotope exchange reactor, where the reaction (5.5) proceeds on the catalyst. As a result of the reaction, a change in the relation between various molecular species of carbon oxide (II) occurs, which can be seen from the following data:

| Molecular species | $^{12}C^{16}O$ | $^{13}C^{16}O$ | $^{12}C^{18}O$ | $^{13}C^{18}O$ |
|---|---|---|---|---|
| Concentration in CO before isotope exchange, % | 5.6 | 91.6 | 2.4 | 0.4 |
| Concentration in CO after isotope exchange, % | 7.50 | 80.52 | 0.48 | 5.50 |

After the reactor 3 (see Figure 5.5), the gas is cooled in the heat exchangers 2, 4, and 5, and fed to the rectification column 6 for the final concentrating. The column consists of a depletion and an enrichment section with a total packing bed height of 20m (SPP 2.5mm $\times$ 2.5mm $\times$ 0.2mm). The column diameter is 20mm. The $^{13}C$ concentration in the waste flow is 33at.%. The waste is returned in the M-10 or M-5 modules for the last column inflow. The liquid nitrogen consumption for the column condenser cooling accounts for 0.4kg per hour. The evaporator is electrically heated.

The waste flows of the M-10 or M-5 modules representing the pure carbon oxide depleted with $^{13}C$ isotope are partly used for the production of the light isotope $^{12}C$. The CO remainder can be chemically processed (e.g. for the production of metal carbonyls, phosgene, spirits, aldehydes, etc.). Otherwise the waste flow can be discharged to the atmosphere in $CO_2$ form, following combustion with air.

The cryogenic rectification column (see Figure 5.6) is adopted for the production of the light isotope $^{12}C$ with a concentration of 99.99at.%. The column operates on a scheme with a head reservoir, in a mode with complete condensate return and periodic product withdrawal. The ML module represents a closed system where vapor–liquid counter-current movement is permanently performed during supply of the heat medium to the still, and of

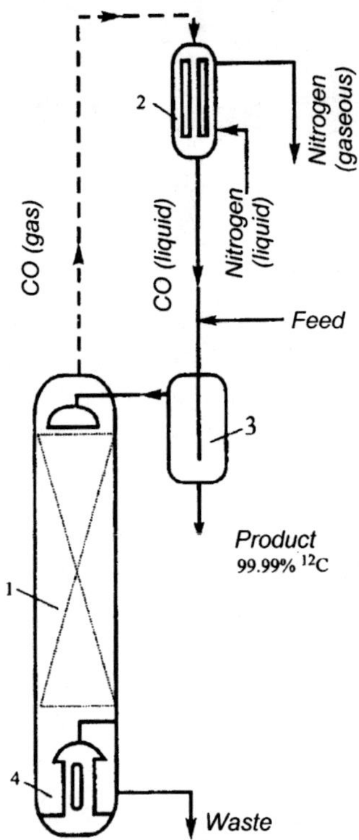


**Figure 5.6**   Scheme of $^{12}C$ production in ML module [22]: 1, rectification column; 2, condenser, (refluxer); 3, head reservoir; 4, column evaporating still.

coolant to the condenser. The still is electrically heated. Because of small-scale flows, it seems inexpedient to organize the liquid hydrogen circulation like in the M-5 or M-10 modules. The rectification conditions ($P$ and $T$) are similar to those in the rectification columns of the above-mentioned modules. The head reservoir size is determined by the amount of liquid carbon oxide in it, equal to 20kg. The rectification column of 50mm diameter is filled with SPP with an element size of 3mm $\times$ 3mm $\times$ 0.2mm. The packing bed height is 20m. The CO flow rate in the column is 5kg per hour.

With the indicated head reservoir volume, the estimated process period (i.e. one operational cycle) accounts for 650hours. With 12 operational cycles per year, the module annual capacity is 240kg of carbon oxide with 99.99at% concentration of $^{12}C$, which is equivalent to 100kg of $^{12}C$ per year. The liquid nitrogen consumption rate is 6kg/h.

The innate of the column, still accounting for about 130g of carbon oxide with an expected concentration of $^{13}C$ in carbon equal to 33at.%, can be used for the replenishment of the last stage of the M-5 or M-10 module.

The suggested technique differs beneficially from the plant for carbon isotope separation by the CO cryogenic rectification developed at the I.G. Gverdtziteli Research Institute of Stable Isotopes in Tbilisi (Georgia) [20], as well as from the plant at the Russian Kurchatov Institute Science Centre [21]. The advantages of the technique [26] include a significantly lower height, the possibility of producing highly enriched $^{13}C$, production of a lighter carbon isotope $^{12}C$, and low consumption rates of the cooling agent – liquid nitrogen.

### 5.1.3 Methane rectification

As source material and working substance for the carbon isotope separation by cryogenic rectification, methane has several advantages over carbon oxide (II): the methane molecular weight is nearly half as high as that of the carbon oxide (II); and methane is an order of magnitude less toxic than carbon oxide (II): (MPC ($CH_4$) = 300mg/m$^3$, whereas MPC (CO) = 30mg/m$^3$).

It is well known that methane is the main component of the natural combustible gas. Liquid methane can be conveniently stored and transported, and liquid methane production is a widely used process technology.

Apart from the fact that methane is less expensive than carbon oxide (II), the use of methane as the working substance makes it possible to perform target isotope extraction by the simplest and most efficient transit scheme. In this case, the waste methane flow with a somehow changed isotopic composition, which leaves the rectification plant, returns to the methane source to serve its main function.

There are data available for the only pilot plant for carbon isotope separation by methane rectification, developed by the Tokyo Gas Company, Japan [12]. The general view of the plant is shown in Figure 5.7. Linked to a liquid natural gas terminal, the plant produces $^{13}C$ with a concentration of 99at. %, $^{12}C$ enriched to 99.9at.%, with highly pure methane (99.9999%) as by-product.

According to the plant operation scheme, after the purification unit, liquid methane is fed to a three-stage system of parallel rectification columns. The first stage consists of seven

**Figure 5.7**  Carbon isotope separation plant by the methane rectification method of Tokyo Gas Company, Japan [12].

columns of 30m height operating in parallel. The second stage incorporates one enrichment column of 30m height and three depletion columns of 6m height each. The third stage comprises one column of 30m height. The scheme includes as well one column of 36m height for the production of methane $^{12}CH_4$. All rectification columns are filled with wire mesh packing. Work is now in progress on the optimization of the rectification process operational conditions with the aim of minimizing production cost and achieving maximum capacity.

More detailed information on the plant process technology parameters, construction, and performance is not yet available.

Sufficiently detailed data on the methane rectification hydrodynamics and mass transfer were obtained from a laboratory-scale plant at the D. Mendeleev University of Chemical Technology of Russia (MUCTR) [26]. The study was done in two columns of 40mm and 50mm diameter with a packing bed height of 1400 and 2000mm. The following four types of packing material were tested in the columns: random (dump) packing materials; SPP packing made from steel wire and from polymeric thread (SPP-P), and MMR wire mesh rings; clone of Dixon rings; and regular structured packing (RSP) made from corrugated metal gauze with a 45° corrugation angle. Characteristics of the packing are presented in Table 5.8.

**Table 5.8**

Characteristics of tested packing materials for rectification columns

| Num-ber | Packing type | Element size, mm | Unit volume mass, kg/m$^3$ | Void Space $V_S$ (m$^3$/m$^3$) | Specific surface $a$, m$^2$/m$^3$ | Equivalent diameter $d_E$, mm |
|---|---|---|---|---|---|---|
| 1 | SPP[a] | $3 \times 3 \times 0.2$ | 1100 | 0.86 | 2580 | 1.33 |
| 2 | SPP-P[b] | $3 \times 3 \times 0.22$ | 185 | 0.83 | 2440 | 1.38 |
| 3 | MMR[c] | $5.2 \times 5.2$ | 600 | 0.93 | 1050 | 3.6 |
| 4 | RSP[d] | $4.0^{e}$ | 700 | 0.92 | 800 | 4.5 |

[a] Spiral-prismatic packing (Levin's packing).
[b] Polymeric spiral-prismatic packing.
[c] Metal mesh rings (clone of Dixon rings).
[d] Regular structured packing.
[e] Crimp height (channel width).

The methane rectification was performed with complete condensate reversion, i.e. in non-withdrawal mode with varying flow rate and temperature (pressure).

The tests consisted of determining hydrodynamic parameters (capacity $L^*_{SP}$ or $G^*_{SP}$, specific hydraulic resistance $\Delta P/H$), and mass-transfer characteristics (HTU, mass-transfer volume factor $K_{OY} \cdot a$) for each packing type.

The results of the capacity determination demonstrate that $L^*_{SP}$ ($G^*_{SP}$) $\sim P^{0.5}$ to within $\pm$ (10–15)%, and that the polymeric packing is inferior to the steel one in capacity [26]:

| Packing type | SPP | SPP-P | MMR | RSP |
|---|---|---|---|---|
| $P$ (column head), kPa | 202 | 192 | 160 | 120 |
| $L^*_{SP}$ ($G^*_{SP}$), kg/(m$^2$·h) | 3400 | 2250 | 4800 | 9800 |

The obtained values of the specific hydraulic resistance and mass-transfer characteristics are given in Table 5.9.

As follows from Table 5.9, the HTU values in the methane rectification are lower than the HETP values in the CO rectification with the use of spiral-prismatic packing (see Tables 5.2 and 5.4). For a column of 41mm diameter, as an example, HETP is equal to 2cm and 3.8cm, respectively. For the MMR packing, the HTU values are about 2.5–3 times higher than those for the SPP packing. For the regular packing with a relatively high capacity, the HTU values do not exceed 8cm, and the mass-transfer coefficient is near to $K_{OY} \cdot a$ for the SPP packing. The polymeric packing is intermediate between SPP steel packing and MMR packing (with HTU values close to those of the SPP packing, and the mass-transfer coefficient $K_{OY} \cdot a$ near those of the MMR packing).

As a whole, the mass-transfer and hydrodynamic characteristics of the methane rectification are quite favorable for the practical separation of carbon isotopes. In the process, the regular packing (RSP) with its high capacity and low hydraulic resistance can be used at the first stages of the cascade, and the SPP or SPP-P packing materials with the lowest HTU values at the subsequent stages.

**Table 5.9**

Specific hydraulic resistance and mass-transfer characteristics in the methane rectification [26].

| Num-ber | Packing type[a] | $P^b$,kPa | $L_{SP}/L^*_{SP}$ | $\Delta P/H$, kPa/m | HTU, cm | $K_{OY} \cdot \alpha$, kmol/(m³·s) |
|---|---|---|---|---|---|---|
| 1 | SPP | 202 | 0.61 | 3.5 | 1.8 | 2.00 |
| 2 | SPP | 202 | 0.78 | 6.0 | 2.0 | 2.15 |
| 3 | SPP-P | 102 | 0.50 | 2.8 | 1.8 - 1.9 | 1.02 |
| 4 | MMR | 160 | 0.63 | 7.6 | 5.6 | 0.97 |
| 5 | MMR | 160 | 0.85 | 12.1 | 6.2 | 1.15 |
| 6 | RSP | 120 | 0.60 | 6.1 | 7.1 | 1.44 |
| 7 | RSP | 120 | 0.82 | 10.0 | 7.6 | 1.84 |

[a] see notes to Table 5.8.

[b] pressure in the condenser.

To compensate for the refrigeration loss in the methane rectification, it is preferable to use the cooling cycle by A. Klimenko [27]. The mentioned cycle is used in practice for production of liquid natural gas.

The methane rectification can be used as a sufficiently simple and safe technique of carbon isotope separation. The choice between the rectification of CO and that of $CH_4$ can be based on the estimation of the extent to which the advantages of the methane utilization (low cost, vast resources, and low toxicity) override the drawback of a lower separation factor compared with carbon oxide (II) rectification.

Preliminary estimations show that the methane rectification operating costs determined by the energy input level can be lowered several times despite a more than twofold reduction in $\varepsilon$.


## 5.2   CARBON ISOTOPE SEPARATION BY CHEMICAL EXCHANGE METHOD

### 5.2.1   Isotope equilibrium

Chemical exchange systems that are the most interesting from the point of view of carbon isotope separation, and isotope effects peculiar to these systems, are presented in Table 5.10 based on the literature [28–38].

According to the properties of chemical compounds forming chemical exchange operating systems (Table 5.10), these systems can be subdivided into three groups: systems with the $CN^-$ ion; systems with gaseous carbon oxide (II); and systems with carbon dioxide (IV), with each group comprising two systems.

The first group consists of the systems that form the basis of cyanhydrine and cyanide isotope separation methods. The group is characterized by the highest (at indoor temperature) values of the separation factor: for the diethyl ketone cyanhydrine – potassium cyanide system $\alpha = 1.035$ at 298K; for the cyanide method (gas–liquid system) $\alpha = 1.013$ at the same temperature. The second group comprises a system based on CO dissolved in an organic substance – liquid propane, propylene, or their mixture in equal proportions (absorption method), and a system with a CO molecular complex in the liquid phase

$Cu_2Cl_2 \cdot 8NH_4Cl \cdot CO$ (complex method of the carbon isotope separation). Both systems have close separation factor values $\alpha \approx 1.014$, which, however, is achieved within different temperature ranges.

Practically equal values of $\alpha$ are as well realized in the third group of systems at indoor temperature (1.012 in the bicarbonate system and 1.01 in the carbamate system at 298K). For carbamate systems $(R_2NCO_2NR_{2(sol,\ org.)} - CO_{2\ (gas)})$, high separation factor values are possible with a temperature decrease down to about 250–233K. In $n$-dibutylamine-based systems, the temperature dependence of the separation factor over the 233–333K range is approximated [36] by

$$\alpha = 0.9654 \exp\left(106.4/RT\right), \tag{5.6}$$

and for a system based on diethylamine in toluene, the separation factor can be found [38] from

$$\ln\alpha = (19.795/T) - 0.058. \tag{5.7}$$

In chemical isotope exchange reactions, higher separation (enrichment) factor values are attained compared with the phase isotope exchange (Figure 5.8).

As follows from Figure 5.8, the carbon isotope separation processes in general are characterized by two groups of methods or systems for which close values of $\varepsilon$ are realized in different temperature regions. In the case of the phase isotope effect (rectification processes), the enrichment factor at a level of $\varepsilon = 0.01–0.012$ is obtained at a sufficiently lower temperature than the same $\varepsilon$ values in the chemical isotope exchange reactions.

From the point of view of $\alpha$ or $\varepsilon$ values, it is the carbamate system that has major advantages over all other liquid–vapour or liquid–gas systems. At a near indoor temperature, the system is characterized by higher enrichment degree factor values compared with the CO or methane rectification. A temperature decrease to about 250K allows the enrichment factor values typical for such chemical exchange systems as the cyanide system, or systems based on the molecular complex $Cu_2Cl_2 \cdot 8NH_4Cl \cdot CO$, to be achieved. As applied to carbon isotope production, the highest separation degree value is a necessary, but not yet sufficient, condition for the use of one or another system as a production process basis. Here, proper account must be taken of the working substance resources, an interphase exchange rate (HETP), and consumption of electric power and/or chemicals, as well as of possible and required flow conversion completeness in the chemical isotope exchange processes. To perform a more detailed analysis, it is necessary to choose from the chemical isotope exchange systems presented in Table 5.10 those operating systems that make it possible to realize the most inexpensive and technologically appropriate flow conversion technique – the thermal method. At close separation factor values, it is hardly probable that the chemical flow conversion technique allows good process performance characteristics (cyanide and bicarbonate separation methods). In addition, efficient catalysts capable of improving the mass-transfer characteristics of the latter method have not yet been found.

Systems with thermal flow conversion include absorption, cyanhydrine, complex and carbamate methods of carbon isotope separation. Despite relatively high enrichment factor values, it is unlikely that the absorption process will have evident advantages over CO or

**Table 5.10**

Isotope effect in carbon isotope separation by chemical isotope exchange method at 0.1 MPa pressure

| Method | Operating system | Isotope exchange reaction | Conditions | | $\alpha$ | Reference |
|---|---|---|---|---|---|---|
| | | | $T$, K | $C^a$ | | |
| Cyanhydrine[b] | $R_2C(OH)CN_{(sol, org.)} - KCN_{(sol, aq)}$<br><br>R – ethyl (-$C_2H_5$); solvent - xylene | $R_2C(OH)^{12}CN_{(sol, org.)} + K^{13}CN_{(sol, aq)} \leftrightarrow$<br>$\leftrightarrow R_2C(OH)^{13}CN_{(sol, org.)} + K^{12}CN_{(sol, aq)}$ | 298 | | 1.035 | [28] |
| Cyanide | $HCN_{(gas)} - NaCN_{(sol, aq)}$ | $H^{12}CN_{(gas)} + Na^{13}CN_{(sol, aq)} \leftrightarrow$<br>$\leftrightarrow H^{13}CN_{(gas)} + Na^{12}CN_{(sol, aq)}$ | 298 | 4.5M | 1.013 | [29, 30] |
| Absorption | $R \cdot CO_{(lq)} - CO_{(gas)}$<br>R – propane + propylene | $R \cdot {}^{12}CO_{(lq)} + {}^{13}CO_{(gas)} \leftrightarrow$<br>$\leftrightarrow R \cdot {}^{13}CO_{(lq)} + {}^{12}CO_{(gas)}$ | 77.4<br>90.2<br>100.8 | 1:1[c] | 1.029<br>1.014<br>1.011 | [31] |
| Complex[d] | $Cu_2Cl_2 \cdot 8NH_4Cl \cdot CO_{(sol, aq)} - CO_{(gas)}$ | $Cu_2Cl_2 \cdot 8NH_4Cl \cdot {}^{12}CO_{(sol, aq)} + {}^{13}CO_{(gas)} \leftrightarrow$<br>$\leftrightarrow Cu_2Cl_2 \cdot 8NH_4Cl \cdot {}^{13}CO_{(sol, aq)} + {}^{12}CO_{(gas)}$ | 293<br>297 | | 1.016<br>1.014 | [32, 33] |
| Bicarbonate | $HCO_3^-{}_{(sol, aq)} - CO_{2 (gas)}$ | $H^{12}CO_3^-{}_{(sol, aq)} + {}^{13}CO_{2 (gas)} \leftrightarrow$<br>$\leftrightarrow H^{13}CO_3^-{}_{(sol, aq)} + {}^{12}CO_{2 (gas)}$<br>${}^{12}CO_{2 (sol, aq)} + {}^{13}CO_{2 (gas)} \leftrightarrow$<br>$\leftrightarrow {}^{13}CO_{2 (sol, aq)} + {}^{12}CO_{2 (gas)}$ | 298<br><br>313 | | 1.012<br><br>1.008 | [34, 35] |
| Carbamate | $R_2NCOO^-{}_{(sol, org.)} - CO_{2(gas)}$<br>R- $n$-butyl (-$C_4H_9$); solvent – octane ($C_8H_{18}$)<br><br><br>R-ethyl (-$C_2H_5$); solvent – toluene ($C_6H_5OH$) | $R_2N^{12}COO^-{}_{(sol, org)} + {}^{13}CO_{2 (gas)} \leftrightarrow$<br>$\leftrightarrow R_2N^{13}COO^-{}_{(sol, org)} + {}^{12}CO_{2 (gas)}$ | 233<br>293<br>298<br>318<br>333<br>285<br>295<br>308 | 2M<br><br>3M<br><br><br>3M | 1.02<br>1.01<br>1.01<br>1.0075<br>1.006<br>1.011<br>1.0097<br>1.006 | [36]<br><br>[37]<br><br><br><br>[38] |

[a] Component concentration in the solution.

[b] In the literature: CYANEX process or system.

[c] Propane : propylene ratio.

[d] In the literature: COCO process or system.

[e] Carbamate – anion.

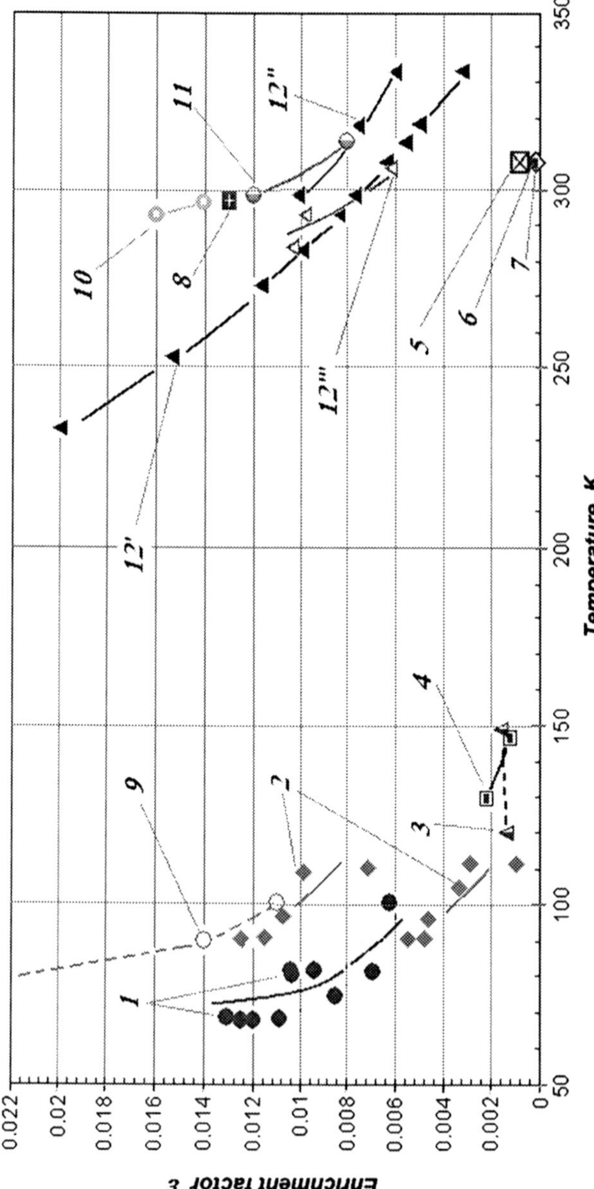

**Figure 5.8** Carbon isotope enrichment factor in rectification and chemical isotope exchange [1–9, 29–38] (enrichment factor for CYANEX system [28] is not shown): 1, $CO_{(lq)}$, $CO_{(vap)}$; 2, $CH_{4(lq)}$, $CH_{4(vap)}$; 3, $C_2H_{4(lq)}$, $C_2H_{4(vap)}$; 4, $C_2H_{6\,(lq)}$, $C_2H_{6(vap)}$; 5, $CH_3Cl_{(lq)}$, $CH_3Cl_{(vap)}$; 6, $C_6H_{6(lq)}$, $C_6H_{6(vap)}$; 7, $CH_3OH_{(lq)}$, $CH_3OH_{(vap)}$; 8, $HCN_{(gas)}$, $NaCN_{(sol.\ water.)}$; 9, $R{\cdot}CO_{(lq)}$, $CO_{(gas)}$, where R is propane $+$ propylene; 10, $Cu_2Cl_2{\cdot}8NH_4Cl{\cdot}CO_{(sol.\ water.)}$, $CO_{(gas)}$; 11, $HCO_3{}^-_{(sol.\ water.)}$ - $CO_{2(gas)}$; 12′, $R_2NCOO^-_{(sol.\ org.)}$ - $CO_{2(gas)}$, where R- n- butyl (-$C_4H_9$), solvent – octane ($C_8H_{18}$), 2 M solution; 12″, $R_2NCOO^-_{(sol.\ org.)}$ - $CO_{2\,(gas)}$, where R- n- butyl (-$C_4H_9$), solvent – octane ($C_8H_{18}$), 3 M solution; 12‴, $R_2NCOO^-_{(sol.\ org.)}$ - $CO_{2\,(gas)}$, where R- ethyl (-$C_2H_5$), solvent – toluene ($C_6H_5OH$), 3M solution.

methane rectification. The point is that the separation factor values given in Table 5.9 were obtained at a relatively low pressure. Besides, these values were found without regard to the variations in oxygen isotope composition [31]. Because of this, it is appropriate to consider three carbon isotope separation methods: the cyanhydrine technique; a process based on the molecular complex $Cu_2Cl_2 \cdot 8NH_4Cl \cdot CO$; and the carbamate technique.

## 5.2.2  Cyanhydrine and complex methods of carbon isotope separation

In the cyanhydrine method, isotope exchange proceeds in the liquid–liquid system (organic and water phases) and can be realized in pulsating extraction columns or with the use of, for example, centrifugal extractors. L. Broun and J. Drury [28] recommend using diethyl ketone cyanhydrine in xylene and KCN aqueous solution as the operating system. The flow conversion is performed thermally: by heating to 52°C at the column end enriched with $^{13}C$, and by cooling to 8°C at the depleted column end. The method is characterized by the highest separation factor value ($\alpha = 1.035$) and by fast isotope exchange kinetics (1.5s half-exchange time with the use of diethyl ketone cyanhydrine in xylene [28]). Among the drawbacks of the method are the high toxicity of the system, the presence of two liquid phases which can result in a considerable increase in holdup period, and possible flow conversion incompleteness under high concentrations of $^{13}C$ isotope [28]. This is probably why the cyanhydrine method has not yet found practical use.

The method based on the molecular complex $Cu_2Cl_2 \cdot 8NH_4Cl \cdot CO$ is characterized by a complex composition of the solution. Since $^{13}C$ is concentrated in the liquid phase, the flow conversion at the enriched end of the column is performed through the complex dissociation in the liquid heating to about 100°C. Under these conditions, a residual CO content in the solution leaving the flow conversion system at a level of (1–10) ppm was experimentally achieved [32] which, as is evident from calculations, suffices to obtain highly concentrated $^{13}C$. The complex method has as well satisfactory mass-exchange characteristics: in a column of 25mm diameter filled with Helipack packing material, HETP ranges from 2.3 to 11cm with a flow rate varying from 245 to 2,500ml/h [33].

Relatively high separation factor values, acceptable flow conversion completeness and mass-exchange rate, and the possibility of the liquid phase circulating in a closed loop between lower and upper phase flow conversion units [33] can ensure good performance characteristics of this method. But it is hardly probable that this method will be able to compete with carbon oxide (II) rectification, especially at the $^{13}C$ initial concentrating stage owing to the necessity of using a liquid phase of a complex chemical composition (in addition to gaseous CO) and special conditions of the process (the need to maintain an acid environment at a pH level of 1–2 through the addition of HCl, and prevention of the liquid phase contact with air to avoid copper tarnishing).

## 5.2.3  Carbamate method

Despite a lower separation factor value, the carbamate method of carbon isotope separation is more technologically feasible and ecologically friendly than the above chemical exchange processes. The mass-exchange and phase flow conversion completeness issues present no special problems. As an example, in a column of 56mm diameter filled with steel spiral-prismatic packing elements of 2.5mm $\times$ 2.2mm $\times$ 0.2mm in size (250mm bed height), at a reflux density of 1.0–4.0ml/(cm$^2$·min) of 2M $n$-dibutylamine (n-DBA) solution in octane, the HETP value varied from 2.87cm to 5.70cm [37]. With the flow conversion at the $^{13}$C-enriched column end through the liquid phase boiling at 131°C, the $CO_2$ loss did not exceed 5μg/ml [37], which demonstrates the possibility of obtaining highly concentrated (99%) $^{13}$C.

The liquid phase (specifically, $n$-DBA solution in octane) can be used in a closed cycle. Separation plants using the carbamate process can operate continuously for some years with the liquid phase closed loop without changes in the phase properties [39]. It should be noted, that in addition to the amines and solvents presented in Table 5.10 above, it is known that a wide variety of their possible combinations was used in the carbamate method [40].

A major drawback of the method is the necessity of a maintaining low concentration of amine carbamate or carbamic acid in the liquid phase owing to the phase viscosity dependence on the carbamate concentration, and to a low carbamate solubility. This leads to an increase in cross-section of mass-exchange columns, and, as a result, to an increase in capital investment in development of production facilities. As for now, it is the carbamate process that is the only technique of carbon isotope separation by chemical exchange with thermal flow conversion that has an industrial history [39, 41]. For example, a six-stage cascade for the production of the heavy carbon isotope with a 90at.% concentration and a capacity of 7.9 $\times$ 10$^{-3}$mmol $CO_2$/min ($\approx$ 180g of $^{13}CO_2$ per year) has been described by J. Agraval [41]. The main cascade characteristics are presented in Table 5.11, and its scheme in Figure 5.9 (the cascade fourth and fifth stages are not shown).

The flow reduction in the cascade is performed after complete $CO_2$ extraction from the liquid phase in the desorber 2 of each $i$ stage of the cascade (see Figure 5.9) through the supply of the obtained $CO_2$ to the next stage absorber 3 (except for the last stage), where carbon dioxide is absorbed by the amine solution in amounts required for this stage feeding. Unabsorbed $CO_2$, together with carbon dioxide from the $i+1$, column head, is returned to the previous column bottom. The pumps 5 are used only for the organization of the amine solution circulation between flow conversion units. Though it is probable that this scheme increases energy inputs for the carbamate dissociation in the lower flow

**Table 5.11**

Parameters of cascade for $^{13}$C concentrating by carbamate method ($n$-DBA solution in TEA) [41].

| Stage | Diameter, mm | Flow rate, mmol $CO_2$/(cm$^2$·min) | HETP, cm | $^{13}$C concentration, at.% |
|---|---|---|---|---|
| 1 | 64 | 1.987 | 3.1 | 2.5 |
| 2 | 52 | 1.349 | 2.7 | 6.1 |
| 3 | 34 | 1.285 | 2.0 | 18 |
| 4 | 25 | 0.776 | 2.0 | 43 |
| 5 | 15 | 0.847 | 2.0 | 73 |
| 6 | 11.8 | 0.560 | 2.0 | 90 |

conversion units, at the same time it prevents the enrichment product loss and ensures stable operation of the cascade.

Reference [10] discusses variations of the calculation of a plant for the production of highly concentrated $^{13}$C (99%) with the use of $n$-DBA solution in octane as a liquid phase component. The calculations resulted in the selection (from the cascade alternatives studied) of a four-stage cascade (with NTP = 1400) characterized by a minimum volume and a minimum time of the passage to a stable operation mode. It was demonstrated that with the plant design operation time equal to 300days per year, half of the costs fell on the source materials – amine and solvent (octane).

### 5.2.4   Comparative economic analysis of carbon isotope separation techniques

Among the best known data on the comparative analysis of $^{13}$C isotope production costs are those reported in references [42] and [43]. The performed studies justify the economic expediency of the CO rectification use for large-scale $^{13}$C production (Figure 5.10). Compared with a physical isotope separation method – gas diffusion of methane – carbon

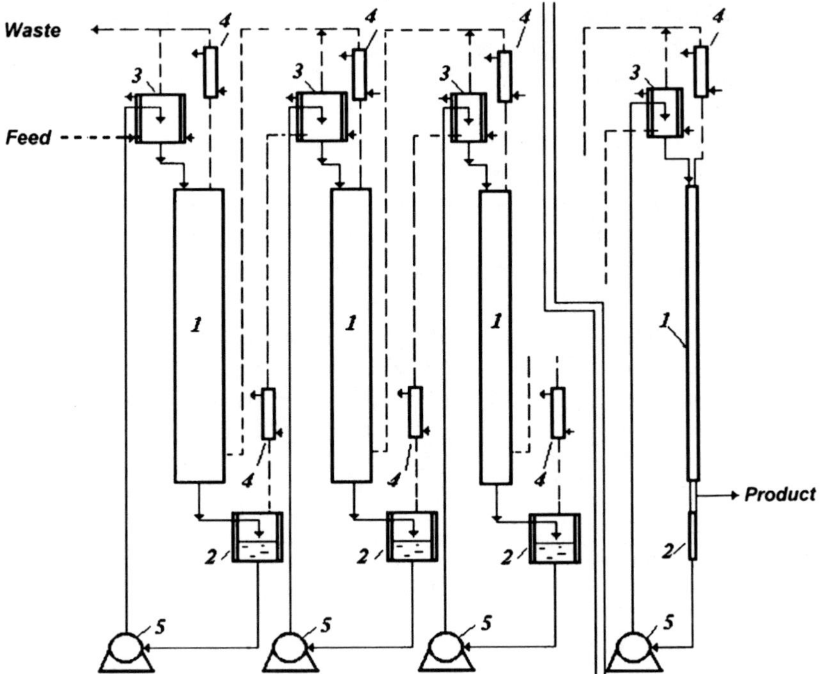


**Figure 5.9**   Scheme of cascade for $^{13}$C concentration [41] by carbamate method: 1, isotope exchange column (columns of stage 4 and Stage 5 are not shown); 2, desorber (lower flow conversion unit); 3, absorber (upper flow conversion unit); 4, cooler; 5, pump.

oxide rectification allows production of carbon-13 at a cost about 1.2–1.4 lower, with an annual production volume at a level of 100kg of $^{13}$C per year.

Worthy of notice is a tendency for a decrease in the ratio between the cost of $^{13}$C production by CO rectification and that by the carbamate method with an increase in the production volume. It is conceivable that with the annual production output increased to about 100–1000kg, the carbamate method would be more efficient.

It should be pointed out that the results of the calculations considered above relate to the carbamate method based on the *n*-DBA solution in triethylamine (TEA). Other variations of the liquid phase composition (*n*-DBA in octane and DEA in toluene) are probably more economically attractive, which has been demonstrated by studies performed in various countries including Georgia [39], Japan [44], Romania [45], and Russia [38, 46, 47].

The CO rectification method was compared as well with other techniques of carbon isotope separation by chemical exchange – cyanhydrine method (CYANEX-process [28]) and the method based on the molecular complex $Cu_2Cl_2 \cdot 8NH_4Cl \cdot CO$ (COCO process [32]). Comparative results are given in Figure 5.11. From the presented data it follows that at an annual production scale of about 10–30kg of $^{13}$C the production costs for different techniques are in close agreement: USD26–30 per gram of $^{13}$C at an annual production scale of 10kg of $^{13}$C, and USD13–16 at an annual production scale of 30kg of $^{13}$C. With an increase in carbon-13 annual production volume to about 150–350kg, it is possible to get over a cost level of USD5 per gram of $^{13}$C.

It might be well to point out that the data presented above are an estimate. It is particularly true for the cyanhydrine and complex methods of carbon isotope separation, which have not found practical use. For the CO rectification and carbamate methods of $^{13}$C production, the comparison results are more reliable.

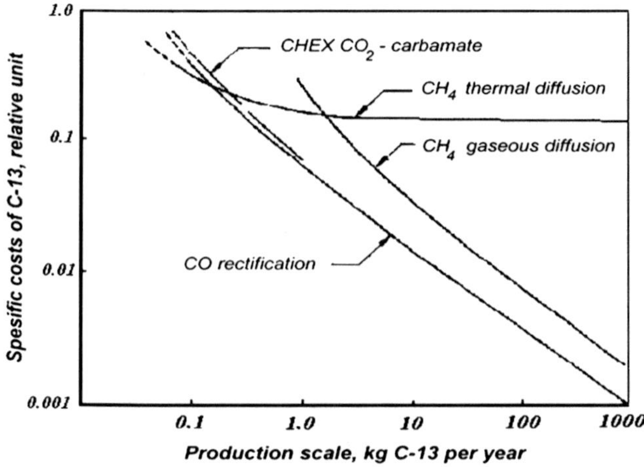


**Figure 5.10**   Specific costs of $^{13}$C production (*in relative units*) for different methods of carbon isotopes separation depending on production scale [5, 43].

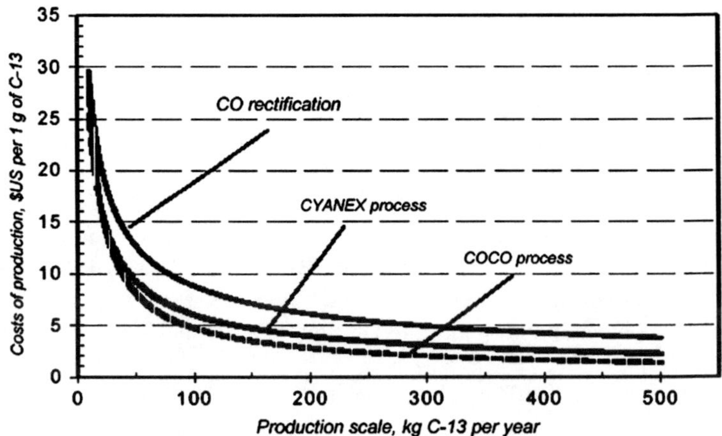


**Figure 5.11**    Comparison between carbon isotope separation by CO rectification and chemical isotope exchange depending on $^{13}C$ production scale (based on data[28, 32]).

## REFERENCES


1.  T. F. Johns In: *Proc. Int. Symp. Isotope Separation*, Amsterdam, North-Holland Pub. Co., **1958**, 74.
2.  E. Ancona, G. Boato, N. Casanova, *Nuova Cimento,* **1962**, 24, 111.
3.  G. G. Devyatykh, A. D. Zorin, N. I. Nikolaev. *Zhurn. Prikl. Khimii*, **1958**. 31, 368.
4.  N. N. Sevriugova, N. M. Zhavoronkov, *Rep. Acad. Sci. U.S.S.R.,* **1960**, 134, 87.
5.  W. Groth, H. Ihle, A. Murrenhaft *Z. Naturforsch. A.*, **1954**, 9, 805.
6.  G. G. Devyatykh, A. D. Zorin, *Zhurn. Phiz. Khimii*, **1956**, 30, 1133.
7.  O. V. Uvarov, N. M. Sokolov, V. V Lyapin, N. M. Zhavoronkov, *Zhurn. Vsesoyuzn. Khim. Ob. Im. D.I. Mendeleeva*, 7, 695.
8.  G. A. Yagodin, O. V. Uvarov, N. M. Zhavoronkov, *Rep. Acad. Sci. U.S.S.R.*, **1956**, 111, 384.
9.  P. Baertschi, W. Kuhn, H. Kuhn, *Z. Naturforsch.*, **1953**, 171, 1018.
10.  E. D. Oziaschvili, A. S. Egiazarov, *Uspekhi Khimii*, **1989**, LVIII, 4, 545.
11.  K. Clusius, F. Endtinger, I. Schleich, *Helv. Chem. Acta*, **1960**, 43, 1267.
12.  K. Haga, H. Soh, In: 5th Int. Symp. on the Synthesis and Applications of Isotopes and isotopically labelled compounds., Strasbourg, France, June 20 – 24, **1994**, PA 026, 144.
13.  M. V. Tikhomirov, N. I. Tunitsky, *Zhurn. Prikl. Khimii*, **1959**, 32, 531.
14.  H. London, In: *Proc. Int. Symp. on Isotope Separation*, Amsterdam, North-Holland Pub. Co., **1958**, 319.
15.  A. O. Edmunds, I. M. Lockhart, In: *Proc. Int. Symp. Isotope Rations as Pollutant Source and Behaviour Indicators*, IAEA, Vienna, **1975**, 279.
16.  W. R. Daniels, A. O. Edmunds, I. M. Lockhart, In: *Proc. Int. Symp.Stable Isotopes in the Life Sciences*, IAEA, Vienna, **1977**, 21.
17.  I. M. Lockhart, *Isotopes: Essential Chemistry and Applications*, **1982**, 7, 1.
18.  T. R. Armstrong, *et al.* In: Stable isotopes, *Proc. Third Intern. Conf.*, New York, Academic Press, **1979**, 177.
19.  Cambridge Isotope Laboratories expands carbon-13 production capacities, CIL Announcement, Andover, 1998.

20. G. Ya. Asatiani, candidate of technological science thesis, Mendeleev Institute of Chemical Technology, **1981**.

21. A. P. Babitchev, I. P. Gnidoy, G. Yu. Grigoryev, K. I. Dyma *et al.*, In: *Proc. VII[th] All-Russian (International) Conference on Physical–Chemical Processes in Selection of Atoms and Molecules*, TsNIIAtominform, **2002**, 290.

22. Ya. D. Zelvenskii, A. V. Khoroshilov, *Khim. Prom.*, **1999**, 4, 25.

23. Ya. D. Zelvenskii, *Khim. Prom.*, **1987**, 7, 425.

24. Ya. D. Zelvenskii, S. A. Arutyunov, V. V. Schitikov, *Proc. Mendeleev Institute of Chemical Technology*, **1989**, 156, 105.

25. Ya. D. Zelvenskii, E. I. Fetisov, A. E. Kovalenko, *Kislorodnaya Prom.*, **1978**, 5, 14.

26. Ya. D. Zelvenskii, A. V. Khoroshilov, N. I. Toropov, In: *Proc. V[th] All-Russian Conference on Physical-Chemical Processes in Selection of Atoms and Molecules*, TsNIIAtominform, **2000**, 160.

27. A. P. Klimenko, *Liquefaction of Hydrocarbon Gases*, Nedra. 1974, 307.

28. L. L. Broun, J. S. Drury, *J. Inorg. Nucl. Chem.*, **1973**, 35, 8, 2898.

29 G. V. Urey. *Isotope Chemistry*, p. 1. Izd. Inostr. Lit, **1948**, 86.

30. E. W. Becker, K. Bier, Z. *Naturforschung*. **1952**, 7a, 651.

31. R. T. Mills *Sep. Sci. Technol.*, **1980**, 15, 3, 475.

32. A. A. Palko, U.S. Pat 3,535,079, (**1970**).

33. A. A. Palko, L. Landau, J. S. Drury *Ind. Eng. Chem. Progr. Des. Develop.*, **1971**, 10, 1, 79.

34. G. C. Urey, A. H. Aten, A. S. Keston, *J. Chem. Phys.*, **1936**, 4, 622.

35. I. A. Semiokhin, A. K. Lykova, A. G. Serenkova, *Vestnik MGU, Chemistry Series*, **1963**, 5, 29.

36. A. Kitamoto, K. Takeshita, In: *World Congress III of Chem.* Engng., Tokyo, **1986**, 1, 719.

37. E. D. Oziaschvili, A. S. Egiazarov, Sh. I. Dzhidzheischvili, N. F. Baschkatova, In: *Stable Isotopes in the Life Sciences*, Vienna, IAEA, **1977**, 29.

38. A. V. Lizunov, A. V. Khoroshilov, S. A. Tcherednitchenko, In: *Proc. VIIth All-Russian (International) Conference on Physical-Chemical Processes in Selection of Atoms and Molecules*, TsNIIAtominform, **2002**, 160.

39. A. S. Egiazarov, G. V. Hatchishvili, T. G. Abzianidze, In: *5th Int. Symp. on the Synthesis and Applications of Isotopes and Isotopically Labelled Compounds,* Strasbourg, France, June 20– 24, **1994**, P028, 146.

40. J. P. Agraval, *Sep. Sci.*, **1971**, 6, 6, 831.

41. M. R. Ghate, T. I. Taylor *Sep. Sci.*, **1978**, 10, 6, 547.

42. R. A. Schwind *Chem. Process Engng*, **1969**, 50, 7, 75.

43. G. Vâsaru, P. Ghete, I. Covaci, M. Atanasiu, *Stable Isotopes in the Life Sciences*, IAEA, Vienna, **1977**, 39.

44. A. Kitamoto, K. Takeshita In: *Proc. Int. Symp. on Isotope Separation and Chemical Exchange Uranium Enrichment, (Bull. Res. Lab. for Nuclear Reactors*, special issue 1), **1992**, 376.

45. D. Axente, A. Bâldea, M. Abbrudean, *ibid.*, 357.

46. A. V. Khoroshilov, S. A. Tcherednitchenko, A. V. Lizunov. Theses of reports at VIth International Conference Zvenigorod, **2001**, 71.

47. A. V. Khoroshilov, A. V. Lisunov, S. A. Tcherednichenko, *Khim. Prom. Segodnia*, **2004**, 5, 30.

# – 6 –

# Nitrogen Isotope Separation

---

## 6.1 NITROGEN ISOTOPE SEPARATION BY RECTIFICATION

### 6.1.1 Isotope effect and properties of operating substances

Relatively few chemical compounds are suitable for the nitrogen isotope separation by rectification. Among these can be listed ammonia, molecular nitrogen, and nitric oxides NO and $NO_2$ ($N_2O_4$). The separation factor values for the above operating substances are presented in Table 6.1 [1–7].

As can be seen from Table 6.1, the highest single-stage separation factor values are characteristic for nitric oxide (II) NO, which is probably the only nitrogen chemical compound used industrially for the separation of $^{14}N$ and $^{15}N$ isotopes by low-temperature (or cryogenic) rectification.

Nitric oxide (II) NO has the critical temperature $T_{cr} = 180.2K$, normal boiling point $T_{b(n)} = 121.39K$ and melting point $T_{m(n)} = 109.5K$. The liquid NO density at $T_{b(n)}\rho_l = 1270$ kg/m$^3$. The heat of evaporation $\Delta H_{ev} = 13.83$kJ/mol.

The most reliable determination of the nitrogen isotope single stage separation factor was performed by K. Clusius [1]. Over a relatively wide temperature range (110–118K), for $^{14}N^{16}O$ and $^{15}N^{16}O$ molecules:

$$\lg\alpha = 3.0230/T - 13.40 \times 10^{-3}. \tag{6.1}$$

As is seen in Table 6.1, the separation factor at the nitric oxide (II) normal boiling point $T_{b(n)}$ is equal to 1.027 (nitrogen heavy isotope $^{15}N$ is concentrated in the liquid phase).

Notice that considerably lower separation factor values are observed in two cases: first, for molecules ($NH_3$ and $N_2$) that are lighter than NO, and secondly, at a lower temperature (for molecular nitrogen) (see Table 6.1):

For ammonia, $\alpha = 1.0025$ (molecules $^{15}NH_3$ and $^{14}NH_3$);

For molecular nitrogen ($^{15}N^{14}N$ and $^{14}N_2$), $\alpha = 1.004$.

An anomalously large isotope effect is typical of the NO rectification. In the opinion of K. Clusius [4–6], that this is due to the presence of $N_2O_2$ dimer (in addition to NO molecules proper) in the liquid phase. The dimerization reaction equilibrium

$$2\,NO \leftrightarrow (NO)_2 \tag{6.2}$$

247

**Table 6.1**

Isotope effect in nitrogen isotope separation by rectification [a]

| Operating substance | Molecules | Temperature, K | $\alpha$ | Approximation of temperature dependence of isotope effect | Reference |
|---|---|---|---|---|---|
| Ammonia NH$_3$ | $^{14}$NH$_3$ – $^{15}$NH$_3$ | 240 198–240 | 1.0025 | lg $\alpha$ = 0.9503/T − 0.00325 | [1] [2] |
| Nitrogen N$_2$ | $^{14}$N$^{14}$N – $^{14}$N$^{15}$N | 63.2−78 77.4[b] | 1.004 | lg $\alpha$ = 0.3985/T − 3.43 $\times$ 10$^{-3}$ | [1] |
| | | 78 80.8 84.7 86.0 90.7 97.5 108.0 | 1.004 1.0036 1.0031 1.0029 1.0025 1.002 1.0014 | ln $\alpha$ = 0.846/T − 6.9 $\times$ 10$^{-3}$ | [3] |
| | $^{14}$N$^{14}$N− $^{15}$N$^{15}$N | 63.2−78 77 | 1.008 | lg $\alpha$ = 0.7974/T − 6.91 $\times$ 10$^{-3}$ | [1] |
| Nitric oxide (II) NO | $^{14}$N$^{16}$O− $^{15}$N$^{16}$O | 110−118 111.8 112.4 119.6 120 121.39[b] | 1.031 1.031 1.028 1.0275 1.027 | lg $\alpha$ = 3.0230/T − 13.40 $\times$ 10$^{-3}$ | [4] [4, 5] [6] |
| | $^{14}$N$^{16}$O− $^{15}$N$^{18}$O | 110−118 | | lg $\alpha$ = 3.042/T − 13.53 $\times$ 10$^{-3}$ | [4] |
| Nitric di (tetra) oxide (IV) N$_2$O$_4$ | $^{14}$N$_2$ $^{16}$O$_4$ − $^{15}$N$^{14}$N $^{16}$O$_4$ | 295 | 1.00313[c] 1.0042[d] | | [7] |

[a] Heavy isotope $^{15}$N is concentrated in the liquid phase in all operating systems.

[b] $T_{b(n)}$.

[c] Single-stage equilibration.

[d] Rayleigh distillation.

is left displaced in the vapor phase, and right displaced in the liquid. The association of heavier molecules (NO)$_2$ is stronger than that of lighter molecules. This effect superimposes on the isotope effect of vapor pressure of monomeric isotopic forms ($^{14}$N$^{16}$O and $^{15}$N$^{16}$O), and the NO rectification thus represents a chemical-exchange rectification process (CHEXR: an isotope separation technique performed by rectification method based on a combination of isotope effects of chemical and phase isotope exchange).

These are anomalously high separation factor values in the NO rectification, which have determined the practical application of this process despite the mass production and readier availability of other working substances (NH$_3$, N$_2$, N$_2$O$_4$), as well as more preferential temperature conditions in the rectification of ammonia or N$_2$O$_4$.

## 6.1.2   Nitrogen isotope separation by NO rectification

When the potentialities of NO cryogenic (or low-temperature) rectification for isotope separation were demonstrated, the chemical isotope exchange method had already been practically applied [4, 8] (see section 6.2). This is due to the peculiar features of cryogenic rectification and, first of all, to the necessity of using special-purpose cryogenic equipment and to the strict requirements on the operating substance (NO) purity.

The need for thorough prepurification of nitric oxide (II) is dictated by the fact that at the NO boiling point, other nitric oxides and water contained in NO transform into solid state which results in the clogging of equipment and upsetting of the required operational mode. Moreover for pure nitric oxide the chemical reaction of NO disproportionation runs with the formation of other nitrogen oxides [9]

$$4\,NO \rightarrow N_2O + N_2O_3, \tag{6.3}$$

which requires special measures to be taken in the purification process. A closeness of the NO melting and boiling points (see section 6.1.1) necessitates using intermediate cooling agents in the condensers of rectification columns, which complicates both their construction and the isotope separation process control system.

It should be noted that in the NO rectification the concentration both of $^{15}N^{16}O$ and of $^{14}N^{18}O$ molecules is performed. But at low temperatures, owing to a very low rate of the HMEX reaction

$$^{15}N^{18}O + {}^{14}N^{16}O \leftrightarrow {}^{14}N^{18}O + {}^{15}N^{16}O, \tag{6.4}$$

simultaneous production of highly enriched $^{15}N$ and $^{18}O$ isotopes does not occur (similar to the CO low temperature rectification: see chapter 5). It should be noted that the presence of trace nitric dioxide ($NO_2$) in NO catalyzes HMEX in the NO rectification process [6, 10].

Despite the above problems, the NO cryogenic rectification process for nitrogen and oxygen isotope production was realized at the Los Alamos National Laboratory in the U.S.A. [11, 12], and at Tbilisi NIISI in the U.S.S.R. [10, 13], with the NIISI plant used primarily for the production of $^{18}O$.

A major distinctive feature of the NIISI plant is the use of the waste flow of nitric oxides from the nitric acid – nitric oxides (Nitrox) cascade for $^{15}N$ production as the NIISI plant feed flow, owing to the limited resources of NO [14] (see section 6.2.4). The NO cryogenic rectification plant at NIISI is considered in more detail in chapter 7.

Brought into operation in 1976, the Los Alamos plant was designed to simultaneously produce both light and heavy nitrogen and oxygen isotopes [12]. The plant scheme is shown in Figure 6.1. From the transport tank 16, nitric oxide, through a two-stage system of after purification by cryogenic adsorption 9 and 10, arrives at the bottom cross-section of the plant unit serving as the depletion section in respect to the less volatile component. The unit comprises six columns 11 operating in parallel, of 51mm diameter and 30m height each. The condenser 5 is cooled by liquid oxygen boiling at 120K under a helium pressure of about 1.1MPa. The NO condensation heat through the intermediate cooling agent (oxygen) is transferred to the liquid nitrogen 2 used as a source of cold. Placed below the intermediate evaporator 15, the enrichment column 17 is 51mm in diameter and 18m in height. Underneath the column evaporator 17, a small column of 25mm diameter and 1.7m height is installed. The column is meant for the complete evaporation of higher nitric oxides accumulating during the rectification process. The columns are filled with random packing material with grate-shaped elements of $4 \times 4$mm in size. The packing HETP is 6cm.

Issuing from the column head 11 (Fig. 6.1), the oxide $^{14}N^{16}O$ flow is hydrated over a platinum catalyst to $H_2^{16}O$ and $^{14}NH_3$ by

$$2NO + 5H_2 \rightarrow 2NH_3 + 2H_2O \tag{6.5}$$

for which purpose NO is mixed in the mixer 6 with oxygen arriving from the tank 3, and passed through the catalytic reactor where the reaction (6.4) proceeds. The ammonia–water mixture formed in the reactor, together with unreacted hydrogen, is delivered to the condenser 12, where $H_2^{16}O$ vapor is condensed and the liquid is collected in the separator 13. Issuing from the separator, the gas mixture arrives at the system of cooled

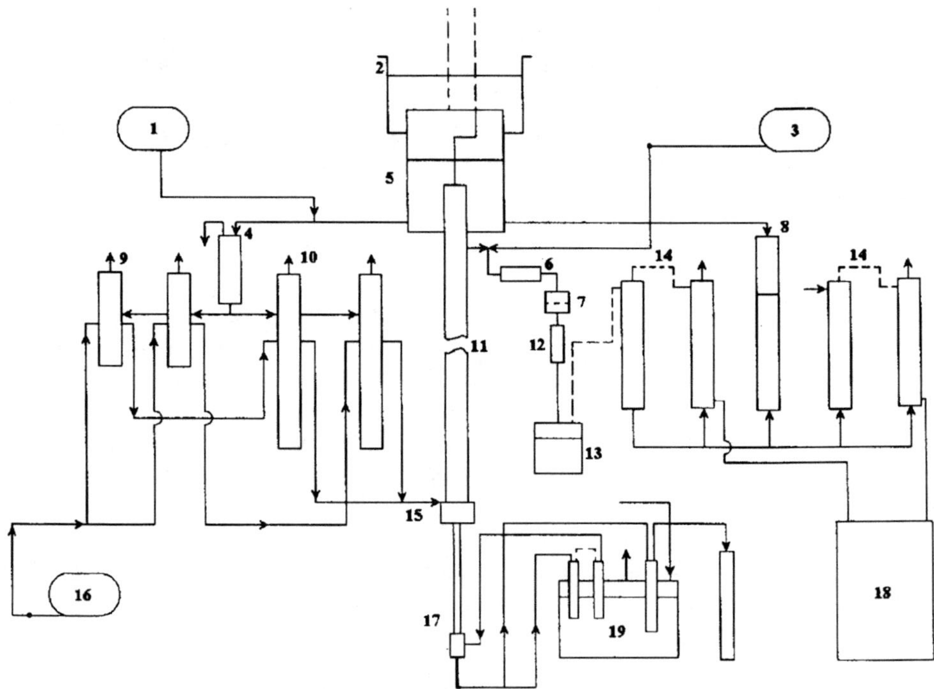

**Figure 6.1**   Scheme of nitric oxide (II) rectification at Los Alamos National Laboratory, USA: 1, liquid nitrogen tank; 2, oxygen condenser cooled by liquid nitrogen; 3, hydrogen tank; 4, 8, Dewar flasks with liquid nitrogen; 5, NO condenser cooled by oxygen; 6, nitric oxide-hydrogen mixer; 7, NO hydrogenation reactor; 9, prepurification adsorbers; 10, second-stage purification adsorbers; 11, depletion rectification columns (concentrating by more volatile component); 12, condenser; 13, $H_2^{16}O$ separator–accumulator; 14, ammonia $^{14}NH_3$ trap; 15, depletion section vaporizers; 16, source nitric oxide reservoir (transport tank); 17, enrichment rectification; 18, $^{14}NH_3$ accumulator; 19, storage of nitric oxide enriched with heavy isotopes.

traps 14 where $^{14}NH_3$ is freezed out from the hydrogen flow. The released ammonia is then bypassed to the $^{14}NH_3$ collector 18.

The main end products of the plant are $^{14}N$ isotope with a concentration of 99.99at.% and $^{16}O$ isotope with a concentration of 99.995at.%, respectively. The withdrawal from the small column evaporator is frozen out in the low product collector 19 (Fig. 6.1). The product isotope concentration is 40at.% of $^{15}N$ and 20at.% of $^{18}O$.

The minimum liquid nitrogen consumption for the production of 1g of $^{15}N$ by cryogenic rectification is estimated, according to D. Stashevski [5], as equal to 49kg.

## 6.2   NITROGEN ISOTOPE SEPARATION BY CHEMICAL ISOTOPE EXCHANGE

### 6.2.1   Isotope effect in the chemical exchange reactions

The nitrogen isotope separation factor has been determined for many operating systems representing two-phase (gas–liquid, liquid–solid, and gas–solid) chemical exchange systems. Based on the data reported in the literature [15–29], Table 6.2 gives single-stage separation factor values at atmospheric pressure for the main types of system.

All chemical exchange systems for which one or the other nitrogen isotope separation method is named (see Table 6.2) can be subdivided into two groups:

- systems using gaseous ammonia or ammonium ion;
- systems using nitric oxides (NO, $N_2O_3$, $NO_2$, $N_2O_4$).

For the first group comprising ammonium (including ammonium carbonate), ion-exchange, and ammonia-complex techniques, the separation factor value does not practically exceed $1.025 - 1.03$, with the lowest values ($\alpha \approx 1.01 - 1.004$) typical for the ammonia-complex technique.

For the second group comprising nitric acid – nitric oxides (or Nitrox system in the literature) in various forms, oxide ($N_2O_3 - NO$, $NO_2$), nitrous ($NO[HSO_4] - NO$, $NO_2$), and oxide–(dioxide)-complex (see Table 6.2) techniques, the separation factor values are generally significantly higher. The minimum values are at a level of $\alpha \approx 1.03$, and the maximum ones (with respect to the enrichment factor $\varepsilon$) are two to three times higher.

In general, from Table 6.2 it follows that the highest nitrogen isotope separation factor values are typical for the Nitrox technique in various forms. The highest value $\alpha \approx 1.1$ was attained at a temperature of $244 - 255K$ in the system with nitric acid applied on granulated silica gel [25] (sorption variant of Nitrox process). In this case, unfortunately, from the large-scale nitrogen isotope production standpoint, an increase in the separation factor does not lead to a reduction in the apparatus size compared with the classic variant of the $HNO_3 - NO$, $NO_2$ method (or Nitrox process) because of a very low flow rate capacity of the columns with solid phase.

For the classical nitrogen isotope separation method (Nitrox process), the single stage separation factor values can be approximated by an equation of the type [30]:

$$\overline{\alpha} = A\exp(B/T), \tag{6.6}$$

**Table 6.2**

Isotope effect in nitrogen isotope separation by chemical exchange method

| Method | Operating system | Main isotope exchange reaction(s) | Conditions | | $\bar{\alpha}$ | Reference |
|---|---|---|---|---|---|---|
| | | | $T$, K | $C^{\text{a}}$ | | |
| 1 | 2 | 3 | 4 | 5 | 6 | 7 |
| | | Systems with gaseous ammonia and/or ammonium ion | | | | |
| Ammonium | $NH_4{}^+{}_{\text{(sol, aq)}} - NH_{3\text{(gas)}}$ | $^{14}NH_4NO_{3\text{(sol, aq)}} + {}^{15}NH_{3\text{(gas)}} \leftrightarrow$ | 298 | 50.1%[b] | 1.018 | [15] |
| | | $\leftrightarrow {}^{15}NH_4NO_{3\text{(sol, aq)}} + {}^{14}NH_{3\text{(gas)}}{}^{\text{ j}}$ | 298 | 54.9%[b] | 1.023 | |
| | | | 293 | 50%[b, c] | 1.020[d] | [16] |
| | | | 303 | | 1.024[d] | |
| | | | 318 | | 1.025[d] | |
| | | | 328 | | 1.022[d] | |
| | | | 293 | 50%[b, c] | 1.0148 | [17] |
| | | | 353 | | 1.0208 | |
| | | $^{14}NH_4{}^{14}NH_4SO_{4\text{(sol, aq)}} + {}^{15}NH_{3\text{(gas)}} \leftrightarrow$ | 293 | 31%[b] | 1.021 | [18] |
| | | $\leftrightarrow {}^{15}NH_4{}^{14}NH_4SO_{4\text{(sol, aq)}} + {}^{14}NH_{3\text{(gas)}}{}^{\text{ j}}$ | 293 | 34%[b] | 1.022 | |
| Ammonium-carbonate | | $^{14}NH_4{}^{14}NH_4CO_{3\text{(sol, aq)}} + {}^{15}NH_{3\text{(gas)}} \leftrightarrow$ | 288.8 | 18.10– | 1.015 | [19] |
| | | $\leftrightarrow {}^{15}NH_4{}^{14}NH_4CO_{3\text{(sol, aq)}} + {}^{14}NH_{3\text{(gas)}}$ | 289.0 | 20.10[e]M | 1.019 | |
| | | $^{14}NH_4COO^-{}_{\text{(sol, aq)}} + {}^{15}NH_{3\text{(gas)}} \leftrightarrow$ | 309.1 | | 1.021 | |
| | | $\leftrightarrow {}^{15}NH_4COO^-{}_{\text{(sol, aq)}} + {}^{14}NH_{3\text{(gas)}}{}^{\text{ j, k}}$ | 317.5 | | 1.022 | |
| Ammonia-complex | $NH_3 \cdot D_{\text{(sol, org)}} - NH_{3\text{(gas)}};$ | $^{14}NH_3 \cdot D_{\text{(sol, org)}} + {}^{15}NH_{3\text{(gas)}} \leftrightarrow$ | | | | |
| | D–methanol | $\leftrightarrow {}^{15}NH_3 \cdot D_{\text{(sol, org)}} + {}^{14}NH_{3\text{(gas)}}$ | 290 | | | [20] |
| | D–ethanol | | | | 1. 011 | |
| | D–propanol | | | | 1.008 | |
| | D–butanol | | | | 1.007 | |
| | D–isopropanol | | | | 1.007 | |
| | D–isoamyl alcohol | | | | 1.007 | |
| | | | | | 1.004 | |

| Method | System | Reaction | $T$, K | Concentration | $\alpha$ | Ref. |
|---|---|---|---|---|---|---|
| Ion-exchange | $NH_4^+R_{(sd)}$ – $NH_4OH_{(sol,\,aq)}$ R-cation exchange resin (Dowex 50 × 12) | $^{14}NH_4^+R_{(sd)} + {}^{15}NH_4OH_{(sol,\,aq)} \leftrightarrow \leftrightarrow {}^{15}NH_4^+R_{(sd)} + 14NH_4OH_{(sol,\,aq)}$ | 293 298 313 328 343 | 0.6M | 1.0257 1.0254 1.0230 1.0212 1.0192 | [21] [22] |

Systems with nitric oxides

| Method | System | Reaction | $T$, K | Concentration | $\alpha$ | Ref. |
|---|---|---|---|---|---|---|
| Nitric acid – nitric oxides (Nitrox), classical | $HNO_{3(sol,\,aq)}$ – $NO, NO_{2(gas)}$ | $H^{14}NO_{3(sol,\,aq)} + {}^{15}NO_{(gas)} \leftrightarrow \leftrightarrow H^{15}NO_{3(sol,\,aq)} + {}^{14}NO_{(gas)};$ $H^{14}NO_{3\,(sol,\,aq)} + {}^{15}NO_{2\,(gas)} \leftrightarrow \leftrightarrow H^{15}NO_{3\,(sol,\,aq)} + {}^{14}NO_{2(gas)}$ | 298 298 298 313 313 313 328 328 328 343 343 343 | 5.7M 9.5M 11.3M 5.8M 9.6M 11.4M 5.8M 9.6M 11.4M 5.8M 9.6M 11.5M | 1.063 1.057 1.047 1.062 1.053 1.043 1.059 1.047 1.037 1.054 1.040 1.031 | [23] |
| Nitrox, cryogenic | $HNO_3 / N_2O_{3(sol)}$ – $NO_{(gas)}$ [f] | $H^{14}NO_{3\,(sol)} + {}^{15}NO_{(gas)} \leftrightarrow \leftrightarrow H^{15}NO_{3\,(sol)} + {}^{14}NO_{(gas)};$ $^{14}N^{14}NO_{3\,(sol)} + {}^{15}NO_{(gas)} \leftrightarrow \leftrightarrow {}^{15}N^{14}NO_{3\,(sol)} + {}^{14}NO_{(gas)}$ | 248 | 14M | 1.057 | [24] |
| Nitrox, sorption | $HNO_{3\,(lq/sd)}$ [g] – $NO, NO_{2\,(gas)}$ | $H^{14}NO_{3\,(lq/sd)} + {}^{15}NO_{(gas)} \leftrightarrow \leftrightarrow H^{15}NO_{3\,(lq/sd)} + 14NO_{(gas)};$ $H^{14}NO_{3\,(lq/sd)} + {}^{15}NO_{2\,(gas)} \leftrightarrow \leftrightarrow H^{15}NO_{3\,(lq/sd)} + {}^{14}NO_{2(gas)}$ | 244 255 291 318 291 | 10M 10M 10M 10M 16M | 1.11 1.10 1.06 1.05 1.05 | [25] |
| Oxide | $N_2O_{3\,(lq)}$ – $NO, NO_{2\,(gas)}$ | $^{14}N^{14}NO_{3\,(lq)} + {}^{15}NO_{(gas)} \leftrightarrow$ $^{15}N^{14}NO_{3\,(lq)} + {}^{14}NO_{(gas)};$ $^{14}N^{14}NO_{3\,(lq)} + {}^{15}NO_{2(gas)} \leftrightarrow \leftrightarrow {}^{15}N^{14}NO_{3\,(lq)} + {}^{14}NO_{2(gas)}$ | 250 296 296 | | 1.034 1.017 [h] 1.030 [i] | [26] |

*(Continued)*

**Table 6.2 (*Continued*)**

| Method | Operating system | Main isotope exchange reaction(s) | Conditions | | $\bar{\alpha}$ | Reference |
| --- | --- | --- | --- | --- | --- | --- |
| | | | $T$, K | $C^a$ | | |
| 1 | 2 | 3 | 4 | 5 | 6 | 7 |
| Nitrose | $NO[HSO_4]_{(sol,\ aq)} - NO,\ NO_{2\,(gas)}$ | $^{14}NO[HSO_4]_{(sol,\ aq)} + {}^{15}NO_{(gas)} \leftrightarrow {}^{15}NO[HSO_4]_{(sol,\ aq)} + {}^{14}NO_{(gas)}$; $^{14}NO[HSO_4]_{(sol,\ aq)} + {}^{15}NO_{2(gas)} \leftrightarrow {}^{15}NO[HSO_4]_{(sol,\ aq)} + {}^{14}NO_{2(gas)}$ | 298 | 1 M | 1.033 | [27] |
| Oxide-complex | $NO\cdot CHClF_{2\,(sol,\ org)} - NO_{(gas)}$ | $^{14}NO\cdot CHClF_{2(sol,\ org)} + {}^{15}NO_{(gas)} \leftrightarrow {}^{15}NO\cdot CHClF_{2\,(sol,\ org)} + {}^{14}NO_{(gas)}$ | 112 | | 1.031 | [28] |
| Dioxide-complex | $N_2O_4\cdot D_{(sol,\ org)} - N_2O_{4\,(gas)}$; D–dimethylamine D–trimethyl phosphate D–tributyl phosphate D–dimethyl sulfoxide | $^{14}N^{14}NO_4\cdot D_{(sol,\ org)} + {}^{14}N^{15}NO_{4\,(gas)} \leftrightarrow {}^{15}N^{14}NO_4\cdot D_{(sol,\ org)} + {}^{14}N^{14}NO_{4\,(gas)}$ | 298 298 298 314 | | 1.030 1.026 1.026 1.014 | [29] |

[a] Component concentration in the solution.
[b] Weight %.
[c] Concentration of the source ammonium nitrate solution.
[d] At a pressure of 0.013MPa.
[e] Concentration of total nitrogen in solution.
[f] Two liquid phases: in the first aqueous solution, mainly $HNO_3$, and in the second, $N_2O_3$ prevails.
[g] Applied on the C-3 silica gel with granulation $(0.25 - 0.50)$ mm in an amount of $((0.6 - 0.5)$mmol $HNO_3$ per gram.
[h] At a pressure of 0.21MPa.
[i] At a pressure of 0.075MPa.
[j] The reaction $^{14}NH_{3\,(sol,\ aq)} + {}^{15}NH_{3(gas)} \leftrightarrow {}^{15}NH_{3(sol,\ aq)} + {}^{14}NH_{3(gas)}$ takes place too.
[k] The reaction $^{14}NH_{4\,(sol,\ aq)}^{+} + {}^{15}NH_{3(gas)} \leftrightarrow {}^{15}NH_{4(sol,\ aq)}^{+} + {}^{14}NH_{3(gas)}$ takes place too.

with $A$ and $B$ being dependent on the $HNO_3$ concentration:

| $C$, M | 7.5 | 10.0 | 12.5 |
|---|---|---|---|
| $A$ | 0.964 | 0.936 | 0.947 |
| $B$ | 29.63 | 35.90 | 27.45 |

## 6.2.2  Comparison of isotope effects in chemical and phase exchange

The enrichment factor ($\varepsilon = \alpha - 1$) values presented in Tables 6.1 and 6.2 are compared in Figure 6.2. Based on $\varepsilon$ values, all two-phase systems and corresponding separation techniques, associated both with rectification methods (Table 6.1) and chemical isotope exchange ones (Table 6.2), can be conventionally subdivided into four groups:

- with maximum values ($\varepsilon \geqslant 0{,}05$);
- with high values ($0.03 \leqslant \varepsilon < 0.05$);
- with medium values ($0.01 \leqslant \varepsilon < 0.03$);
- and with low values ($\varepsilon < 0.01$).

The maximum $\varepsilon$ values are typical only for chemical-exchange systems with nitric oxides, namely, $HNO_{3\,(lq/sd)}-NO$, $NO_{2\,(gas)}$; $HNO_{3\,(sol,\,aq)}-NO$, $NO_{2\,(gas)}$, and $N_2O_{3\,(lq)}-NO$, $NO_{2\,(gas)}$.

Systems with high and medium $\varepsilon$ values are the largest in number. The highest enrichment factor values are characteristic as well of systems with nitric oxides. The low-temperature NO rectification, oxide, nitrose, and oxide–(dioxide)-complex chemical-exchange techniques of isotope separation are characterized by similar $\varepsilon$ values.

The group with medium enrichment factor values comprises generally the chemical-exchange systems with ammonia or ammonium ion.

The smallest isotope effect (the group with low $\varepsilon$ values) is achieved in the rectification of $NH_3$, $N_2$, $N_2O_4$, and in systems with ammonia molecular complex ($NH_3 \cdot D_{(sol,\,org)}-NH_{3\,(gas)}$).

Despite the diversity of systems characterized by large isotope effects, only some of them have been practically applied for nitrogen isotope separation. The point is that the practical industrial realization of a separation technique depends on a variety of factors comprising, aside from the separation factor value, process operating conditions (temperature, pressure, concentration), method of phase flow conversion (for chemical-exchange systems, including the completeness of a chemical element conversion from one phase into the other), acceptable mass-exchange parameters, availability of source materials, and collected experimental data.

## 6.2.3  Main production technologies

Chemical-exchange techniques of nitrogen isotope separation with an industrial record comprise three processes: ammonium, Nitrox (in the classical form), and ion-exchange, with all three having acceptable mass-exchange characteristics and requiring chemical methods of the phase flow conversion (flow reflux).

Notice that for the ammonium technique, the flow reflux completeness can be insufficient for the production of highly enriched $^{15}N$ ($> 90\%$), and the low-temperature rectification of nitrogen oxide technique has its own specific features (see section 6.1.2).

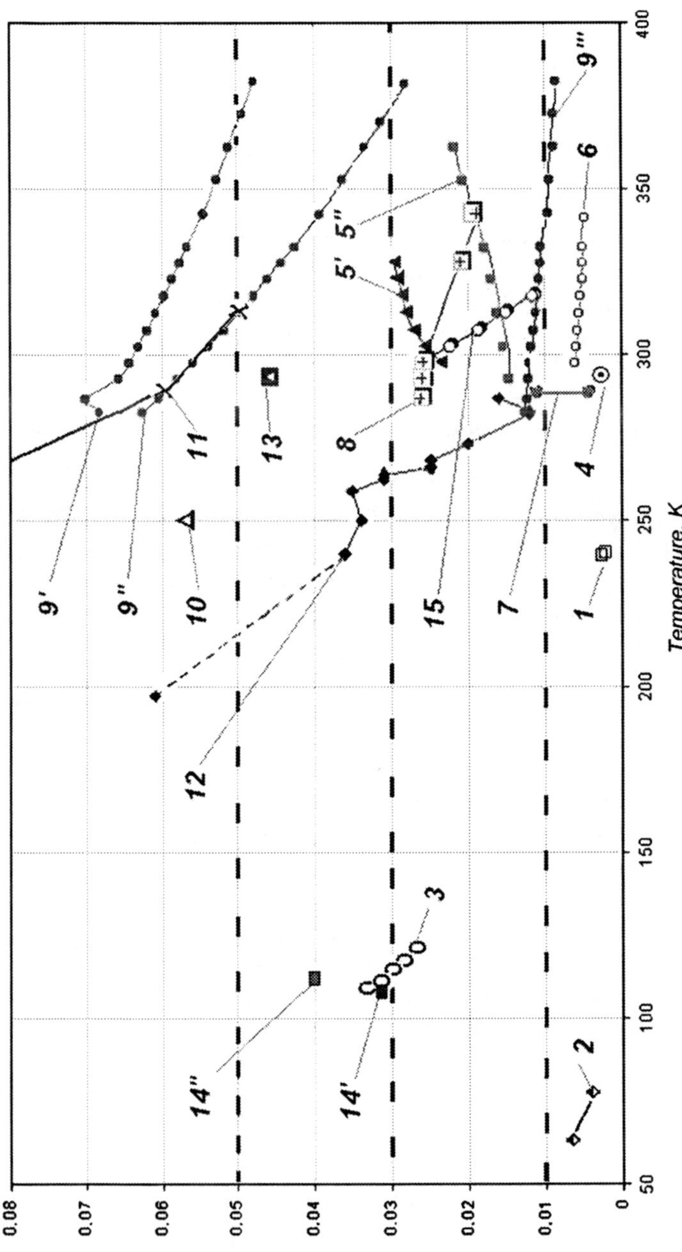



**Figure 6.2** Enrichment factor in nitrogen isotope separation processes using two-phase systems [1–7, 15–29]: 1, $NH_{3\,(lq)} - NH_{3\,(vap)}$; 2, $N_{2\,(lq)} - N_{2\,(vap)}$; 3, $NO_{(lq)} - NO_{(vap)}$; 4, $N_2O_{4\,(lq)} - N_2O_{4\,(vap)}$; 5', $NH_4NO_{3\,(sol,\,aq)} - NH_{3\,(gas)}$ ($P = 0.013$ MPa); 5'', $NH_4NO_{3\,(sol,\,aq)} - NH_{3\,(gas)}$ ($P = 0.1$ MPa); 6, $NH_4OH_{(lq,\,aq)} - NH_{3\,(gas)}$; 7, $NH_3 \cdot D_{(sol,\,org)} - NH_{3\,(gas)}$; 8, $NH_4R_{(sd)} - NH_4OH_{(sol,\,aq)}$; 9', $HNO_{3\,(sol,\,aq)} - NO, NO_{2\,(gas)}$; 9'', $HNO_{3\,(sol,\,aq)}$ (10M) $- NO, NO_{2\,(gas)}$; 9''', $HNO_{3\,(sol,\,aq)}$ (15M) $- NO, NO_{2\,(gas)}$; 10, $HNO_3$ (14M)/$N_2O_{3\,(sol)} - NO, NO_{2\,(gas)}$; 11, $HNO_{3\,(sol,\,aq)}/$ silica gel$_{(sd)} - NO, NO_{2\,(gas)}$; 12, $N_2O_{3\,(lq)} - NO, NO_{2\,(gas)}$; 13, $NO[HSO4]_{(sol,\,aq)} - NO, NO_{2\,(gas)}$; 14', $NO \cdot CHClF_{2\,(sol,\,org)} - NO_{(gas)}$; 14'', $NO \cdot HCl, CHClF_{2\,(sol,\,org)} - NO_{(gas)}$; 15, $- N_2O_4 \cdot D$ (sol, org) $- N_2O_{4\,(gas)}$.

Though more technologically suitable, chemical exchange methods with thermal flow reflux have, for the most part, low $\varepsilon$ values, which is the main obstacle to their practical use. In the dioxide-complex technique of nitrogen isotope separation [29], for one, with its acceptable separation factor values, the flow reflux completeness required for $^{15}$N production is not achieved.

## 6.2.4   Ammonium technique of nitrogen isotope separation

The ammonium technique of nitrogen isotope separation is the first-ever process based on the chemical isotope exchange method (see Table 6.2) and implemented for stable isotope production in general and $^{15}$N production in particular (H. Thode and H. Urey [8]).

The effective separation factor value (Table 6.2) depends on the dissolved ammonium (ammonia) fraction in the liquid phase and varies with a change in conditions affecting the $NH_3$ content in the solution (ammonium salt concentration, pressure, temperature). For 50% ammonium nitrate, as an example, at a temperature of 293K and a pressure of 0.1MPa, $\bar{\alpha} = 1.0148$. A pressure reduction to about 0.013MPa brings about an increase in the effective separation factor up to $\bar{\alpha} = 1.020$. A similar influence is exerted by the temperature change at atmospheric pressure (see Table 6.2).

The ammonium system $NH_4NO_{3\,(sol,\,aq)} - NH_{3\,(gas)}$ belongs to so-called salt systems characterized by a high gas solubility and high isotope exchange rate in the liquid. Mass exchange in such systems is limited by the gas diffusion, with the optimum pressure (from the minimum column height standpoint) being lower than atmospheric pressure at indoor temperature. For the $NH_4NO_3 - NH_3$ system, $P_{opt} \approx 6$–$8$ kPa [31].

The lower flow reflux (at the column end enriched with $^{15}$N) is performed by the chemical method through the reaction of the ammonium salt decomposition by alkali:

$$NH_4NO_3 + MeOH \rightarrow NH_3\uparrow + MeNO_3 + H_2O, \tag{6.7}$$

where Me is an alkaline metal (Na or K).

To enhance the flow conversion completeness, efforts are made to more completely liberate ammonia from the forming $MeNO_3$ salt solution, for which purpose the solution is subjected to boiling.

The upper flow reflux (at the column end depleted from $^{15}$N) can be realized, if need be, by the ammonia – nitric acid interaction reaction:

$$NH_3 + HNO_3 \rightarrow NH_4NO_3. \tag{6.8}$$

For the first time, the ammonium technique was used for $^{15}$N concentrating in late 1930s in the U.S.A. [32, 33]. The plant for nitrogen isotope separation by the ammonium technique represented a three-stage cascade, and is shown in Figure 6.3.

A 60% (by mass) ammonium nitrate solution (flow $L$) is used for the cascade feed. In the column C1, the $^{15}$N concentration increases by a factor of 7. One-seventh of the flow $L$ is delivered to the column C2 where the target isotope concentration further increases by a factor of 9. A similar flow division is performed in passing to the third cascade stage

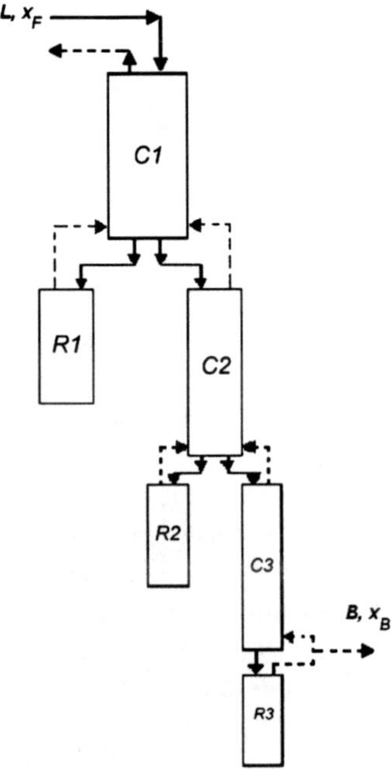



**Figure 6.3**  Scheme of first cascade for nitrogen isotope separation by the ammonium method: C1–C3, isotope exchange columns; R1–R3, reflux units.

where the $^{15}N$ concentration enhances by a factor of 11. The overall separation degree makes up about 800. The flow conversion in the units (refluxers) R1–R3 is performed with the use of caustic soda by eq. (6.7).

Based on the ammonium technique of the nitrogen isotope separation, several cascades of various capacity and type were developed in the U.S.S.R. An example is a three-stage laboratory-scale cascade constructed in 1950 at the Institute of Biological and Medical Chemistry of the Academy of Medical Sciences of the U.S.S.R. [34] and operated for several years. The cascade had a low capacity – only 0.5g of nitrogen per day ($\approx$ 180g/year) with a concentration of $\approx$ 40 at.% of $^{15}N$. Nevertheless, the cascade development required of the developers not only knowledge of the technological features of the process but also the manufacturing of the packing, special-purpose pumps, feeders, automatic adjustment system, etc. Each cascade stage represented a single column. The diameter of the first cascade column was 62mm, and that of the last column 10mm. The columns were filled with fine stainless-steel wire coils with a size of 4.5 mm $\times$ 4 mm $\times$ 0.2 mm to 2 mm $\times$ 2 mm $\times$ 0.2 mm.

The second example is a two-stage industrial cascade developed in 1950s by S. Babkov and N. Zhavoronkov [35], which operated for a long time at the Novomoskosky Chemical Industrial Complex, Tula Region, U.S.S.R. The cascade annual capacity accounted for 10kg of 50% of $^{15}N$. The first stage comprised two rotary-type horizontal (inclined) apparatuses (of 270mm diameter and 2.7m length) providing for $^{15}N$ concentration up to 7.15 at.%. A packed column of 50mm diameter and 11.5m height filled with SPP packing material with elements of 3mm $\times$ 3mm $\times$ 0.2mm in size served as the second stage.

The cascade was designed to operate at a temperature of 298K and absolute pressure of 16.3kPa, which allowed for a sufficiently high value of $\bar{\alpha} = 1.026$ with an ammonium nitrate solution concentration of 60% (by weight).

Later, the evolution of other nitrogen isotope separation techniques resulted in the exclusion of the ammonium technique which, over many years, had not attracted the attention of scientists and technologists whose interests were generally focused on the nitric acid – nitric oxides (or Nitrox) system (see 6.2.4). But in recent years, the scientists of MUCTR suggested performing the ammonium process at atmospheric pressure using elevated temperatures to reduce the ammonia water solubility and, hence, to increase the effective single-stage separation factor [17, 36].

From the standpoint of the flow conversion completeness and mass exchange, these conditions are more stringent. This is why the studies were in the first place devoted to these problems. The D. Mendeleev University researchers have developed a pilot plant for nitrogen isotope separation by the ammonium technique providing for the possibility of investigating the flow conversion, hydrodynamics, and mass exchange at atmospheric pressure and at a temperature of $323 - 363K$ [37]. The plant comprises an isotope exchange column of 18mm diameter filled with SPP packing of which the elements are $2mm \times 2mm \times 0.2mm$ in size, with a packing bed height of 80cm. The studies demonstrated that the residual nitrogen concentration in the sodium nitrate waste flow depends on the excess alkali, on temperature, and on the reacting mixture residence time in the lower flow reflux system [38]. On the basis of the obtained data, the reaction (6.7) regime was determined from the allowable loss level in the production of $^{15}N$ with a concentration of $10 - 50at.\%$. The attainment of a higher $^{15}N$ concentration is associated with a considerable increase in the consumption of alkali, which is produced by electroplating techniques and, hence, is an expensive product. The use of $NaNO_3$, obtained as the $^{15}N$ production waste, reduces the separation costs only slightly [36, 39]. It was suggested replacing NaOH for $Ca(OH)_2$, which is far cheaper than the caustic soda. In this case, a by-product – calcium nitrate, $Ca(NO_3)_2$ – can be used as fertilizer as well.

A study of the dependence of the mass exchange on the relative flow rate and temperature was performed for 50% ammonium nitrate solution [37]. It was found that within the relative flow rate region of $\varphi \approx 0.3 - 0.8$ and over a temperature range of $323 - 353K$, the HETP value remains practically constant and equals $3.1 - 3.4cm$. At a temperature of 363K, HETP increases almost twofold, which tells us about the ammonia diffusion as the limiting factor in the gas phase under indicated conditions. The study results, on the basis of the maximum specific capacity per unit volume of the plant, made it possible to recommend new conditions for the realization of nitrogen isotope separation allowing for a significant decrease in the separation section of the plant compared with operation under vacuum.

To compare the experimental data on mass exchange, and obviate the need for the use of alkali, experiments with aqua ammonia (or ammonia spirit) ($NH_4OH_{(sol,\ aq)} - NH_{3(gas)}$ system), and, hence, with thermal flow reflux [37] (similar to CHEX in $NH_3 \cdot D_{(sol,\ org)} - NH_{3(gas)}$ system [20] see Table 6.2), were performed at the plant. The height equivalent for the theoretical plate of separation (HETP) accounted for 2.8cm and the flow conversion completeness proved to be no less than that in the case of the ammonium nitrate use. Owing to a low value of $\alpha = 1.005$, a practical use of the system is, of course, highly conjectural, but it is premature to completely rule out the possibility.

## 6.2.5   Nitrox technique of nitrogen isotope separation

The $HNO_3 - NO$, $NO_2$ technique or Nitrox process, developed by W. Spindel and I. Taylor [40] in 1955 in the U.S.A., is based on the isotope exchange reactions between aqueous solutions of nitric acid and nitric oxides, performed with $^{15}N$ concentration in the liquid phase (see Table 6.2). At $T = 298K$, the single-stage separation factor for the first and second reactions equals, respectively, 1.115 and 1.094 (for non-dissociated nitric acid) [30].

What actually happens is that several tens of isotope-exchange reactions occur simultaneously in the nitric acid – nitric oxides system, owing to a multicomponent composition of the gas (NO, $NO_2$, $N_2O_4$, $N_2O_3$, $HNO_3$ vapors), and liquid ($HNO_3$, $NO_3^-$, dissolved nitric oxides) phases, which depends on the nitric acid concentration and temperature [30].

The resultant value of the reactions is the effective separation factor $\bar{\alpha}$ equal, as an example, for 10M $HNO_3$ to 1.055 at indoor temperature [41]. Both an enhancement in the nitric acid concentration and an increase in temperature lead to a decrease in $\bar{\alpha}$ (see Table 6.2).

In the $HNO_3 - NO$, $NO_2$ system, isotope exchange proceeds with a lower rate than that in the ammonium separation technique. The main resistance to the mass-exchange is exhibited by the liquid phase, and the process is generally limited by the isotope exchange reaction, which is demonstrated by a rather steep HETP dependence on the nitric acid flow rate (see Figure 6.4) and on the temperature (Figure 6.5) [39, 42].

At near-indoor temperature, the HETP values for laboratory-scale columns fall generally within a range of 5–10cm. Table 6.3 shows the dependence of the mass-exchange characteristics on $HNO_3$ concentration, type of packing, flow rate and other parameters of the nitrogen isotope separation process in the Nitrox system.

The lower flow reflux in the nitric acid – nitric oxides process is performed by chemical methods through the nitric acid reduction by gaseous sulfur dioxide:

$$HNO_3 + 1.5\,SO_2 + H_2O \rightarrow NO\uparrow + 1.5\,H_2SO_4; \tag{6.9}$$

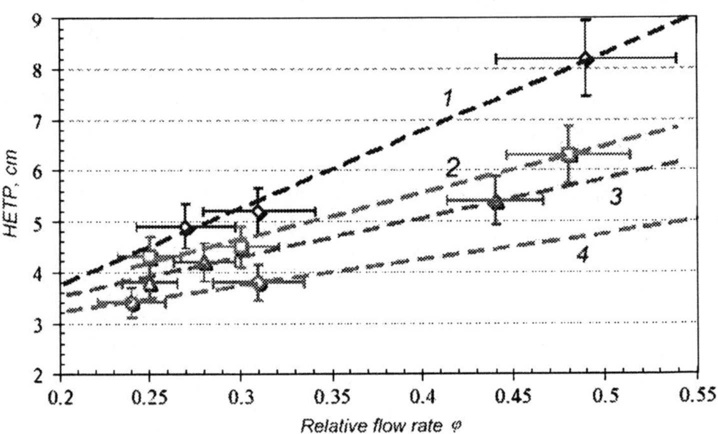


**Figure 6.4**   Relative flow rate $\varphi$ influence on HETP for $10.0 \pm 0.1M$ nitric acid at various process operation temperatures: $\Diamond$, 283K; $\square$, 293K; $\triangle$, 303K; $\bigcirc$, 313K.

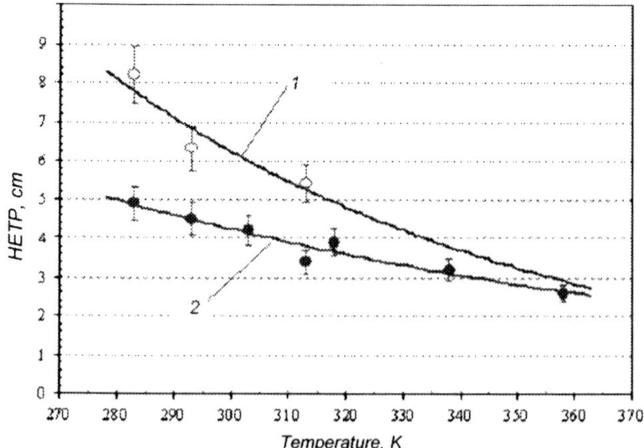



**Figure 6.5**  Influence of temperature and nitric acid flow rate on HETP for $10.0 \pm 0.1$M $HNO_3$ [39]: 1, $G = 33 - 36$mg-atom N·(cm$^2$ min)$^{-1}$, $\varphi = 0.5$; 2, $G = 19 - 22$mg-atom N·(cm$^2$ min)$^{-1}$, $\varphi = 0.3$.

$$HNO_3 + 0.5\,SO_2 \rightarrow NO_2\uparrow + 0.5\,H_2SO_4. \qquad (6.10)$$

Reactions (6.9 and 6.10) proceed with abundant heat release: 207.4kJ/mol of $HNO_3$ and 45.7kJ/mol of $HNO_3$, respectively.

In general, with regard to the dependence of the gas phase composition on nitric acid concentration and on temperature, the reaction of $HNO_3$ reduction by sulfur dioxide to the equilibrium gas phase [45] is as follows:

$$HNO_3 + X\,SO_2 + (X - 0.5)\,H_2O \rightarrow NO_{(2.5-X)}\uparrow + H_2SO_4, \qquad (6.11)$$

where $X$ is specific sulfur dioxide consumption (or molar ratio), mol $SO_2$·mol$^{-1}$ $HNO_3$.

With an increase in the nitric acid concentration and in the separation process temperature, the molar ratio $X$ decreases, which is illustrated in Figure 6.6.

The upper flow conversion is based on the NO deterioration by oxygen (atmospheric oxygen)

$$NO + \tfrac{1}{2}O_2 \leftrightarrow NO_2, \qquad (6.12)$$

and the absorption of nitric dioxide by water with the formation of nitric acid

$$3NO_2 + H_2O \leftrightarrow NO + 2HNO_3. \qquad (6.13)$$

The highly enriched $^{15}$N was first produced by the nitric acid – nitric oxides technique at a two-stage laboratory-scale cascade in the U.S.A. [41]. The achieved concentration amounted to 99.8at.%. The columns were startingly low, 5.2m and 5.5m, and the process itself was considerably simpler than the ammonia one and did not require a reduced pressure.

**Table 6.3**

HETP and mass-transfer volume factor dependence on nitric acid concentration and specific flow rate at indoor temperature 293–298 K

| Column $D/H$, mm/cm | Packing Type, material | Element size, mm | $C_{HNO_3}$, M | $L_{SP}$, g-at N $\times$ $(m^{-2} \cdot s^{-1})$ | $K_m$ | HETP, cm | $K_{OY} \cdot a$, g-at N $\times$ $(m^{-3} \cdot c^{-1})$ | Reference |
|---|---|---|---|---|---|---|---|---|
| 10/150 | coils, glass | 1.6 | 2 | 0.29 | 1.2 | 50 | 0.58 | [41] |
| | | | 6 | 1.41 | 2.0 | 13 | 10.8 | |
| | | | 8 | 2.18 | 2.3 | 11 | 19.8 | |
| | | | 10 | 2.72 | 2.5 | 8.8 | 30.9 | |
| | | | 10 | 2.72 | 2.4 | 9.2 | 29.6 | |
| | | | 10 | 2.90 | 2.6 | 8.4 | 34.5 | |
| | | | 10 | 5.80 | 1.9 | 12.5 | 46.4 | |
| | | | 12 | 2.83 | 1.7–2.0 | 10 | 28.3 | |
| 25/95 | Helipack, steel | 1.3 $\times$ 2.5 $\times$ 2.5 | 10 | 2.90 | 6.2 | 2.8 | 104 | |
| | | | | 5.44 | 3.6 | 4.0 | 136 | |
| | | | | 9.07 | 2.4 | 5.8 | 156 | |
| 11/152 | SPP, Stainless steel | 2.0 $\times$ 2.0 $\times$ 0.25 | 10 | 3.63 | | 5.3 | 68.5 | [14] |
| | | | | 4.35 | | 5.8 | 75.0 | |
| | | | | 4.90 | | 6.5 | 75.4 | |
| | | | | 5.44 | | 7.0 | 77.7 | |
| | | | | 6.53 | | 7.4 | 88.2 | |
| 11 | SPP, Stainless steel | 2.0 $\times$ 2.0 $\times$ 0.25 | 10 | 1.63 | | 2.7 | 60.4 | [43] |
| | | | | 2.36 | | 3.1 | 76.1 | |
| | | | | 3.63 | | 4.1 | 88.5 | |
| | | | | 4.71 | | 5.0 | 94.2 | |
| 35/300 | SPN, Stainless steel | 2.0 $\times$ 2.0 $\times$ 0.25 | 5.3 | 2.43 | | 11.7 | 20.8 | [14] |
| | | | 7.2 | 3.33 | 3.4 | 8.2 | 40.6 | |
| | | | 10 | 4.66 | | 7.0 | 66.6 | |
| 11/280 | SPP, Stainless steel | 2.0 $\times$ 2.0 $\times$ 0.25 | 14.5 | 2.05 | | 4.1 | 50 | [44] |
| | | | | 4.35 | | 6.6 | 65.9 | |
| | | | | 5.80 | | 4.3 | 135 | |
| | | | | 6.40 | | 9.3 | 68.8 | |

These virtues, and sulfuric acid as a more valuable by-product compared with sodium salts, together with water and air instead of nitric acid (in the upper flow reflux realization), gave a powerful impetus to a rapid worldwide distribution of the novel technique. Apart from the U.S.A., pilot and industrial plants were developed in the U.S.S.R. [14, 46], Great Britain [47], France [48], Germany (the formes GDR) [49, 50], Romania [51], and others.

In the U.S.S.R., the plants for the nitrogen isotope separation by the nitric acid – nitric oxides method were developed by many organizations including Novomoskosky Chemical Industrial Complex (Tula Region), L.Ya. Karpov NIPhKhI (Moscow), GIPKh (Leningrad Region), and PhTI (Sukhumi, Georgia) [14].

In the early 1960s, a cascade of two columns (48mm and 14mm in diameter, and 9m and 8m in height, respectively) with a rather low annual capacity (250g of $^{15}$N with a concentration of 98at.%) was developed by GIPKh.

At the same time, several pilot cascades were developed by PhTI in Sukhumi. After that, a then large two-stage cascade was designed and constructed at the pilot factory of the

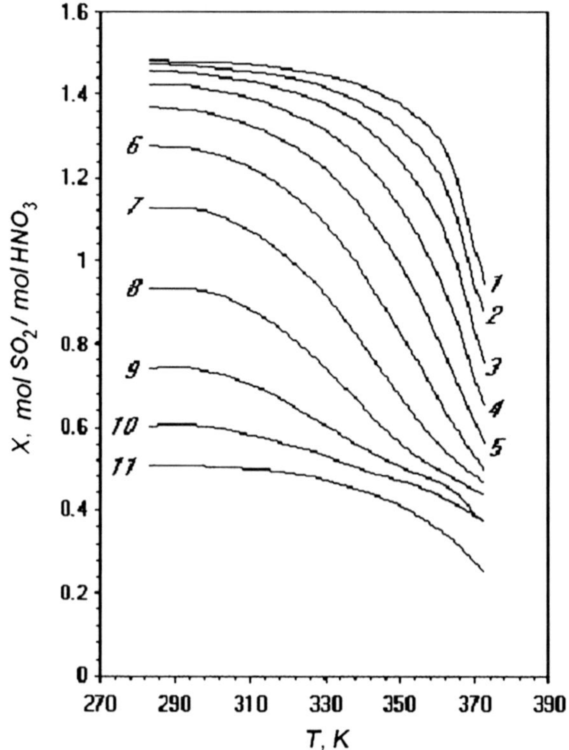

**Figure 6.6** Temperature dependence of molar ratio $X$ at various nitric acid concentrations (weight per cent) [45]: 1, 25%; 2, 30%; 3, 35%, 4, 40%; 5, 45%; 6, 50%; 7, 55%; 8, 60 %; 9, 65%; 10, 70%; 11, 80%.

Stable Isotopes Research Institute (NIISI, Tbilisi). The cascade, with an annual capacity of 4.5kg of nitrogen, was located in a 15-storey tower [14] (Fig. 6.7). The heights of the packed section of the cascade columns are for 17.1m and 17.8m, with the first-stage column being 120mm in diameter, and the second-stage column 26mm. The columns were filled with a fine spiral-prismatic packing material and arranged one above the other in a production building 60m high. During the cascade's industrial operation, the annual capacity varied from 3 to 4 kg of nitrogen with a concentration of 95–99% (AISI plant) [14, 52]. The cascade operated at indoor temperature and was fed by 10M nitric acid. The main characteristics of the AISI plant are presented in Table 6.4.

Worthy of notice is the experience of the $HNO_3$ – NO, $NO_2$ technique implementation for [15]N production (up to 10kg per year, with a concentration of 99%) at the Bitterfeld Chemical Complex in Wolfen (the formes GDR) [49, 50]. A two-stage cascade was deployed for the initial [15]N concentration, which obviated the need for construction of additional disposal systems for working and by-product flows and to decrease transportation expenditures for source material delivery and waste removal.

The first-stage column of the cascade was 200mm in diameter (enrichment of up to 1at.% of [15]N) and the second-stage column 150mm, with a height of 11.5m each. The

**Table 6.4**

Main characteristics of AISI plant for $^{15}$N production by the nitro-acid method (based on the data of sources [14, 52])

| Stage | Column | | | HNO$_3$[a] flow, kg·h$^{-1}$ | HETP, cm | $^{15}$N concentration, at.% |
|---|---|---|---|---|---|---|
|  | Diameter, mm | Height, m | SPP packing, mm |  |  |  |
| I | 120 | 17.1 | 2.5 × 2.5 × 0.25 | 31.44 | ≈ 28 | 3.5 |
| II | 26 | 17.8 | 2.2 × 2.2 × 0.25 | 1.65 | ≈ 8 | ≤ 99 |

[a] 10M nitric acid.

**Figure 6.7**  NIISI, Tbilisi, Georgia: 1, industrial building; 2, laboratory building; 3, warehouse for acids and sulfur dioxide.

columns were made of glass tubes which were filled by SPP packing. The HETP values are 20cm and 15cm for the first-stage and second-stage columns, respectively. The cascade produced up to 300kg of nitric acid with a $^{15}$N concentration of 10at.%.

The final concentrating (up to 99% of $^{15}$N) was performed in a two-column cascade housed in a 40-storey tower of the Isotope and Radiation Research Institute (I&RI) of the

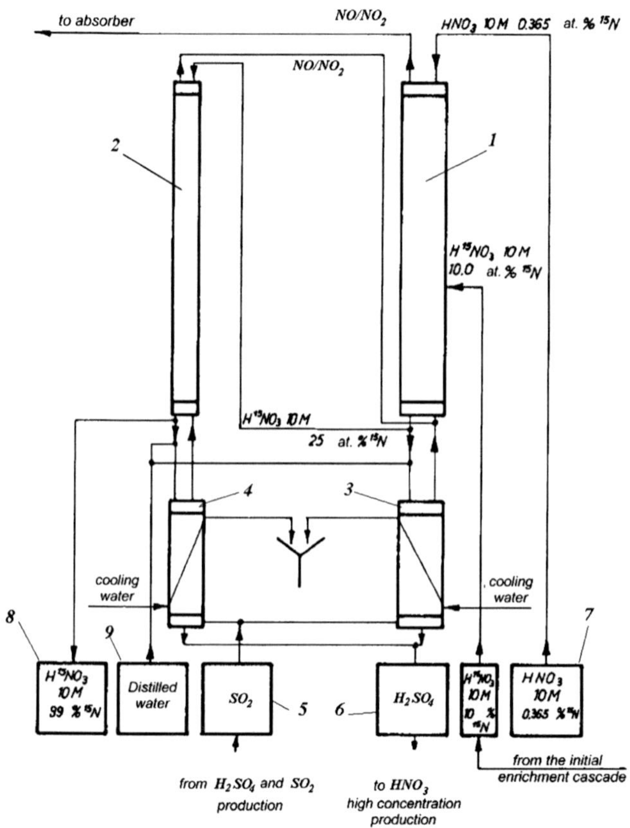


**Figure 6.8** Scheme of two-stage cascade for $^{15}$N concentrating from 10at.% up to 99at.% at Isotope and Radiation Research Institute (I&RRI) of Academy of Sciences of the former German Democratic Republic (Leipzig) [50]: 1, 2, first and second stage columns, respectively; 3, 4, phase flow conversion reactor (nitric acid reduction by sulfur dioxide); 5, source of $SO_2$; 6, nitric acid receiver; 7, nitric acid tank (natural abundance of heavier nitrogen isotope); 8, product tank; 9, distilled water tank.

Academy of Sciences of the former GDR in Leipzig. The cascade is shown in Figure 6.8. The first-stage column 1 is 56mm in diameter, the second-stage column 2 – 27mm. Both columns are 10.5m in height. The columns are made from X10CrNiTi stainless steel. The HETP values are 7cm and 5cm for the first-stage and second-stage columns, respectively. The cascade annual capacity is 25kg of $H^{15}NO_3$ with a 99% concentration [50]. The main plant characteristics are given in Table 6.5.

To remove the heat of reactions (6.9 and 6.10 or 6.11) each reactor for the nitric acid reduction is provided with an efficient cooling system (3, 4 in Figure 6.8). A quantity of distilled water is delivered for the reflux reactor (9 in Figure 6.8) to ensure the specified composition of the oxide mixture formed in the reflux unit. The flow reflux in the Nitrox process is performed very effectively and residual nitrogen concentration in a formed sulphuric acid (see Eqs. (6.9 and 6.10 or 6.11)) does not limit production of highly enriched $^{15}$N (more than 99at. %).

The cost items of the production of 1 kg of $^{15}$N are presented in Table 6.6 [50]. It should be noted that the raw material costs will be significantly lower because of the utilization,

**Table 6.5**

Main characteristics of the plant for $^{15}N$ production by the Nitrox method in I&RI
(based on the data of sources [49, 50])

| Stage | Diameter, | Column Height, m | SPP packing, mm | HETP, cm | $^{15}N$ concentration, at.% |
|---|---|---|---|---|---|
| | | | Initial enrichment | | |
| I | 200 | 11.5 | $2 \times 2 \times 0.2$ | 20 | 1.0 |
| II | 150 | 11.5 | | 15 | 10.0 |
| | | | Final enrichment | | |
| I | 56 | 10.5 | $2 \times 2 \times 0.2$ | 7 | 25 |
| II | 27 | 10.5 | | 5 | 99 |

**Table 6.6**

Main cost items of production of 1kg of $^{15}N$ by the Nitrox process
(based on the data of source [50])

| Position | Cost item | Unit measure | Consumption |
|---|---|---|---|
| | Source materials | | |
| 1 | 10 M $HNO_3$ | ton | 68.27 |
| 2 | $SO_2$ | ton | 44.0 |
| 3 | $H_2O$ in the reactor | ton | 11.73 |
| | Energy inputs | | |
| 4 | Electric power | kWh | 4,000 |
| 5 | Cooling water | $m^3$ | 2,666.7 |

in one way or another, of corresponding amounts of by-products (nitric oxides, sulfuric acid) of the production process [53].

The waste flow of nitric oxides leaving the nitric acid – nitric oxides plant can be utilized, for example, as source flow for the nitrogen and oxygen isotope separation by NO cryogenic rectification [10, 13] or by the chemical exchange method [51] (see chapter 7).

To compensate for the expenditure for sulfur dioxide, it is essential to utilize the resultant sulfuric acid (by increasing its concentration to a marketable level), integrate separation plants into the structure of appropriate production facilities, develop techniques for the reduction of the $SO_2$ specific consumption (i.e. through optimum temperature conditions for the isotope separation process), create combined process flow sheets with the use of thermal flow reflux techniques for the $^{15}N$ initial enrichment, and, finally, to reduce $H_2SO_4$ to $SO_2$ for the realization of a near-closed cycle of reagents in the flow reflux units [54, 55].

To realize the last measure, it is suggested using the process of thermo catalytic reduction of sulfuric acid by methane

$$3CH_4 + 11H_2SO_4 \rightarrow 11SO_2 + 2CO_2 + 2CO + 17H_2O, \qquad (6.14)$$

followed by the $SO_2$ removal from the formed vapor-gaseous reaction mixture [54].

Experimental studies on the reaction (6.14) were performed for the $H_2SO_4$ solutions generally with a concentration of 50–65% (by mass), formed in the $^{15}N$ production process (i.e. at NIISI plant). The studies were performed with the use of an extra-pure silica gel S-3 modified with ferric oxide or vanadium admixtures ($\approx$ 1.5% by mass); and industrial silica gels KSK-2 and KSK-G [54]. The reduction process was done over a temperature range 770–870K. The experiments demonstrated that this is a catalyst modified with vanadium oxide (C-3/$V_2O_5$) that possesses the highest catalytic reactivity. The studies reported in reference [55] were performed on a vanadium catalyst applied on KSM silica gel with the reduction carried out at 580°C. As was seen from the studies, the method under consideration can allow return of up to 90% of sulphur dioxide to the flow reflux system.

A four-section scheme (Figure 6.9) [54] is suggested for the $SO_2$ extraction from the vapour-gas mixture formed by reaction (6.14). The scheme operates as follows: from the sulfuric acid reduction reactor, the gas mixture flow arrives at the first adsorber I, where the $SO_2$ sorption takes place at a temperature of 300K, with the remaining gas phase components being displaced to the second adsorber II.

After the adsorber I is saturated with sulfur dioxide, this latter is delivered to the adsorber II, where the temperature increases. It is registered by a heat-sensing device, and the gas mixture flow is completely changed over to the adsorber, and adsorber I is linked to the liquid sulfur dioxide receiver V and heated to 450–480K. As a result of a pressure increase to about 0.3MPa, sulfur dioxide condenses in the receiver V. The adsorber I is disconnected from the $SO_2$ receiver and cooled to 300K. The adsorber I is ready for a new cycle. The same operations are successively performed with the second, third, and fourth adsorbers.

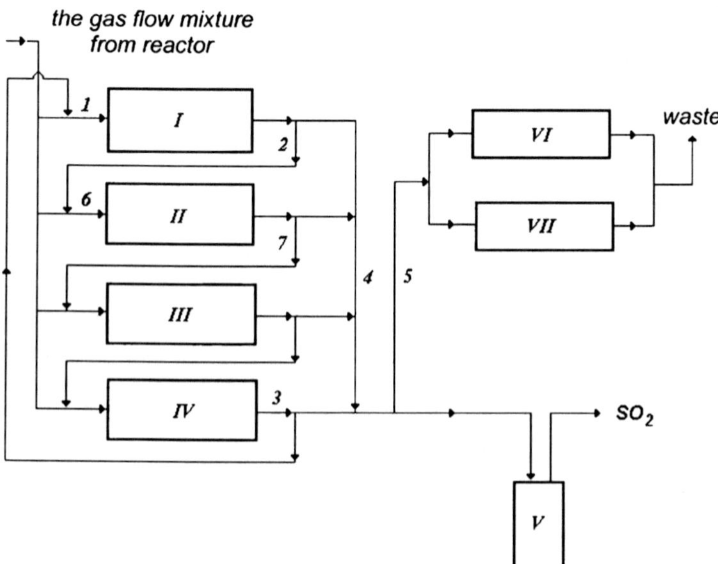


**Figure 6.9**   Scheme of adsorption unit for $SO_2$ extraction from vapor-gas mixture following catalytic thermal reduction of sulphuric acid: I, II, III, IV, adsorber with SKT-6 carbon; V, liquid sulfur dioxide receiver; VI, VII, environmental protection units [54].

To utilize $SO_2$ obtained with the use of the above scheme, sulfur dioxide must be subjected to further dehumidification. For this purpose it is suggested using silica gel with applied sulfuric acid. The capacity of this sorbent accounts for about 5mol of $H_2O$ per kilogram at 300K and 2.5mol of $H_2O$ at 480K. As has been reported in literature [54], the sulfur dioxide itself is practically not adsorbed.

### 6.2.6  Nitrogen isotope separation by ion exchange

The ion exchange is characterized by single-stage separation factor values close to those of the ammonium method (see Table 6.2), and by lower HETP values. The method is based on the reaction

$$^{14}NH_4^+R_{(sd)} +^{15}NH_4OH_{(sol)} \leftrightarrow {}^{15}NH_4^+R_{(sd)} + {}^{14}NH_4OH_{(sol)}, \qquad (6.15)$$

where R is ion-exchange resin (cationite) where the heavier nitrogen isotope is concentrated.

At a concentration of ammonium solution in water equal to 0.1–0.6 M for cationites Dowex 50 $\times$ 4, Dowex 50 $\times$ 12, Dowex 50 $\times$ 16, and Amberlite IR-120, the separation factor values lie within a range 1.0202 – 1.028 [21, 22], and for cationite KU 2 $\times$ 9 at a $NH_3$ concentration of 0.25–2.68 M, $\alpha = 1.022 - 1.017$ [56].

The F. Spedding stationary zone method [21], with the use of which 100g of $^{15}N$ with a concentration of 99.7% was produced [22], remains the main technique of ion-exchange method implementation. As applied to large-scale isotope production, the ion exchange attractiveness is due to lower HETP values compared with liquid–gas chemical isotope exchange systems. But a relatively low flow rate capacity of the ion-exchange columns ($\approx 1mg$-at $N\cdot(cm^2\cdot min)^{-1}$) determines their large diameter.

The ion exchange process realized under conditions of complete phase flow reflux is similar to the organization of the chemical exchange separation method in the liquid–gas system and, hence, is characterized by associated alkali and acid consumption.

For the above-mentioned reasons, the ion exchange method has not gained wide acceptance for relatively large-scale nitrogen isotope separation. The method falls into the category of processes that are under development [56–59]. The papers concerning this matter generally have to do with the development of mathematical models for steady and unsteady separation modes. For the first case, as an example, a novel theoretical approach and numerical method have been suggested to solve the problem of the description of separation process in an ion-exchange plant operating by the Spedding method. The design and experimental data are in a satisfactory agreement which, as is pointed out by the authors of some papers [56–59], allows use of the developed method for the assessment of the ion exchange applicability for $^{15}N$ large-scale production.

### 6.3  COMPARISON OF NITROGEN ISOTOPE SEPARATION TECHNIQUES

The production scale plays an important part in the selection of separation technique. A comparative economic assessment of nitrogen isotope separation techniques subject to the production scale has been made on the basis of the determination of components of specific costs (per $^{15}$N unit) [53, 60]:

$$C_{SP} = (C_V + C_L + C_C)/B, \tag{6.16}$$

where $C_V$ and $C_L$ are separation costs proportional to the apparatus volume and operation flow, respectively; $C_C$ is other costs depending or slightly depending on the volume and flow rate.

The specific costs were determined for two production scale ranges (5–50kg and 50–1000kg of $^{15}$N per year) with the use of physical–chemical methods [60]. Notice that, at present, physical separation methods (i.e. gas-centrifugal) are not competitive [61].

Among chemical exchange separation methods (Figure 6.10), nitric acid – nitric oxides and oxide methods have the best characteristics, with the latter featuring lower sulfur consumption per unit of product. The ammonium method in its classical form is significantly inferior to the above-mentioned nitrogen isotope separation processes.

For the thermal flow conversion systems (Figure 6.11), the NO cryogenic rectification dominates over the whole range of the production scale variation. In the chemical exchange methanol-based system, the costs per $^{15}$N units are higher by about twofold. The chemical exchange rectification of nitric di(tetra)oxide and ammonia rectification are considerably inferior to the first two methods.

A comparison between the two groups shows that the NO cryogenic rectification is somewhat less attractive than the Nitrox technique, especially for small-scale (several kilograms of $^{15}$N per year) production plants. With an increase in production scale the costs become closer, and at an annual capacity of 20–50kg $^{15}$N, the nitric oxide low temperature rectification can be more economically expedient (without regard to the costs of the homomolecular isotope exchange (HMEX) reaction – see (6.4)).

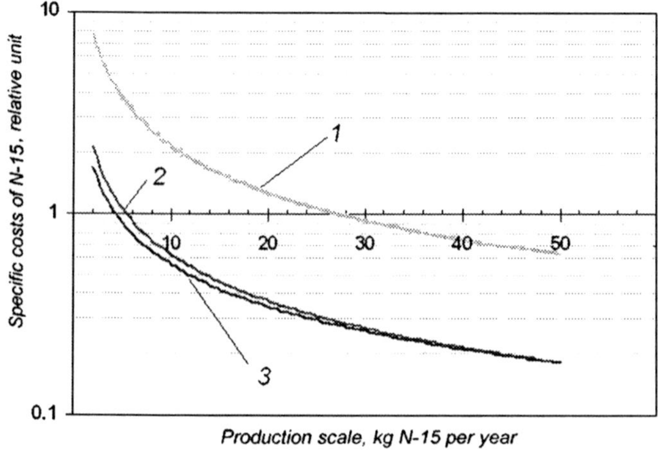


**Figure 6.10**  $^{15}$N production cost (in relative units) at annual production scale of 5 – 50kg for separation techniques with chemical flow reflux [60]: 1, ammonium method ($P = 0.024$MPa); 2, nitric acid (Nitrox) method ($T = 298$K); 3, oxide method ($T = 264$K).

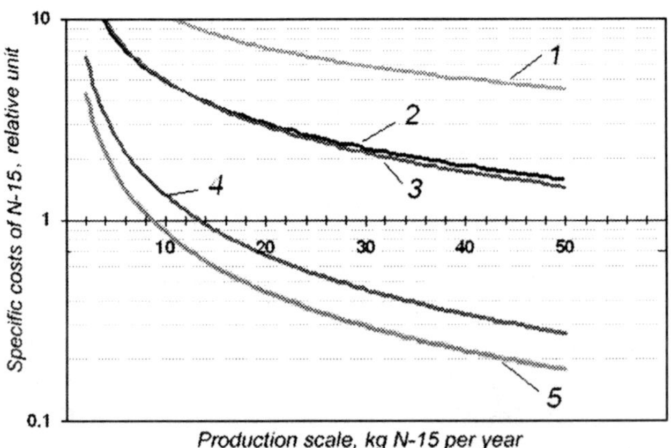


**Figure 6.11**   $^{15}$N production cost (in relative units) at annual production scale of 5 – 50kg for separation techniques with thermal flow reflux [60]: 1, ammonia rectification; 2, molecular nitrogen rectification; 3, nitric dioxide (tetra-oxide) chemical exchange rectification; 4, chemical exchange in ammonia complex in ethanol–ammonia; 5, nitric oxide NO low-temperature rectification.

## 6.4   LARGE-SCALE PRODUCTION CHARACTERISTICS

Fifteen or twenty years ago, an annual capacity of a nitrogen-15 concentration plant (production complex) accounted for hundreds of grams or several kilograms, and, in the best case, up to 10kg of $^{15}$N [50]. As for $^{13}$C production, where the world's level was estimated as not exceeding a $^{13}$C annual production equal to 10kg or even less, by now a threshold of over 100 kg has been passed. The required production scale of $^{15}$N can turn out to be all the more sizable.

The results of the specific cost calculations for large-scale $^{15}$N production by the eight nitrogen isotope separation methods are given in Figure 6.12. For this case, three groups of techniques can be distinguished. In the first group (I) (an annual production scale of about 250kg of $^{15}$N) the following can be observed: the nitric dioxide (tetra-oxide) chemical exchange rectification (CHEXR) becomes more attractive then the ammonium technique.

The second group (II) indicates that specific costs of the Nitrox and oxide methods are practically equal.

And finally, in the third group (III), the costs of the chemical exchange in the $NH_3 \cdot CH_3OH_{(lq)}-NH_{3(gas)}$ system become commensurate with those for the nitric oxide chemical exchange rectification. Of course, in a specific context, the costs ratio can displace over the production scale in one or another direction. But, nonetheless, the obtained data allow a different consideration of, for example, the ammonia-complex method.

First of all, some different technological solutions are required to organize large-scale production. The issues of source material provision, disposal of the by-products (if at all), or by-product recovery with their return to the processing cycle, and ecological friendliness are brought to the forefront. The organization of transit schemes and, as far as possible, of closed cycles of main process flows, hence, becomes a necessity.

As a case in point, the scheme of the nitrogen isotope separation for the $^{15}$N initial concentration with the use of either low temperature NO rectification, or chemical exchange

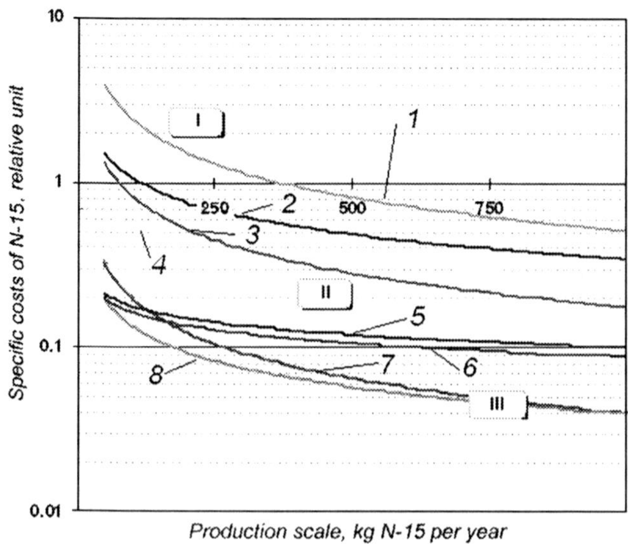



**Figure 6.12**  Comparison of costs per unit of $^{15}N$ production methods (in relative units) for large-scale production [60]: 1, ammonia rectification; 2, molecular nitrogen rectification; 3, nitric dioxide (tetra-oxide) chemical exchange rectification; 4, ammonium method ($P$ = 0.024MPa); 5, oxide method ($T$ = 264K); 6, nitric acid (Nitrox) method ($T$ = 298K); 7, chemical exchange in ammonia complex in ethanol–ammonia; 8, nitric oxide NO low-temperature rectification.

in the oxide system, put forward by the scientists of MUCTR, can be cited (Figure 6.13) [60]. The main idea of the scheme is the creation of several closed loops of source flows, withdrawal flows, working substance waste flows (in the case of rectification), or of the working system components (in the case of chemical isotope exchange).

The flows of operating substances circulating in the separation column and in the flow reflux units form the first inner processing loop.

Owing to the normalization of the isotopic composition of product and waste flows by chemical exchange ($E_D$ and $E_C$ units) with nitric acid (in the case under consideration), the flows of feed F′, F″, of product B′, B″ and of waste W′, W″ form the second inner loop. To ensure the phase equilibrium conditions in the above units, nitric acid is previously saturated with nitric oxides in the $A_D$ and $A_C$ units through the absorption of oxides of the gas flow inert to $HNO_3$ and nitric oxides from the standpoint of both chemical and isotopic interaction (i.e. $N_2$). The last is preliminarily saturated with nitric oxides (as well as with water and nitric acid vapours) in the desorption units $D_D$ and $D_C$, respectively. The nitric acid leaving the desorption units must contain small amounts of dissolved nitrogen oxides. The carrier gas circulation is assured by blowers.

The absorption units – isotope exchange unit – desorption unit form the third loop. It should be noted that when the product and waste flows are covariant, two gas-carrier loops can be integrated into one. This scheme makes it possible to isolate the operating system used for the separation.

If for the chemical isotope exchange it is planned to use chemical flow conversion methods, additional units $U_D$ and $U_C$, linked to the flow conversion units $R_D$ and $R_C$

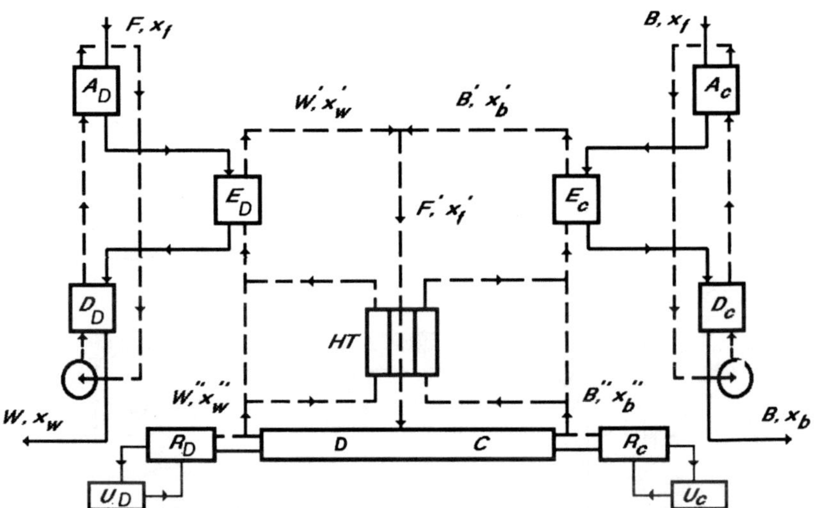


**Figure 6.13** Concept of transit scheme organization at $^{15}$N initial concentration stage using, for example, nitric acid (Nitrox) or oxide separation methods, or NO low-temperature rectification [60]: D, C, separation column depletion and enrichment sections, respectively; $R_D$, $R_C$, flow conversion units of corresponding column sections; $U_S$, $U_C$, units for recovery of reagents used in corresponding flow reflux units; HT, recuperative heat exchanger; $D_D$, $D_C$, desorption units; $A_D$, $A_C$, adsorption units; $D_D$, $E_C$, isotope exchange units; F, F', external and internal feed flows, respectively; B, B', B'', external, depleted internal, and enriched internal product flows, respectively; W, W', W'', external, enriched internal, and depleted internal waste flows, respectively; x, x', x'', extracted component concentration in corresponding flows.

respectively, are required. For the Nitrox or oxide methods, the unit $U_C$, for one, can represent a system of the sulfuric acid thermo-catalytic reduction (see Figure 6.9).

The scheme presented in Figure 6.13 can be realized not only for the oxide methods of nitrogen isotope separation but also for the chemical exchange systems with the use of ammonia. Under conditions of gaseous ammonia deficiency and ammonium salt sufficiency, for example, the $NH_3$ closed loop (the second process circuit) can be organized through the isotope exchange between ammonia and $NH_4^+$ salt solution in the units $E_D$ and $E_C$, and the separation process as such realized with the use of the ammonia-complex method of nitrogen isotope separation (the first process circuit) with the methanol closed loop between the units $R_D$ and $R_C$. In this case, of course, the additional units $U_D$ and $U_C$ cease to have significance. By and large, the above solution will probably be the most attractive if there is a need to produce $^{15}$N in the ammonium form as such.

With sufficient resources of gaseous ammonia, and with the use of the ammonia-complex method, the transit circuits in the scheme under consideration are made unnecessary because in this case the process is realized in accordance with the classical transit scheme and with the closed loop of the complex-forming material. In so doing, it is advisable to perform the final concentrating by the ammonium technique with the formation of a combined (by the separation methods) process scheme, similar to the combination of the NO rectification and Nitrox method when the use of rectification at the initial concentration stage (up to $\approx$ 3.6at.% of $^{15}N$), together with the nitric acid – nitric oxides nitrogen isotope separation method, can allow reduction in the processing component of the costs down to a few dollars (\$4–\$5) per gram of $^{15}N$).

## REFERENCES


1. K. Clusius, K. Schleich, *Helv. Chim. Acta*, **1958**, 41, 5, 1342.
2. H. G. Thode, *J. Am. Chem. Soc.*, **1940**, 62, 581.
3. E. Ancona, G. Boato, *Casanova Nuovo Cimento*, **1962**, 24, 111.
4. K. Clusius, M. Vecchi, *Helv. Chim. Acta*, **1959**, 42, 6, 1921.
5. D. Staschevski, *Chem. Technol.*, **1975**, 4, 8, 269.
6. K. Clusius, K. Schleich, M. Vecchi, *Helv. Chim. Acta*, **1959**, 42, 7, 2654.
7. G. M. Begun, *J. Chem. Phys.*, **1956**, 25, 6, 1279.
8. H. G. Thode, H. C. Urey, *J. Chem. Phys.*, **1939**, 7, 34.
9. K. Clusius, K. Schleich, In : *Proc. Second UN Int. Conf. on the Peaceful Uses of Atomic Energy*, UN, Geneva, **1958**, 4, 485.
10. P. Ya. Asatiani, I. P. Giorgadze, G. L Partsakhaschvili, G. A. Tevzadze *et al.*, In: *Stable Isotopes in the Life Science*,. IAEA, Vienna, **1977,** 75.
11. B. B. McInteer, R. M. Potter, *Ind. Eng. Chem.*, **1965**, 4, 35.
12. D. E. Armstrong, B. B. McInteer, T. R. Mills, J. G. Montoya, In: *Stable Isotopes. Proc. Third Intern. Conf.*, New York, Academic Press, **1979**, 175.
13. P. Ya. Asatiani, author's abstract, Ph.D. thesis, Mendeleev University of Chemical Technology of Russia, **1981**, 20.
14. Yu. V. Nikolaev, Ph. D. thesis, Sukhumi, PhTI, **1964**, 110.
15. H. C. Urey, M. Fox, J. R. Huffman, H. G. Thode, *J. Am. Chem. Soc.*, **1937**, 59, 1407.
16. V. Yu. Orlov, N. N. Tunitsky, *Zhurn. Phiz. Khimii*, **1956**, 30, 9, 2085.
17. D. V. Peshtchenko, E. A. Tyupina, A. V. Khoroshilov, *Uspekhi Khimii i Khim. Technol.* ?IV, 4, Mendeleev University of Chemical Technology of Russia, **2000**, 13.
18. H. C. Urey, J. R. Huffman, H. G. Thode, M. Fox, *J. Chem. Phys.*, **1937**, 5, 11, 856.
19. G. M. Begun, A. A. Palko, L. L. Brouwn, *J. Chem. Phys.*, **1956**, 60, 48.
20. A. I. Kuznetsov, G. M. Pantchenkov, *Zhurn. Phiz. Khimii*, **1970**, 44, 7, 1802.
21. F. H. Spedding, G. E. Powell, H. J. Svec *J. Am. Chem. Soc.*, **1955**, 77, 5, 1393.
22. D. Carrillo, M. Urgell, J. Iglesias, *Anales de Quimica*, **1968**, 64, 9–10, 841.
23. M. Stern, L. Kauder, W. Spindel, *J. Chem. Phys.*, **1962**, 36, 764.
24. V. I. Gorshkov *et al.*, report 77033366, M. V. Lomonosov Moscow State University, **1979**, 61.
25. V. T Bayramov *et al.*, A. C. SU 1443259 A1, bulletin 6, **1996**.
26. E. U. Mouse, L. N. Kauder, W. Spindel, *Z. Naturforschg.*, **1963**, 18a, 235.
27. G. Stiehl, *Isotopentechnik*, **1960/61**, 1, 4, 118.
28. R. Nakane, T. Watanabe, *Isotopes and Radiation*, **1960**, 3, 6, 506.
29. S. G. Katalnikov, I. A. Myshletsov, A. V. Khoroshilov, In: *Proc. Int. Symp. on Isotope Separation and Chemical Exchange Uranium Enrichment*, Oct. 29 – Nov. 1, 1990, Tokyo, Japan, (Bull. Research Lab. for Nuclear Reactors), **1992**, 368.
30. S. G. Katalnikov, A. V. Khoroshilov, M. M. Tchelyak, *Atomnaya Energia*, **1986**, 60, 2, 109.
31. G. K. Boreskov, S. G. Katalnikov, *Technology of Chemical Isotope Exchange Processes*, Mendeleev University of Chemical Technology of Russia, **1974**, 220.
32. H. G. Thode, H. C. Urey, *J. Chem. Phys.*, **1939**, 7, 34.
33. H. G. Thode, H. C. Urey, *J. Chem. Phys.*, **1940**, 8, 904.

34. B. V. Ottesen, M. E. Aerov, *Zhurn. Phiz. Khimii*, **1956**, 30, 6, 1356.
35. S. I. Babkov, N. M. Zhavoronkov, *Khim. Prom.*, **1955**, 7, 388.
36. S. G. Katalnikov, B. M. Andreev, E. A. Tyupina, A. V. Khoroshilov, In: Proc. *2nd All-Russian Conference on Physical-Chemical Processes in Selection of Atoms and Molecules*, TsNIIatominform, **1997**, 117.
37. E. A. Tyupina, A. V. Khoroshilov, In: *Scientific-Innovative Cooperation, IInd Scientific-Technological Conference*, **2003**, 4.1, 65.
38. A. V. Lizunov, E. A. Tyupina, A. V. Khoroshilov, *Uspekhi Khimii i Khim. Technol.* XIII, 3, Mendeleev University of Chemical Technology of Russia, **1999**, 69.
39. A. V. Khoroshilov, In *Proc. 5th All-Russian Conference on Physica–Chemical Processes in Selection of Atoms and Molecules*, TsNIIatominform, Izd. Trovant, **2000**, 166.
40. W. Spindel, T. I. Taylor, *J. Chem. Phys*, **1955**, 23, 5, 981.
41. T. I. Taylor, W. Spindel, In: *Proc. Intern. Symp. on Isotope Separation*, Amsterdam, North-Holland Pub. Co, **1958**, 158.
42. A. V. Khoroshilov, *Proc. Mendeleev University of Chemical Technology of Russia*, 119, 138.
43. V. A. Kaminsky, *Atomnaya Energia*, **1963**, 14, 6, 586.
44. M. S. Safonov, V. I. Gorschkov, N. E. Tamm, V. A. Ivanov, In: *Proc.* 4th *All-Russian Conference on Physical-Chemical Processes in Selection of Atoms and Molecules*, TsNIIatominform, **1999**, 168.
45. A. V. Khoroshilov, S. G. Katalnikov, *Isotopenpraxis*, **1989**, 5, 12, 546.
46. I. G. Gverdtziteli, Yu. V. Nikolaev, E. D. Oziaschvili, K. G. Ordzhonokidze *et al., Atomnaya Energiya*, **1961**, 5, 10, 487.
47. A. O. Edmunds, I. M. Lockhart, In: *Proc. Symp. Isotope Rations as Pollutant Source and Behaviour Indicators*, IAEA, Vienna, **1975**, 279.
48. J. Mahenc, G. Pompidor, *Chimie et Industrie. Genie Chimique*, **1968**, 99, 1137.
49. E. Krell, *Isotopenpraxis*, **1976**, 12, 5, 188.
50. E. Krell, C. Jonas, In: *Stable Isotopes in the Life Sciences*, **IAEA**, Vienna, **1977**, 59.
51. D. Axente, A. Baldea, M. Abbrudean, In: *Proc. Int. Symp. on Isotope Separation and Chemical Exchange Uranium Enrichment* (Bull. Res. Lab. for Nuclear Reactors, special issue 1), **1992**, 357.
52. A. B. Bakhtadze, G. A. Tevzadze, E. D. Oziaschvili *et al., Isotopes in the USSR*, **1980**, 59, 54.
53. A. V. Khoroshilov, S. G. Katalnikov, In: *Proc. Mendeleev University of Chemical Technology of Russia*, **1984**, 130, 18.
54. A. A. Razmadze, author's abstract, Ph. D. thesis, M. V. Lomonosov Moscow State University, **1999**, 21.
55. S. L. Vasilyak, S. E. Salnikov, A. V. Khoroshilov, *Uspekhi Khimii i Khim, Technol. Collection of Scientific Proceedings*, XVI, 5 (22), Mendeleev University of Chemical Technology of Russia, **2002**, 56.
56. E. Aoki, T. Kai, Y. Fujii, *J. Nucl. Sci. Technol.*, **1997**, 34, 3, 277.
57. Y. Fujii, M. Aida, M. Okamoto, *Sep. Sci. Technol.*, **1985**, 20, 377.
58. T. Kai, E. Aoki, Y. Fujii, *J. Nucl. Sci. Technol.*, **1999**, 36, 4, 371.
59. M. Ohwaki, Y. Fujii, K. Morita, K. Takeda *Sep. Sci. Technol.*, **1998**, 33, 1, 31.
60. V. D. Borisevich, A. V. Khoroshilov, V. I. Gorshkov, V. A. Ivanov, In: *Proc. 7th Workshop on SPLG*, Moscow, Russia, **2000**, 218.
61. A. V. Khoroshilov, *Khim. Prom.*, **1999**, 4, 37.

# – 7 –

# Oxygen Isotope Separation

## 7.1  OXYGEN ISOTOPE SEPARATION BY RECTIFICATION

### 7.1.1  Isotope effect and properties of operating substances

Rectification is the main technique in oxygen isotope production. The most important substances are nitric oxide (NO), molecular oxygen ($O_2$), and water ($H_2O$). The single-stage separation factor values typical for the process of rectification of these operating substances are given in Table 7.1 [1–12]. Table 7.1 presents as well the values of $\alpha$ for carbon monoxide (CO) by the rectification of which $^{13}C$ enrichment is performed (see chapter 5) with $^{18}O$ as by-product. All substances shown in Table 7.1, hence, are of practical significance for oxygen isotope separation.

Nitric oxide (II) NO (properties are given in chapter 6) is characterized by the maximum values of the oxygen isotope separation factor, with the heaviest isotope $^{18}O$ concentrating in the liquid phase.

According to K. Clusius [1], the separation factor for the molecules $^{14}N^{16}O$–$^{14}N^{18}O$ over a temperature range of $110 - 118K$ is approximated by

$$\lg \alpha = 4.4684/T - 21.03 \times 10^{-2} \tag{7.1}$$

and at the NO normal boiling point $\alpha = 1.037$. This value exceeds the isotope effect in the nitrogen isotope separation under the same conditions (see chapter 6) for the same reason of a high value of $\alpha$.

Oxygen $O_2$ has the critical temperature $T_{cr} = 154.77K$, normal boiling point $T_{b\,(n)} = 90.17K$ and melting point $T_{m(n)} = 54.36K$. Liquid $O_2$ density at $T_{b(n)}$ $\rho_1 = 1142kg/m^3$. The heat of evaporation $\Delta H_{ev} = 6.81kJ/mol$.

According to A. Kovalenko and Ya. Zelvenskii [5], the isotope effect in molecular oxygen rectification over a temperature range of $76.8–90.2K$ in the region of $^{18}O$ initial concentration (for molecules $^{16}O^{16}O$–$^{16}O^{18}O$) is described by

$$\ln \alpha = 43.4/T^2 - 0.0626/T. \tag{7.2}$$

## Table 7.1

Isotope effect in oxygen isotope separation ($^{16}O$, $^{18}O$) by rectification (experimental and calculation data)

| Operating substance | Molecules | Temperature K | $\alpha$ | Approximation of temperature dependence of isotope effect | Reference |
|---|---|---|---|---|---|
| Nitric oxide NO | $^{14}N^{16}O - $ $^{14}N^{18}O$ | 110–118 111 120 | 1.043 1.037 | $\lg \alpha = 4.4684/T - 21.03 \cdot 10^{-2}$ | [1] |
| | $^{14}N^{16}O - $ $^{15}N^{18}O$ | 110–118 120 | 1.0632 | $\lg \alpha = 7.3230/T - 32.93 \cdot 10^{-3}$ | [2] |
| Carbon monoxide CO | $^{12}C^{16}O - $ $^{12}C^{18}O$ | 69–77 69 77 | 1.008 1.0059 | $\lg \alpha = 25.3/T^2 - 0.132$ | [3] |
| | | 76.8–88 | | $\ln \alpha = 45.7/T^2 - 0.088/T$ | |
| | | 76.8–90.2 | | $\ln \alpha = 43.4/T^2 - 0.0626/T$ | |
| Oxygen $O_2$ | $^{16}O^{16}O - $ $^{16}O^{18}O$ | 76.8 78.8 80.4 82.2 84.8 86.4 88.0 83.2 90.6 98.5 | 1.0066 1.0062 1.0059 1.0055 1.0053 1.0052 1.0049 1.0063 1.0052 1.004 | | [4, 5] [6] |
| | $^{16}O^{16}O - $ $^{18}O^{18}O$ | 63.5–90 | | $\ln \alpha = 57.285/T^2 - 0.145/T - 0.661 \cdot 10^{-3}$ | [7] |
| Water $H_2O$ | $H_2^{16}O - $ $H_2^{18}O$ | 273–403 | | $\ln \alpha = 1991.1/T^2 - 4.1887/T + 0.001197$ [a] | [8] |
| | | 293–443 | | $\ln \alpha = 5.869/T - 0.0108$ | [9] |
| | | 294–483 | | $\lg \alpha = 3.300/T - 7.22 \times 10^{-3}$ | [10,11] |
| | | 323 333 343 353 363 373 | 1.0078 1.0068 1.0063 1.0058 1.0053 1.004 | | [12] [10] |

[a] Calculation.

At the normal $O_2$ boiling point $\alpha \approx 1.0047$ and the enrichment factor is significantly lower than that in the NO rectification (by a factor of 7–8).

Water $H_2O$ has the critical temperature $T_{cr} = 647.27$K, normal boiling point $T_{b(n)} = 373.15$K and melting point $T_{m(n)} = 273.15$K. Liquid water density at $T_{b(n)}$ $\rho_1 = 1000$kg/m$^3$. The heat of evaporation $\Delta H_{ev} = 125.56$kJ/mol.

In the region of low concentration of the heavy oxygen isotope $^{18}O$ over a temperature range of 293–443K, the separation factor for the molecules $H_2^{16}O$–$H_2^{18}O$ [9] is approximated by

$$\ln\alpha = 5.869/T - 0.0108. \tag{7.3}$$

At $T_{b\,(n)}$ the separation $\alpha = 1.0038$–$1.004$ [10] and close to that for molecular oxygen rectification at atmospheric pressure.

Experimental values of the separation factor [9, 12] agree well with the calculation data by A. Van Hook [8] with the best agreement at 340–345K.

For the molecules $H_2^{16}O$–$H_2^{17}O$, the separation factor is considerably lower: 1.0033 and 1.0026 at 331K and 340K, respectively [10].

In the substitution of protium atoms by deuterium atoms in water, the oxygen isotope separation factor, according to D. Staschewski [13], decreases:

$$(\alpha_D - 1)/(\alpha_H - 1) = 0.825 \pm 0.022, \tag{7.4}$$

where $\alpha_D$ is the separation factor for molecules $D_2^{16}O$–$D_2^{18}O$; $\alpha_H$ is the separation factor for $H_2^{16}O$–$H_2^{18}O$ at the same temperature.

For water with medium oxygen isotope $^{17}O$, the separation factor is, naturally, lower than that for $H_2^{18}O$. According to data [13] it can be taken that:

$$(\alpha_{17} - 1)/(\alpha_{18} - 1) = 0.564 \pm 0.014, \tag{7.5}$$

where $\alpha_{17}$ is the separation factor for molecules $H_2^{16}O$–$H_2^{17}O$; $\alpha_{18}$ is the separation factor for $H_2^{16}O$–$H_2^{18}O$ at the same temperature.

The above isotope effect values for four operating substances (NO, CO, $O_2$, and $H_2O$) count in favor of nitric oxide (II) cryogenic rectification. Nevertheless, all rectification processes with the use of the above substances have an industrial record and, at present, can be considered as competitive (and, first of all, NO, $O_2$, and $H_2O$). Each process has its own advantages and disadvantages, and the selection of one or the other operating system is determined by many factors. These factors include the working substance availability, requirements on its purity, corresponding hardware implementation, and costs per unit product (consumption coefficients).

From the chronological point of view, the first industrial plant for highly enriched heavy oxygen isotope $^{18}O$ production was based on water rectification.

## 7.1.2  Heavy oxygen isotope production by water rectification

The advantages of water rectification as an isotope separation method were noted in chapter 2 when considering the issues of heavy water $D_2O$ production. The negative characteristics of the process, associated with high energy inputs and the necessity to process large water flows, are not of decisive importance for the $H_2^{18}O$ production since the natural concentration of $^{18}O$ is an order of magnitude higher than the deuterium content, and the $H_2^{18}O$ production scale is incommensurably smaller. For these reasons, water rectification was and still remains a main oxygen isotope separation method. Both from the process organization standpoint, and from the historical retrospective point of view, it is the scheme for $H_2^{18}O$ and $H_2^{17}O$ production from natural water (Figure 7.1) put forward by I. Dostrovsky [14] that is deserving of more attention.

In accordance to the adopted engineering solutions, an eightfold pre-concentration of the oxygen heavy isotope is performed in ten columns of 100mm diameter and with a packing bed 10m in height, operating in parallel independently of one another on the open scheme (without depletion) with withdrawal of first kind at a 35kPa pressure in the column head (water vacuum rectification). In the column, the mean separation factor value for molecules $H_2^{16}O$–$H_2^{18}O$ is taken as $\alpha \approx 1.006$.

The use of packed columns of 100mm diameter allows avoidance of HETP deterioration through the scale factor. In all ten columns, the flow rate accounts for 7.292kg of $H_2O$ per hour, with the withdrawal rate equal to 3.33ml of 1.6% $H_2^{18}O$ and 0.12% $H_2^{17}O$ per hour. The obtained concentrate at a rate of 3.33ml/h is fed to a three-stage cascade with the fourth column as a depletion section.

In the depletion column D (Figure 7.1), the waste flow of the stage EI is depleted to the natural content of $^{18}O$, which obviates the need for the feedback coupling of the cascade (EI–EIII) with pre-concentration columns. The coupling is very difficult to realize since it is necessary to divide the return flow into ten parts so the withdrawal of second kind corresponds to the actual efficiency of each pre-concentration column.

Owing to the adopted scheme, each preconcentration column, not coupled with the following columns by the return flow, operates independently. The concentrate delivery from the first to the second stage, and from the second to the third stage, is performed without

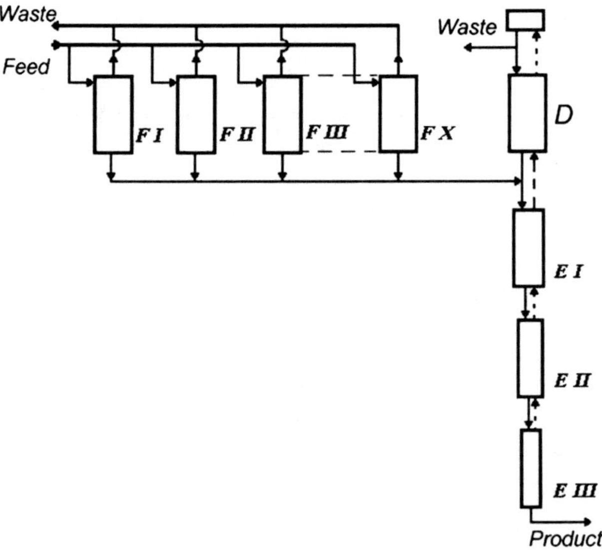


**Figure 7.1**  Scheme of water vacuum rectification for oxygen isotope separation by I. Dostrovsky [14]: FI–FX pre-concentration columns; EI, EII and EIII, columns of cascade enrichment section, namely, first, second, and third stages, respectively; D, depletion section of column.

pumps in the vapor phase with following vapor condensation. The condensate arrives at the tank mounted at a level ensuring a hydraulic drop required for the next stage refluxing. An example of the column-to-column linkage is given in Figure 7.2.

This technique of liquid flow displacement through the separation cascade stages, though associated with an additional heating steam consumption (which is of no significance for small flows), makes it possible to avoid constructing very high production facilities due to the necessity of mounting the columns vertically (one above the other) to provide for the gravity flow of the phases.

As indicated in reference [10], under conditions of prolonged continuous operation of separation cascades with small flow rates, the flow displacement through the cascade stages by pumps does not work well.

The above plant annual capacity is about 4kg of heavy-oxygen water $H_2^{18}O$ with a concentration of over 90%. The plant characteristics are presented in Table 7.2.

To increase the heavy oxygen isotope concentration and to enhance the plant capacity, the cascade has been improved over the course of 25 years. The modified plant comprises 40 packed columns of 20–150mm diameter and produces up to 6kg of $H_2^{18}O$ with a concentration of 98–99% and $H_2^{17}O$ enriched to 40% (the $^{17}O$ concentrating to 90% was performed by the thermal diffusion method).

From the engineering point of view, the plant represented two cascades coupled by direct and return flows. The first cascade was designed to achieve the optimal (with respect

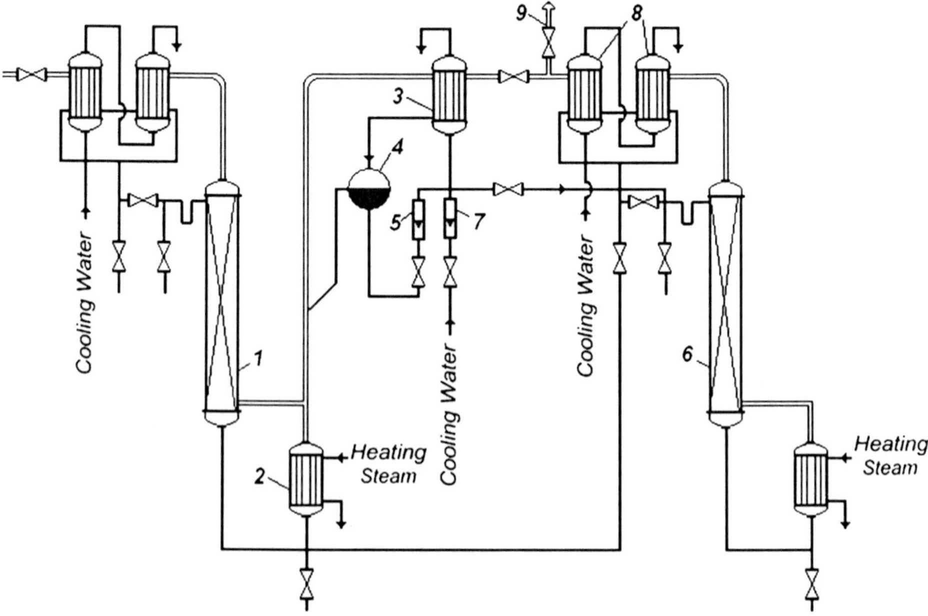


**Figure 7.2** Scheme of links between cascade columns in water vacuum rectification – scheme with intermediate condensation [14]: 1, first-stage column; 2, evaporator; 3, condenser; 4, hotwell with constant level; 5, second-stage column feed flow rotameter; 6, second-stage column; 7, cooling water rotameter; 8, condensers of second-stage vapor; 9, to vacuum system.

**Table 7.2**

Characteristics of columns of I. Dostrovsky cascade [14]

| Column[a] | | | Packing[b], mm | Flow rate, kg/h | $x_B$, at.% | |
|---|---|---|---|---|---|---|
| Designation | Diameter, mm | Packing bed height, m | | | $^{18}O$ | $^{17}O$ |
| FI–FX | 100 | 9.5 | 3.2 × 3.2 | 7.29 | 1.6 | 0.12 |
| D | 100 | 9.5 | 3.2 × 3.2 | 7.08 | 0.1 | 0.035 |
| EI | 100 | 9.5 | 3.2 × 3.2 | 7.08 | 1.4 | 0.15 |
| EII | 76 | 9.3 | 3.2 × 3.2 | 7.08 | 3.3 | 0.39 |
| EIII | 76 | 7 | 1.6 × 1.6 | 2.08 | 70 | 1.78 |
| EIV | 25 | — | 1.6 × 1.6 | 0.21 | 90 | 0.97 |

[a] Made of copper.

[b] Dixon gauze (100 mesh) rings made of phosphor bronze.

to the subsequent enrichment) values of $^{18}O$ and $^{17}O$ concentrations. The second cascade was meant for the subsequent enrichment of the oxygen heavy isotopes.

The first cascade consisted of two stages. The first stage (or the pre-enrichment stage) included (similar to Figure 7.1) 20 columns of 150mm and 100mm diameter operating in parallel, which performed the oxygen isotope concentrating from the natural level to 2% of $^{18}O$ and 0.2% of $^{17}O$. The stage feed flow rate accounted for 57.3kg of $H_2O$ per hour. The withdrawal flow of the stage equal to 40ml/h arrived at the second stage comprising two pairs of columns operating in series-parallel.

At the outlet of the second stage enrichment section, the concentration accounted for 5–6% of $^{18}O$ and 0.5% of $^{17}O$.

The second stage withdrawal flow was delivered for the feeding of the cascade third stage incorporating two columns in the enrichment section and one column in the depletion section. The oxygen isotope concentration in the enrichment section increased up to 48% of $^{18}O$ and 3.7% of $^{17}O$ (as reported in reference [14], the maximum enrichment with $^{17}O$ is achieved over the region of intermediate concentration of the heavy oxygen isotope), and the withdrawal flow equal to 1.46ml/h was fed to the plant second cascade. The second cascade, comprising 11 columns linked together to form two branches, was used for the enrichment of oxygen heavy isotopes: up to 98–99% of $^{18}O$, and up to 40% of $^{17}O$. The plant withdrawal flow rate in general accounted for about 0.625ml/h and 0.1ml/h of water enriched with $^{18}O$ and $^{17}O$ isotopes, respectively.

Several plants for heavy-oxygen water ($H_2^{18}O$ and $H_2^{17}O$) production are at present operable in Israel. Their overall annual capacity is estimated at 60–80kg of $H_2^{18}O$.

A water rectification plant for heavy-oxygen water production was developed by B. Zamansky, and has operated until now in Gorlovka, Ukraine (at present Concern Stirol Corporation, Stirolpharm Company). The plant includes 34 packed columns of 160 to 25mm diameter filled with cylindrical packing material made of stainless steel wire with elements of 2mm size. Rectification is performed under vacuum at a pressure of 20kPa. The plant operates with the complete condensate return through the head tank. Once a day, the liquid volume of each column still is displaced to the head tank of the next stage, with the content of the head tank displaced to the previous column still. From the above total column number, 14 columns with an overall flow rate of 73l/h (with the source flow consisting of condensate with natural isotope content) form the first stage. Six columns form the second stage with a 28l/h overall flow rate. The third and the fourth stages comprise two columns each, with the total number of stages equal to eleven. The plant produces monthly 170g of heavy-oxygen water with a concentration of up to 80–85% of $^{18}O$, i.e. below the product concentration level equal 95%.

Since the plant was developed progressively over a period of years without adequate design calculations, it is far short of optimum. At present, the plant is being reconstructed to achieve an expected annual capacity of 4–5kg of $H_2^{18}O$ with a 95% concentration.

A plant for concentrating heavy-oxygen water by water rectification at atmospheric pressure operated successfully in Moscow [15]. The plant incorporated four columns filled with spiral-prismatic packing made of stainless-steel wire of 0.2mm diameter, with elements of 2.5mm $\times$ 3mm and 1.5mm $\times$ 2.0mm. In each column, the packing bed height was 12m. The columns operated on the closed-loop principle with the upper tank in the unsteady mode, complete condensate return, and periodic withdrawal from the still as the required concentration is obtained. The upper tank of a stage was filled with the concentrate produced in the still of the previous stage, which provided for $^{18}O$ sequential concentrating on a discrete cascade principle.

Two columns of 122mm diameter each served as the first stage with a flow rate of 10.2l/h in each column. The second and third stages were represented by single columns of 70 and 32mm diameter, respectively.

The head tanks of the first stage columns were filled with water or condensate (production waste from the heavy-water plant in Aleksin, Tula Region, U.S.S.R.) with a somewhat heightened concentration of $^{18}O$ oxygen isotope (initially $\approx$ 0.9% of $H_2^{18}O$). The final product concentration achieved at the above plant did not exceed 60–70% of $H_2^{18}O$.

The Moscow plant operated successfully and profitably for over 10 years satisfying, via V/O Isotop, domestic demand for heavy-oxygen water. It should be noted that the plant's performance could be considerably improved with operation under vacuum.

The plant was shut down after the pilot plant for $^{18}O$ isotope production by NO rectification at the NIISI (Tbilisi, Georgia) had been put into service and commercialized [16].

A vacuum rectification plant for the extraction of $^{18}O$ and $^{17}O$ heavy isotopes from heavy water was developed at the Nuclear Centre in Karlsruhe, Germany [17]. The use of heavy water as source material is determined by the fact that in the process of deuterium concentrating by water electrolysis, an increase in the heavy oxygen isotope concentration incidentally occurrs. In the case under consideration, heavy water, supplied by Norsk Hydro, a Norwegian company, containing 1.4at.% of $^{18}O$ and 0.12at.% of $^{17}O$ was used. A sevenfold increase in the $^{18}O$ initial concentration in the source material, as compared with a natural one, compensates for a decrease in the separation factor for $D_2O$ as against that for $H_2O$ in accordance with eq. (7.4).

The Karlsruhe plant, comprising 42 packed columns, is shown in Figure 7.3. Initially, the plant was rated at an annual capacity of about 4kg of $H_2^{17}O$ with a concentration of no less than 99.8%, and 0.2kg of $H_2^{17}O$ with a 99% concentration.

In all columns, the packing bed height is 12m, with wire mesh rings (Dixon rings) as packing material.

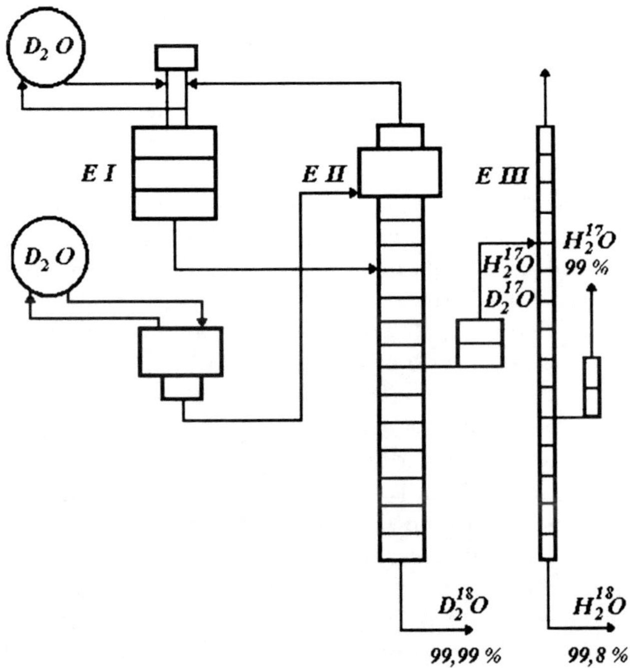


**Figure 7.3**  Scheme of plant for $H_2^{18}O$ and $H_2^{17}O$ production by heavy water rectification at Nuclear Centre in Karlsruhe, Germany [17].

From the above source material at a flow rate of 7–8.3l/h, the initial concentration stage incorporating four columns EI (Figure 7.3) with packing elements of 3mm × 3mm height produces 2.08ml/h of concentrate containing 18at.% of $^{18}O$ and 0.9at.% of $^{17}O$.

Further concentrating to up to 99.99% of $^{18}O$ is performed in a branched series EII consisting of 16 columns of 34mm diameter (2mm × 2mm packing element size) with a constant flow rate of 0.7–0.83l/h. The fraction enriched with $^{17}O$ to 10at.% is withdrawn from the eleventh column head.

After $D_2^{17}O$ conversion into $H_2^{17}O$ for separation factor enhancement, which is achieved by water electrolysis followed by oxygen combustion in hydrogen, the fraction is supplied to the third series EIII comprising 18 columns, each of 12mm diameter with a flow rate of 0.1l/h for $H_2^{17}O$ final concentrating.

The initial concentration section operates in periodic mode, with complete condensate return through the feed water tanks. Over a period of several months, the feed water is depleted from 1.4at.% to 0.3–0.4at.% of $^{18}O$, and then returned to be used as heavy water, and the feeding tank is filled with another source material load. Two small regenerating columns are designed lest the deuterium content in the water decreases.

A pressure of 26–27kPa is maintained in each column head. A pressure drop in the columns accounts for about 44–48kPa, with an average rectification temperature equal to 347K. Under these conditions, an average separation factor value is taken to be $\alpha = 1.005$.

Withdrawn from the stage EII (Figure 7.3), the product can be radioactive as a result of the accumulation of tritium contained in heavy water. This is why the water is electrolytically decomposed. Produced in the process, heavy oxygen is immobilized in water or carbon dioxide gas.

In the three years after the plant was put into service, an $^{18}O$ concentration of over 99% was attained, and a steady profile of $^{18}O$ was established in the system under consideration over a period of about ten years. The plant operation was fully automatic, without shift maintenance.

A portion of the annual heavy-oxygen water output could be with a concentration of 99.99at.%, and the remainder in the form of 99.8% product.

Water rectification is used as well in the production of light oxygen $^{16}O$, or more precisely, of water purified from heavy oxygen isotopes $^{18}O$ and $^{17}O$. In this case, the specific energy consumption per unit product is 250 times lower than that in the production of $^{18}O$ heavy isotope. At the Karlsruhe plant, a single column operating in periodic mode produced annually 2kg of water containing 99.99at.% of $H_2^{16}O$ [17].

An enlargement of the rectification plant included an increase in annual capacity of up to 11kg of $^{18}O$ and 0.7kg of $^{17}O$. But because of the termination of heavy water production by the electrolytic method, and corresponding cessation of source material supply, heavy-oxygen water production from $D_2O$ at Karlsruhe was suspended.

In 1997, a relatively large-scale plant for heavy-oxygen water production by water vacuum rectification was developed and put into operation in Sosnovy Bor, Leningrad region, Russia [18]. The project was funded by the Global Scientific Technologies Corporation.

Characteristics of two separation cascades with an annual design capacity of 9kg of $^{18}O$ with a concentration of 92–95at.% and 1.5kg with a concentration of 36at.% are reported in reference [18].

The large cascade includes four stages. The first stage incorporates eight packed columns of 100mm diameter and with a packing bed height of 16m, operating in parallel. The second stage consists of four columns of 80mm diameter and a packing bed height of 12m. Two columns of 80mm diameter and 12m height form the third stage. The fourth stage comprises a single column of 30mm diameter and of the same height.

All columns are filled with trihedral spiral-prismatic packing material made of stainless-steel wire of 0.2mm in diameter. In the first stage, the packing element size is 3.5mm $\times$ 3.5mm, and 2mm $\times$ 2mm in the other stages.

The cascade columns are linked on the basis of a discrete cascade scheme.

The evaporators (stills) of all columns are electrically heated. A closed circuit of the heat medium (water) is used to cool the condensers.

In the heads of the first-stage columns, a pressure of 20kPa is maintained.

The annual capacity of the above cascade is 9kg of oxygen with a 92–95at.% concentration of $^{18}O$ [18]. In the third-stage bottom, an enrichment with $^{17}O$ equal to 12at.% was achieved.

The second small cascade in Sosnovy Bor was initially developed as experimental plant intended to gain information on the design arrangement of the components and to train personnel. The cascade comprises three separation stages, the first of which includes two rectifications columns of 100mm diameter operating in parallel, with the second and third stages consisting of a single column of 80mm diameter each. Each column has a spiral-prismatic packing bed height of 12m, with an element size of 3.5mm $\times$ 3.5mm $\times$ 0.2mm in the first-stage columns, and 2mm $\times$ 2mm $\times$ 0.2mm in the second and third stages.

In a steady withdrawal mode of the small cascade operation, concentrations of 2%, 10% and 36% of $^{18}$O were attained at the bottom ends of the stages, respectively. The cascade annual capacity is 1.5kg of oxygen with a concentration of $^{18}$O in the product equal to 36at.%.

At present, the $H_2^{18}$O production plant of the Global Scientific Technologies Corporation is being modernized. It is planned that the plant's annual capacity will be increased to 20–25kg of heavy-oxygen water.

### 7.1.3  $^{18}$O concentrating by molecular oxygen cryogenic rectification

The separation of oxygen isotopes $^{16}$O, $^{17}$O, and $^{18}$O can be realized through molecular oxygen $O_2$ rectification (the process is characterized by separation factor values similar to those of water rectification – see Table 7.1), but the production of highly concentrated $^{18}$O is troublesome because of the $^{18}$O isotope content in the form of $^{16}O^{18}O$. The reason is that the rate of the homomolecular isotope exchange (HMEX) reaction in liquid oxygen is very low

$$2\left(^{16}O^{18}O\right) \leftrightarrows {}^{16}O_2 + {}^{18}O_2 . \tag{7.6}$$

Nevertheless, oxygen rectification can serve as a simple method of the initial concentration of $^{18}$O featuring such advantages over water rectification (see section 7.1.2) as a somewhat higher value of $\alpha$, and an evaporation heat an order of magnitude lower.

A plant for the cryogenic rectification of $O_2$ was developed at Prochem, British Oxygen Company, in Great Britain [19, 20].

The plant comprised two first-stage columns of 37.5mm diameter and one second-stage column of 25mm diameter packed with wire mesh rings of 2mm $\times$ 2mm in size. In each column, the packing bed height was equal to 10m. Gaseous oxygen containing 25–27at.% of $^{18}$O was continuously withdrawn from the plant bottom, with the composition of gas containing 27at.% of $^{18}$O being:

| $^{16}O^{18}O$ | $^{16}O^{16}O$ | $^{18}O^{18}O$ |
|:---:|:---:|:---:|
| 53.6 % | 46.2 % | 0.2 %. |

A very low content of $^{18}O^{18}O$ molecules confirms the fact that the HMEX reaction (7.6) almost does not proceed at all in liquid oxygen.

Over a period of years, experiments on the oxygen isotope separation by $O_2$ rectification were performed by the Department of the Isotope Technology and High-Pure Substances of the D. Mendeleev University of Chemical Technology of Russia (MUCTR) [4, 21–23]. The results of experiments performed in a column filled with stainless-steel spiral-prismatic packing material with elements of 2mm $\times$ 2mm $\times$ 0.2mm in size demonstrated that a temperature of 79–80K, corresponding to a pressure of 25–30kPa, is the optimum condition for the rectification process. It is convenient to maintain a near-optimum regime by using liquid nitrogen itself for the condenser cooling. In the process, the separation factor for the molecules $^{16}O^{16}O–^{16}O^{18}O$ is equal to 1.0062–1.0060 (Table 7.1).

In a column of 40mm diameter at the indicated values of $T$ and $P$ with a flow rate of 1150 kg/(m$^2$·h), the HETP value was 2.3cm [21].

A plant for heavy oxygen-isotope initial concentration by molecular oxygen rectification was developed on the basis of the studies and experiments performed in MUCTR [22, 23]. The plant is characterized by the following main engineering considerations:

- it is expedient that the heavy oxygen-isotope extraction from liquid air be linked up to an industrial air separation plant (ASP);
- to produce heavy oxygen isotope in annual amounts of kilograms (or even of tens of kilograms), a required rate of the flow intended for processing accounts for some tens of kilograms, that is, it is very small compared with available liquid oxygen resources measured in tons. Hence it is appropriate to adopt the so-called transit scheme without depletion, according to which the waste liquid oxygen from the isotope separation plant with somewhat lowered $^{18}O$ content is returned to ASP to be used for its original purpose, i.e. as common liquid oxygen.

For a fiftyfold initial concentration, i.e. to 10at.% of $^{18}O$, a two-stage cascade of concentrating packed columns is suggested (see Figure 7.4), with a limitation on the packing bed height equal to 20m for the plant to be installed at a common production facility.

It is expected that the subsequent concentrating of the $^{18}O$ isotope, which involves the processing of considerably smaller flows with correspondingly lower energy inputs and smaller equipment size, can be performed by other available separation methods such as water vacuum rectification (see section 7.1.2) after catalytic combustion of oxygen enriched with $^{18}O$ and mixed with hydrogen or natural gas, or the thermal diffusion method, etc. Sorption processes can be as efficient for the final concentrating [24].

It is well known that the thorough removal of acetylenic hydrocarbons is a major condition for preventing explosions during liquid oxygen production. Estimates demonstrate that even if the content of acetylenic hydrocarbons in the industrial liquid oxygen does not exceed $1 \times 10^{-10}$ volume fraction (corresponding to the threshold perception of the used analytical control methods), it is possible that less volatile hydrocarbons have accumulated in the still liquid in dangerous amounts over a prolonged period of hydrogen continuous rectification in the isotope separation plant. This is why, to ensure the process is explosion-proof and to avoid the accumulation of hydrocarbons over a prolonged period of the continuous operation of the oxygen rectification plant, provision should be made for a rectification column on the plant feed flow line that makes it possible to remove hydrocarbon admixtures (and other heavy impurities) upon their accumulation above current standards of permissible content in liquid oxygen.

In a purification column 9 (Fig. 7.4) of small height (compared with isotope separation columns), the oxygen source flow is purified from less volatile trace contaminants, and the purified oxygen from the column head is delivered to the condenser 4, whence it gravitates to the purification column 9 as a reflux, and to the plant first-stage column 1 as feeding.

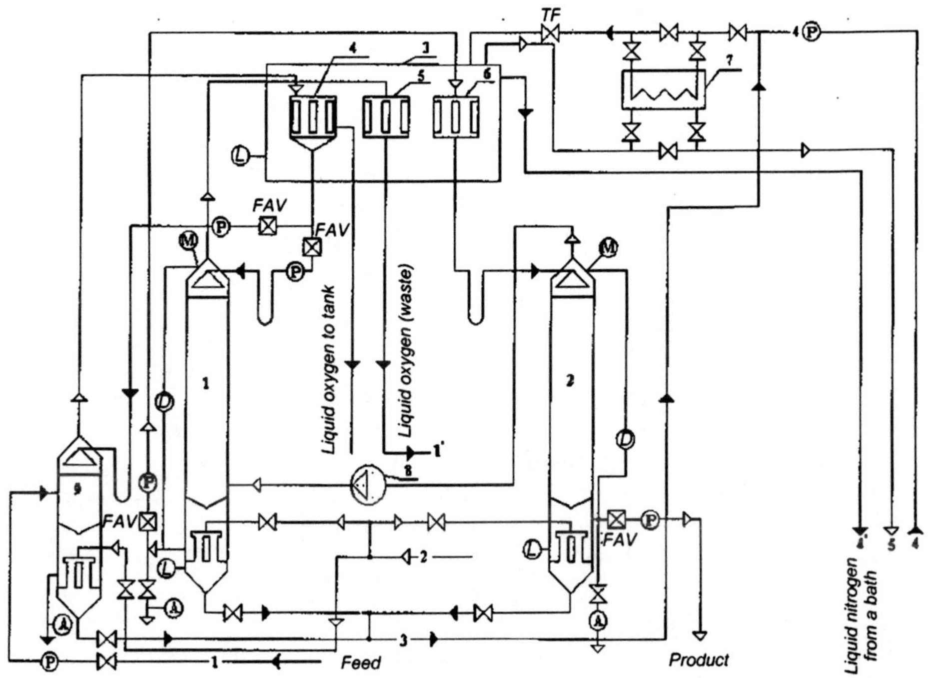

**Figure 7.4**   Scheme of plant for oxygen isotope separation by $O_2$ cryogenic rectification [23] : 1, 2, rectifiction columns of first and second stages, respectively; 3, liquid nitrogen; 4, 5, 6, purified oxygen condensers of first- and second- stage columns, respectively; 7, liquid nitrogen over cooler; 8, blower; 9, purifier; A, sampling for analysis; FAV, fine adjustment valve; TF, throttle flap; D, differential pressure gage; M, manometer; F, flow meter; L, level gauge.

Waste oxygen from the first-stage column head is liquefied in the first-stage condenser 5 and discharged to the commercial oxygen tank for subsequent distribution.

In the first-stage condenser, a pressure of 30kPa ($T = 80K$) is adopted, which approaches the above-mentioned optimum conditions of oxygen rectification. A pressure in the column bottom equals 75kPa (a temperature of 87K).

To avoid a pressure increase and a separation factor decrease, the pressure and temperature in the cascade second stage are similar to those in the first stage, and the vapor return flow delivery from the second stage to the first-stage lower cross-section is done by a mechanical device (rotary glandless blower) 8.

Concentrate withdrawal from the first stage (1at.% of $^{18}O$) is performed in the vapor phase. The withdrawn flow is delivered to condenser 6 by differential pressure, where it is liquefied, and supplied by gravity to the cascade second stage 2 as a reflux.

The oxygen concentrate withdrawn from the second stage bottom is supplied for further processing (to the reactor for combustion with hydrogen if the after-concentrating is performed by water rectification).

At an annual plant capacity of 2kg of $^{18}O$, the design dimensions of the columns are:

- first stage: flow rate 33kg/h, column diameter 100mm, MMR packing in the form of wire mesh rings with elements of 4–5mm size;
- second stage: flow rate 6.6kg/h, column diameter 60mm, MMR packing, or spiral-prismatic packing with elements of 3mm $\times$ 3mm $\times$ 0.2mm size.

All oxygen vapor condensers are immersed in the nitrogen bath 3 (Figure 7.4) and placed at an upper level for reflux of the columns by gravity.

The constancy of reflux flows is ensured by a uniform fluid level in the lower sections of condensers (serving as collectors), which is achieved by an overflow of small amounts of excessive liquid. The solution described above allows for the necessary displacement and stabilization of flows to be performed without pumps and complex automatic control devices.

The evaporators of all three columns are heated by compressed gaseous nitrogen at a pressure of 0.6MPa (input line 2, Figure 7.4). Giving up its heat to the oxygen evaporation, nitrogen liquefies and comes by line 3 through the heat exchanger–overcooler 7 to the liquid nitrogen bath, where it boils at a near atmospheric pressure which allows for cooling of the oxygen vapor condensers immersed in the bath. At the plant start, and when a makeup is required, liquid nitrogen is supplied from the tank by line 4 as refrigerant.

From under the bath lid, gaseous nitrogen is withdrawn through the liquid nitrogen overcooler 7 into the gaseous nitrogen line 5 for compression and reuse.

In the plant described above, the energy input is determined by the consumption for the nitrogen compression to 0.6MPa (only 6kWh at the indicated plant parameters which corresponds to 24,000kWh per kilogram of $^{18}O$).

According to estimates, the nitrogen cycle accounts for 50% of cold consumption, and the remaining 50% is due to the bath makeup with liquid nitrogen from the outside (from ASP) [22, 23]. The same source accounts for the cold consumption for the system cool down and establishment of normal process conditions at the plant start.

## 7.1.4  NO cryogenic rectification

Abnormally high values of the separation factor in the NO rectification compared with those in water and molecular oxygen rectification (Table 7.1) have resulted in the practical use of this isotope separation method despite the lack of reliable sources of NO.

In the U.S.S.R., nitric oxide low-temperature rectification for the production of heavy oxygen isotope $^{18}O$ was performed at NIISI (Tbilisi, Georgia). The cascade's annual capacity was 3.5kg of $^{18}O$ [16, 25]. The waste flow of nitric oxides issuing from the nitric acid – nitric oxide (Nitrox) plant for $^{15}N$ production and containing about 80% (by volume) of nitric oxide (II), served as source flow. The plant is shown in Figure 7.5.

The feed flow of nitric oxides arrives at the packed absorber 1, of 120mm in diameter and 2m in height, provided with a water-cooling jacket and irrigated with water. The absorber is designed to remove the higher nitric oxides ($NO_2$, $N_2O_3$, $N_2O_4$) from the gas flow. Weak nitric acid formed in the process is withdrawn from the absorber. Leaving the

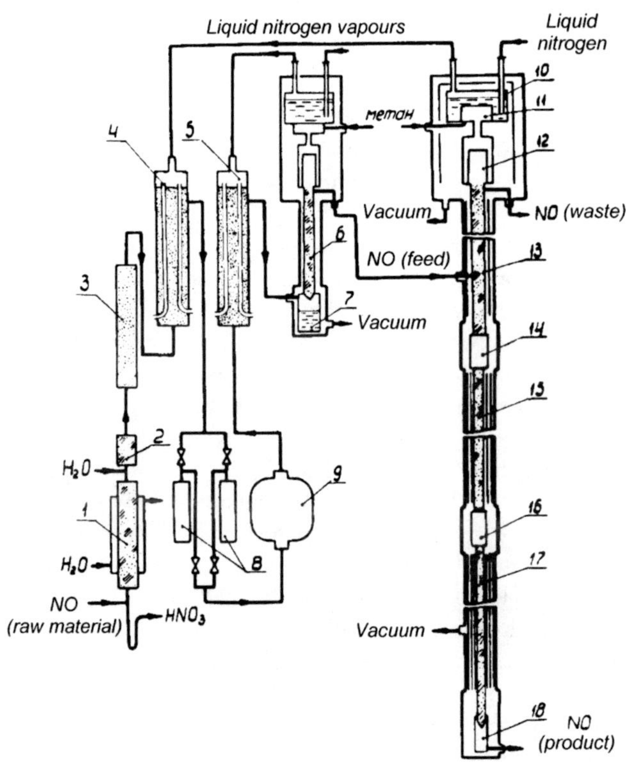



**Figure 7.5**  Scheme of isotope separation by NO cryogenic rectification at NIISI, Tbilisi, Georgia [16]: 1, absorber; 2, mist extractor; 3, 4, 5, adsorber with silica gel; 6, purifying rectification column with still; 7, 8, purified nitrogen oxide trap; 9, receiver; 10, liquid nitrogen bath; 11, methane condenser; 12, nitrogen oxide condenser; 13, 15, 17, rectification columns; 14, 16, 18, column evaporator.

absorber, the gas flow arrives at the mist extractor, and then, for further purification and dehydration, is passed through the adsorbers 3 and 4, of 140mm diameter, each filled with silica gel and cooled by the waste nitrogen vapor.

The adsorber 3 operates at a temperature of 278K, and adsorber 4 at 175–190K. The purified nitric oxide in liquid form is accumulated in a trap 8.

For final, sufficiently fine, cleaning, the nitric oxide is delivered, through the receiver 9 and silica gel adsorber 5 operating at 175K, to the purifying rectification column 6 with the still 7. The purifying rectification column has an internal diameter of 57mm filled to a height of 2m with spiral-prismatic packing material with elements of 3mm × 3mm × 0.2mm. The temperature in the condenser is 125K. The purified NO is withdrawn from the column head and delivered as a feed flow to the three-stage isotope separation cascade 10–18. The cascade comprises three packed columns forming a small depletion section (of 5.6m height), and a three-stage enrichment section.

The columns are placed coaxially one above the other to allow for the displacement of liquid by gravity. The overall height of the columns is 53m. The columns are filled with spiral-prismatic packing made of stainless-steel wire of 0.2mm diameter with elements of 3mm $\times$ 3mm, 2mm $\times$ 2mm and 1.5mm $\times$ 1.5mm (by stages). The internal diameters of the columns are 89mm, 32mm, and 15mm, with packing bed heights by enrichment section stages of 12m, 8m, and 26m, respectively. From the outside, each column is wrapped around with a screen-vacuum heat insulator and placed in a pressurized vacuum casing.

In the NO rectification, HETP is 5.6cm, 3.3cm, and 2.7cm for columns of 57mm, 32mm, and 15mm in diameter, respectively.

Between the first and the second columns, as well as between the second and the third columns, intermediate evaporators – flow dividers 14 and 16 with electric heating elements are installed. The evaporators are fed with such power amounts that are required to evaporate a desired liquid flow portion.

The third-stage column bottom is linked to the evaporator 18 to ensure the complete flow conversion at the enriched end of the cascade. The product is withdrawn from the vapor phase.

Condensate is formed in the condenser 12 cooled with methane boiling at a pressure of 0.22MPa (123K). The pressure control and stabilization are performed with a helium cushion. The methane vapor, in its turn, is condensed in the condenser 11 located above and cooled by liquid nitrogen boiling at atmospheric pressure. The purification column condenser operates in a similar way with methane as intermediate cooling agent.

As described above, from the condensers of rectification columns, the boiling nitrogen vapor is supplied for cooling the purification system adsorbers. The pressure in the nitrogen oxide condenser is 112kPa, with a 157kPa pressure in the lower cross-section of the column.

The waste flow is returned to the absorption system of the plant for $^{15}N$ production by chemical exchange. The heavy isotope concentration in the waste flow accounts for 0.13% which corresponds to an $^{18}O$ extraction degree of 35%.

The consumption coefficients per kilogram of $^{18}O$ are: nitric oxide 1800kg, liquid nitrogen 200tons, electric power 12,000kWh.

The plant described above for $^{18}O$ concentrating by NO rectification was complemented by an auxiliary unit for the production of light isotopes in the form of $^{14}N^{16}O$. The process is implemented with a single packed column of 28mm diameter and 25m height operating with the complete return of condensate and provided with a large lower evaporating tank of 25l volume, with the waste flow from the cascade depletion section serving as the feed flow. The unit's daily capacity is 180–240g of $^{14}N^{16}O$ with a concentration of 99.998%.

As noted above, in the nitric oxide rectification process, together with $^{14}N^{18}O$ concentrating, concentrating of $^{15}N^{16}O$ molecules also takes place, with the maximum concentration of the latter observed somewhere in the middle of the cascade. The concentration of $^{15}N$ in the end product is rather low and accounts for about 5–10at.%. A low level of product enrichment with $^{15}N$ isotope shows that in cryogenic rectification conditions the reaction (6.4), probably, proceeds slowly. According to observational data, the concentration of $^{15}N$ in the product increases when nitric oxide containing admixtures of higher oxides is used [25].

To solve the problem of simultaneous production of $^{18}O$ and $^{15}N$ isotopes by NO rectification, it is suggested supplementing the separation cascade with an isotope exchange stage to perform the HMEX reaction (6.4) with the formation of $^{15}N^{18}O$ molecules (see, for example, reference [26]). The problem of simultaneous production of highly enriched heavy isotopes of oxygen and nitrogen is also considered in refernce [27]. In the above-mentioned papers, in particular, the influence of isotope exchange stages, located at the interfaces between separation stages of the cascade enrichment section, on the concentrations of $^{15}N$ and $^{18}O$ in the product withdrawal flow is studied by calculation for a four-stage cascade of which the parameters correspond to those of the NIISI cascade. The studies demonstrate that with the use of one or two exchange stages it is possible to increase the $^{15}N$ concentration to 90at.%, with the $^{18}O$ concentration remaining not lower than 90at.%. An alternative solution can consist of the development of a more complex double cascade, i.e. an auxiliary separation stage fed by the withdrawal flow of the NIISI cascade.

High separation-factor values attained in the NO rectification are still attracting the attention of production engineers despite serious problems associated with this process realization. A new pilot plant for NO cryogenic rectification, for example, was recently developed at National Institute for Research and Development of Isotopic and Molecular Technologies in Cluj-Napoka, Romania [28].

The column $C_1$ (see Figure 7.6), of 7.1m height and 16mm diameter, is filled with Helipack packing material of which the elements are 1.8mm $\times$ 1.8mm $\times$ 0.2mm in size. The still B, with a volume of 100ml, has a built-in electric heater with a maximum power of 150W. To ensure adiabaticity, the column is placed in a vacuum jacket M evacuated to a residual pressure of $1 \times 10^{-5}$ Torr. The plant is completed with a double-chamber condenser C. As an intermediate cooling agent, molecular oxygen is used (similar to the Los Alamos laboratory [29]). The column operates at atmospheric pressure.

As the NO source flow (similar to the NIISI plant [16, 25]) the flow of nitric oxides issuing from the plant was used for nitrogen isotope separation by the Nitrox method and passed through a special system for the purification to a NO content of 99.99%.

The plant was developed for research purposes, and at a reflux density of 1.2ml/(cm$^2$·min) the HETP measured value was 7.8cm [28] which is almost threefold that for a column of a similar diameter of the NIISI plant.


## 7.2  OXYGEN ISOTOPE SEPARATION BY CHEMICAL EXCHANGE METHOD

### 7.2.1.  Separation factor and operating systems

The effective separation factor values for oxygen isotope separation in operating systems employed in chemical isotope exchange are presented in Table 7.3 [3, 30–34].

From the data given in Table 7.3 it can be seen that the highest separation factor values are observed in the $CO_{2\,(gas)} - H_2O_{(lq)}$ system ($\alpha \approx 1.042$ at $T = 298K$). The isotope effect in the system is commensurate with the oxygen isotope separation factor in NO rectification (Table 7.1).

The NO, $NO_{2\,(gas)} - HNO_{3\,(aq.\,sol)}$ system has a far lower separation factor value compared with the isotope effect in nitric oxide (II) low-temperature rectification. For the nitric acid – nitric oxide technique, for one, the $\varepsilon$ value (6.2M $HNO_3$; 298K) is half as great.

In sulfur dioxide-water exchange, the enrichment factor is about 2.5 times smaller than that in the NO low-temperature rectification.

Notice that for all systems (methods) under consideration, the heavy oxygen isotope $^{18}O$ is concentrated in the gas phase.

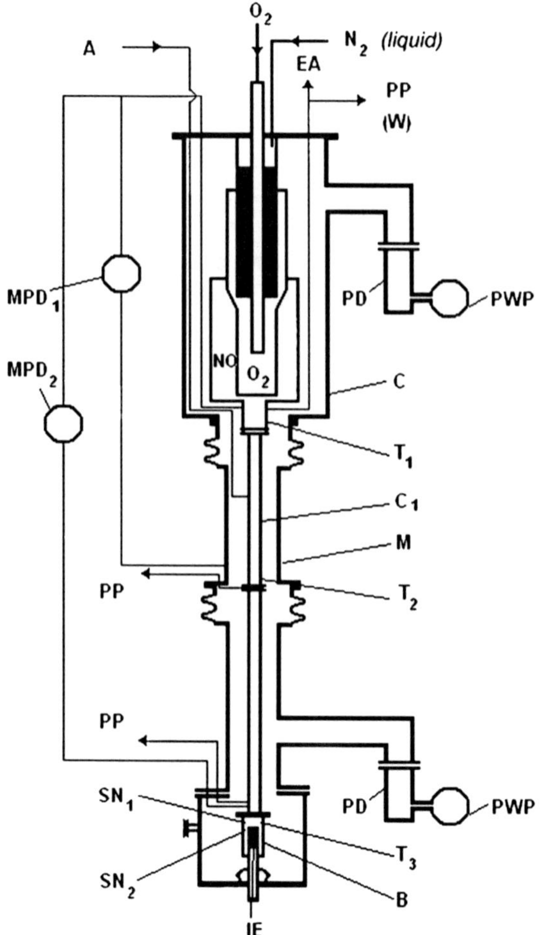


**Figure 7.6**   NO cryogenic rectification column [28]: A, feed inlet; W, NO outlet; EA, NO emergency discharge; PWP, roughing-down pump; PD, diffusion pump; C, condenser; $C_1$, column; $T_1$–$T_3$, temperature sensors; M, jacket; B, still; IE, electric heating unit; $SN_1$, $SN_2$, level gauges; PP, sampling; $MPD_1$, $MPD_2$ – differential pressure gauge.

By this time, only two oxygen isotope separation methods (initial concentration region) have found practical use. They are: chemical isotope exchange between nitric oxides and nitric acid aqueous solution (or the nitric acid – nitric oxide isotope separation method) [35, 36]; and the carbamate isotope separation method [37, 38]. It should be pointed out that both processes are used first of all for concentrating other isotopes, with the former for nitrogen isotope separation (chapter 6), and the latter for carbon isotope separation (chapter 5). Despite a small isotope effect, the carbamate system draws attention by the thermal flow reflux method.

**Table 7.3**

Isotope effect in oxygen isotope separation by the chemical exchange method (at 0.1 MPa pressure)

| Method | Operating system | Basic isotope exchange reaction(s) | Conditions | | $\alpha$ | Reference |
|---|---|---|---|---|---|---|
| | | | $T$, K | $C$ [a] | | |
| 1 | 2 | 3 | 4 | 5 | 6 | 4 |
| Carbon dioxide | $CO_{2\,(gas)} - H_2O_{(lq)}$ | $C\,^{16}O^{16}O_{(gas)} + 2H_2{}^{18}O_{(lq)} \leftrightarrow C\,^{16}O^{18}O_{(gas)} + H_2{}^{16}O_{(lq)}$ | 278<br>298<br>313<br>333<br>353<br>373 | | 1.0465<br>1.0424<br>1.0392<br>1.0349<br>1.0320<br>1.0292 | [30] |
| Nitric acid – nitric oxides | $NO, NO_{2\,(gas)} -$<br>$HNO_{3\,(sol,\,aq)}$ | $N\,^{16}O_{(gas)}, N^{16}O^{16}O_{(gas)} + HN\,^{16}O_2{}^{18}O_{(sol,\,aq)} \leftrightarrow$<br>$\leftrightarrow N\,^{18}O_{(gas)}, N^{16}O^{18}O_{(gas)} + HN^{16}O_{3\,(sol,\,aq)};$<br>$N\,^{16}O_{(gas)}, N^{16}O^{16}O_{(gas)} + H_2{}^{18}O_{(lq)} \leftrightarrow$<br>$\leftrightarrow N\,^{18}O_{(gas)}, N^{16}O^{18}O_{(gas)} + H_2{}^{16}O_{(lq)}$ | <br><br>298 | 4.1M<br>6.2M<br>8.0M<br>9.7M | 1.028<br>1.020<br>1.018<br>1.015 | [31] |
| Sulfite | $SO_{2\,(gas)} - H_2O_{(lq)}$ | $S\,^{16}O^{16}O_{(gas)} + H_2{}^{18}O_{(lq)} \leftrightarrow S\,^{16}O^{18}O_{(gas)} + H_2{}^{16}O_{(lq)}$ | 294 | | 1.016 | [32] |
| Carbamate | $CO_{2\,(gas)} -$<br>$R_2NCOO^{-b}{}_{(sol,\,org)}$<br>R- $n$-butyl<br>$(-C_4H_9)$;<br>solvent – octane<br>$(C_8H_{18})$ | $C^{16}O^{16}O_{(gas)} + R_2NC^{16}O^{18}O^-{}_{(sol,\,org)} \leftrightarrow$<br>$\leftrightarrow C^{16}O^{18}O_{(gas)} + R_2NC^{16}O^{16}O^-{}_{(sol,\,org)}$ | 293 | 2M | 1.013 | [33, 34] |

[a] Component concentration in solution.

[b] $R_2NCOO^-$, carbamate-anion.

In terms of the organization of the isotope separation process, knowledge of the separation factor must be supplemented with information on the availability and quality of source materials, on kinetics and mass exchange, on the flow conversion completeness, as well with other knowledge that makes it possible to more reasonably speak about technological applicability, advantages, and drawbacks of one or other isotope separation method.

## 7.2.2   Characteristic properties of separation processes

*$CO_{2\,(gas)} - H_2O_{(lq)}$ system*

The $CO_{2\,(gas)} - H_2O_{(lq)}$ system (carbon dioxide method) is the most favorable in respect of the chemical properties of its compounds. The required upper flow conversion is performed chemically on the basis of the methanization reaction (7.7) proceeding due to carbon dioxide reduction by hydrogen

$$CO_2 + 4H_2 \rightarrow CH_4 + 2H_2O. \tag{7.7}$$

The process is performed on a nickel catalyst at a temperature of 400°C in two stages with intermediate condensation of water vapor [39]. The process feature is that it requires close control over oxygen-containing admixtures in hydrogen to avoid isotopic dilution of the product ($H_2^{18}O$). It is also necessary to perform a thorough dewatering of methane to ensure against $^{18}O$ losses. The main drawbacks of this method are high hydrogen consumption (4mol per mol of $CO_2$ by stoichiometry only) and a very low mass exchange rate limited by the $CO_2$ hydration. In a column of 25mm diameter packed with stainless-steel coils at a relatively high pressure (1.46MPa), for example, HETP was equal to 7.68m at indoor temperature [39].

A study on the conditions for the mass exchange efficiency increase, performed in MUCTR, demonstrated that HETP can be reduced to 6–10cm through the use of activating additives (the experiments were performed in a column of 20mm diameter packed with trihedral wire coils of 1.5mm × 2.0mm size), such as monoethanolamine together with sodium selenite at a temperature of over 50°C and pressure greater than 1MPa [40, 41]. Nevertheless, despite favorable (from the mass exchange standpoint) results, a practical realization of the conditions found during the study is questionable. The problem of the circulation of additives catalyzing the isotope exchange is very difficult from the technological point of view. The removal of monoethanolamine, and especially selenite, from the water waste flow followed by their introduction into $H_2^{18}O$ (at the column upper end) is difficult to realize.

*$SO_{2(gas)} - H_2O_{(lq)}$ system*

The sulfur dioxide – water system is characterized by the absence of kinetic problems in the exchange of oxygen atoms. But the sulfite method of realization (Table 7.3) is conditioned by the flow conversion problem.

For the organization of flow conversion, it suggested to reducing $SO_2$ by sulfuretted hydrogen to water by

$$SO_2 + 2H_2S \rightarrow 2H_2O + 3S, \tag{7.8}$$

which proceeds readily in the presence of water vapor. For the complete reduction of sulfur dioxide, however, a large excess of sulfuretted hydrogen is required ($H_2S:SO_2 = 2.2:1$). Owing to possible water losses in this case, it is unlikely that the process can be used for the flow reflux for separation of oxygen isotopes.

### $NO, NO_{2\,(gas)} - HNO_{3\,(aq,\,sol)}$ system

By now, only two chemical isotope exchange processes have been realized (experimentally) for oxygen isotope production. These are $NO, NO_2 - HNO_3$, and carbamate CHEX methods [35–37].

The former is based on the isotope exchange reaction between nitric oxides and the aqueous solution of nitric acid (Table 7.3). The isotope exchange rate in this system is fairly high: for laboratory-scale columns, HETP accounts for some centimeters [36, 40] (according to E. Oziashvili and co-workers [40], HETP = 5.6cm).

This method is distinguished by the possibility of integration with a plant for nitrogen isotope separation by the Nitrox method (see chapter 6), both through the gas phase and, probably, through the liquid phase. First and foremost, it allows eliminatation of the problem of the NO source in the $^{18}O$ concentrating (similar to nitric oxide cryogenic rectification, see section 7.1.3).

Such an integrated engineering solution is exemplified by the plant for $^{15}N$ and $^{18}O$ production developed by D. Axente and co-workers [36] at the National Institute for Research and Development of Isotopic and Molecular Technologies in Cluj-Napoka, Romania, for which the scheme is shown in Figure 7.7.

The left-hand side of Figure 7.7 (2, 3, 4) displays a two-stage cascade for $^{15}N$ concentrating linked through the NO flow (column 1 head, purifier 5, column 6 bottom) with a two-stage cascade (columns 6, 13) for $^{18}O$ concentrating.

From the point of view of isotope separation technology with the chemical exchange method, of interest is the solution of the flow conversion problem with respect to $^{18}O$ (Figure 7.7, columns 6, 13 head). Since the liquid phase includes (generally) two oxygen-containing compounds, $H_2O$ and $HNO_3$, it is necessary to obtain these from NO without introducing oxygen compounds from outside.

This rather complicated problem can be solved by partly decomposing $N^{18}O$ to $N_2$ and $^{18}O_2$ molecules either by electric discharge (7, 14 in Figure 7.7) or at the Pt-catalyst at a temperature of over 700°C, as reported by D. Axente and co-workers [36]. After that, $N^{18}O$ is oxidized by oxygen ($^{18}O_2$) with the formation of $N^{18}O_2$. The obtained nitrogen oxide (IV) is then absorbed by water ($H_2^{18}O$) (8, 15), formed from part of the $N^{18}O_2$ through the reaction with hydrogen over chrome–nickel catalyst at a temperature of 750–800°C [40]. Formed in the absorber, the $HN^{18}O_3$ solution in the $H_2^{18}O$ is supplied to the isotope exchange columns 6, 13 as a reflux, and $N_2$ is withdrawn from the plant together with unreacted hydrogen.

To absorb water vapor contained in $N_2$ during the production of concentrated $^{18}O$, it is necessary to use the sorption of water vapor over zeolites (12, 19 in Figure 7.7), because, even

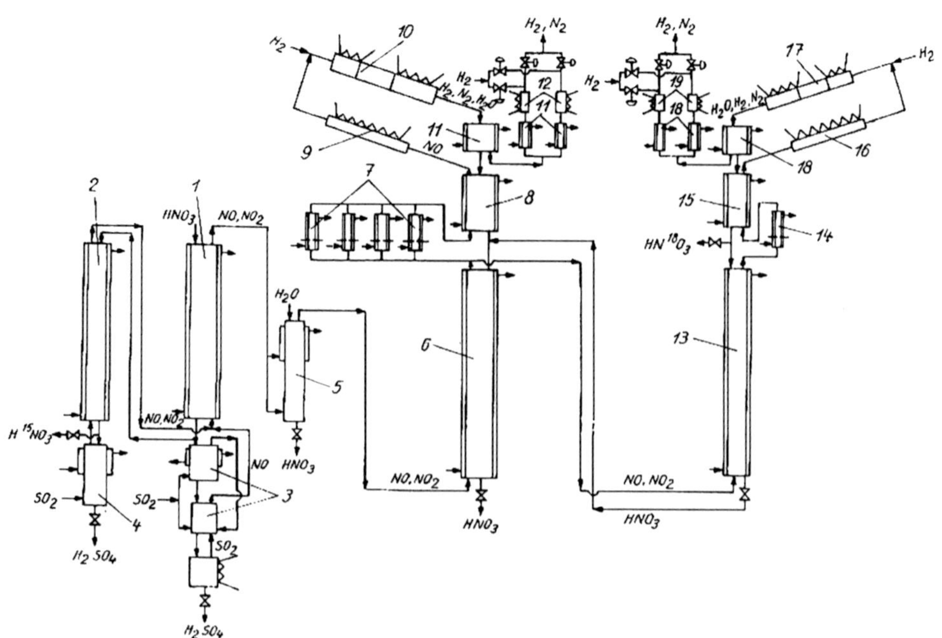

**Figure 7.7**   Scheme of nitrogen and oxygen isotope separation by the nitric acid – nitric oxides method at National Institute for Research and Development of Isotopic and Molecular Technologies in Cluj-Napoka, Romania [36]: 1, 2, $^{15}$N concentration columns of first and second stages, respectively; 3, 4, $HNO_3$ reduction reactor; 5, packed column for NO purification; 6, 13, $^{18}$O concentrating columns; 7, 14, reactor for NO decomposition by electrical discharge; 8, 15, absorber; 9, 16, reaction mixture preheater; 10, 17, $H_2O$ production reactor; 11, 18, heat exchanger; 12, 19, zeolite adsorber.

if the condenser is cooled to 0°C, the content of water vapor in the gas is $\approx 7 \times 10^{-4}$ g/l, as against the permissible value accounting for $2 \times 10^{-6}$ g/l for the production of 70at.% $^{18}$O [40].

$CO_{2\,(gas)} - R_2NCO_2^{-}{}_{(sol,\,org)}$ *system*

Oxygen isotope separation by the carbamate method provides the second example of the application of chemical exchange, but with more attractive thermal methods of flow reflux [33, 34] (Table 7.3).

In this case the flow conversion is performed through the reaction equilibrium displacement with heat application or rejection

$$2R_2NH_{(sol,org)} + CO_{2(gas)} \underset{cooling}{\overset{heating}{\rightleftharpoons}} R_2NCOONR_{2(sol,org)} \underset{cooling}{\overset{heating}{\rightleftharpoons}} R_2NCOO^{-}_{(sol,org)}$$

$$+ R_2NH_2^{+}{}_{(sol,org)}.$$

$$(7.9)$$

For a laboratory-scale column, with the use, for example, of 2M $n$-DBA solution in triethylamine, at a temperature of 25°C, a specific gas flow rate of 0.06–0.24mol $CO_2$/(cm²·h), and with Helipack packing, HETP was equal to about 2–8cm [36]. Somewhat higher HETP values were obtained for spiral-prismatic packing with an element size of 2.3mm × 2.3mm × 0.2mm [36].

Enrichment with $^{18}O$ of more than 200-fold was obtained at an enlarged pilot plant with a 40m packed column (51mm internal diameter). This was intended primarily for $^{13}C$ concentrating, with the use of the carbon dioxide-$n$-DBA carbamate in octane system, simultaneously with $^{13}C$ concentrating [37]. In addition to the interest in this isotope, the result is capable of significantly improving the carbamate method's economic performance for carbon isotope separation (chapter 5). Besides, the process also allows solution of the problem of initial $^{18}O$ concentration for further enrichment with the use of combined process schemes [41], i.e. in combination, for example, with the gas-centrifugal method [42] or with sorbtion on zeolites [24].

# REFERENCES


1. K. Clusius, K. Schleich, *Helv. Chim. Acta.*, **1958**, 41, 6, 1342.
2. K. Clusius, K. Schleich, M. Vechi, *Helv. Chim. Acta.*, **1959**, 42, 6, 2654.
3. T. F. Johns, In: *Proc. Intern. Symp. on Isotope Separation*, Amsterdam, North-Holl and Pub. Co., **1958**, 74.
4. A. E. Kovalenko. author's abstract, Ph.D. thesis, Mendeleev University of Chemical Technology, **1971**, 24.
5. A. E. Kovalenko, Ya. D. Zelvensky, *Isotopenpraxis*, **1969**, 5, 1, 20.
6. E. Ancona, G. Boato, G. Casanova *Nuovo Cimento*, **1962**, 23, 1041.
7. K. Clusius, F. Endtinger, K. Schleich, *Helv. Chim. Acta*, **1961**, 44, 98.
8. A. Van Hook, *J. Phys. Chem.*, **1968**, 72, 234.
9. S. Shapiro, F. Steckel, *Trans. Farad. Soc.*, **1967**, 63, 4, 883.
10. O. V. Uvarov, N. M. Sokolov, N. M. Zhavoronkov, *Zhurn. Phiz. Khimii*, **1962**, 36, 2699.
11. O. V. Uvarov, *Production and Properties of Heavy-Oxygen Water*, L. Ya. Karpov NIPhKhI, **1963**, 44.
12. D. Jickli, D. F. Staschewski, *Chem. Soc. Farad. Soc.*, **1977**, 72, 1505.
13. D. Staschewski, *Ber. Bunsenges. Physik. Chem.*, **1964**, 68, 5, 454.
14. I. Dostrovsky, A. Raviv, *Proc. Int. Symp. Isotope Separation*, Amsterdam, McGraw Hill Book Co., **1958**, 336.
15. L. Streltsov, N. Zhavoronkov, T. Gumeniuk *et al.*, *Khim. Prom.*, **1974**, 3, 221.
16. P. Ya. Asatiani *et al.*, In: *Stable Isotopes in the Life Sciences*, IAEA, Vienna, **1977**, 75.
17. D. Staschewski, In: *Stable Isotopes in the Life Sciences*, IAEA, Vienna, **1977**, 85.
18. A. S. Polevoy, M. N. Polyansky, In: *Proc. 2nd All-Russian Conference on Physical-Chemical Processes in Selection of Atoms and Molecules*, TsNIIatominform, **1997**, 111.
19. A. O. Edmunds, I. M. Lockhart, In: *Isotope Rations as Pollutant Source and Behav. Indic.*, IAEA, Vienna, **1975**, 229.
20. W. R. Daniels, A. O. Edmunds, I. M. Lockhart, In: *Stable Isotopes in the Life Sciences*, IAEA, Vienna, **1977**, 21.
21. A. E. Kovalenko, Ya. D. Zelvensky, E.S. Vaynerman, *Atomnaya Energiya*, **1969**, 27, 541.
22. Ya. D. Zelvensky, *Isotope Separation by Cryogenic Ractification*, Mendeleev University of Chemical Technology of Russia, **1998**, 208.

23. Ya. D. Zelvensky, *Khim. Prom.*, **1999**, 4, 32.

24. B. M. Andreev, E. P. Magomedbekov, I. L. Selivanenko, In: *Proc. 2nd All-Russian Conference on Physical-Chemical Processes in Selection of Atoms and Molecules*, TsNIIatominform, **1997**, 123.

25. G. Ya. Asatiani, candidate of technological science thesis, Mendeleev Institute of Chemical Technology, **1981**, 20.

26. R. I. Sidenko, G. A. Sulaberidze, V. A. Tchuzhinov, V. M. Vetsko, In: *Proc. Mendeleev Institute of Chemical Technology*, **1989**, 156, 57.

27. R. I. Sidenko, G. A. Sulaberidze, V. A. Tchuzhinov, V. M. Vetsko, *Atomnaya Energiya*, **1990**, 69, 4, 255.

28. M. G. Gligan, A. Radoi, S. Dronca, C. Bidian *et al.*, *Revista de Chimie*, **1997**, 48, 335.

29. D. E. Armstrong, B. B. McInteer, T. R. Mills, J. G. Montoya, In: *Stable Isotopes. Proc. Third Intern. Conf.*, New York, Academic Press, **1979**, 175.

30. D. Staschevski, *Chem. Technol.*, **1975**, 4, 8, 269.

31. S. C. Saxena, T. I. Taylor, *J. Phys. Chem.*, **1962**, 66, 8, 1480.

32. L. L. Broun, J. S. Drury, *J. Phys. Chem.*, **1959**, 63, 11, 1885.

33. J. P. Agraval, *Separation Sci.*, **1971**, 6, 6, 819.

34. J. P. Agraval, *Separation Sci.*, **1971**, 6, 6, 831.

35. M. Abrudean, D. Axente, S. Bâldea, *Isotopenpraxis*, **1981**, 17, 11, 377.

36. D. Axente, A. Bâldea, M. Abbrudean, In: *Proc. Int. Symp. on Isotope Separation and Chemical Exchange Uranium Enrichment (Bull. Res. Lab. for Nuclear Reactors*, special issue 1), **1992**, 376.

37. A. S. Egiazarov, G. V. Hatchisvili, T. G. Abzianidze, In: *Proc. 5th Intern. Symp. on the Synthesis and Applications of Isotopes and Isotopically Labelled Compounds*, Strasbourg, France, June 20-24, 1994, P028, 146; T. I. Taylor, *J. Chim. Phys. Phys.-Chim. Biol.*, **1963**, 60, 1–2, 157.

38. B. M. Andreev, T. D. Gumeniuk, Ya. D. Zelvensky, In: *Proc. Mendeleev Inst. Chem. Technol.*, **1970**, LXVII, 100.

39. B. M. Andreev, T. D. Gumeniuk, Ya. D. Zelvensky, A. M. Meretsky, *Isotopenpraxis*, **1971**, 7, 5, 180.

40. E. D. Oziashvili, Yu. V. Nikolaev, N. Ph. Myasoedov, *Rep. Acad. Sci. Georgian SSR*, **1962**, XXIX, 3, 289.

41. A. V. Khoroshilov, In: *Tenth Symp. on Separation Sci. and Tehnol. for Energy Applic.*, Gatlinburg, Tennessee, October 20–24, **1997**, 76.

42. Yu. V. Petrov *et al.*, Patent RU 2092234, C1, 10. 10. 1997, *Inventions*, **1997**, 28, 199.

# Subject Index